Photocatalysis: Principles and Applications

Photocatalysis: Principles and Applications

Edited by
Jensen Gilbert

www.willfordpress.com

Published by Willford Press,
118-35 Queens Blvd., Suite 400,
Forest Hills, NY 11375, USA

ISBN: 978-1-68285-793-9

Cataloging-in-Publication Data

Photocatalysis : principles and applications / edited by Jensen Gilbert.
    p. cm.
Includes bibliographical references and index.
ISBN 978-1-68285-793-9
1. Photocatalysis. 2. Catalysis. I. Gilbert, Jensen.
QD716.P45 P46 2020
541.395--dc23

For information on all Willford Press publications
visit our website at www.willfordpress.com

Contents

Preface

Photocatalysis is the acceleration of a photoreaction in the presence of a catalyst. Light is absorbed by an adsorbed substance in catalyzed photolysis. However, in photogenerated catalysis, the photocatalytic activity is based on the ability of the catalyst to generate electron-hole pairs which create free radicals. Homogeneous photocatalysis and heterogeneous photocatalysis are the two primary categories of photocatalysis. In homogeneous photocatalysis, the photocatalysts and the reactants exist in the same phase. Whereas, in heterogeneous photocatalysis reactants are in a different phase. Heterogeneous photocatalysis includes a large variety of reactions, such as mild or total oxidations, dehydrogenation, hydrogen transfer, and deuterium-alkane isotopic exchange, etc. This book brings forth some of the most innovative concepts and elucidates the unexplored aspects of photocatalysis. Different approaches, evaluations, methodologies and advanced studies on photocatalysis have been included in this book. This book is appropriate for students seeking detailed information in this area as well as for experts.

This book unites the global concepts and researches in an organized manner for a comprehensive understanding of the subject. It is a ripe text for all researchers, students, scientists or anyone else who is interested in acquiring a better knowledge of this dynamic field.

I extend my sincere thanks to the contributors for such eloquent research chapters. Finally, I thank my family for being a source of support and help.

Editor

Photocatalytic Oxygenation by Water-Soluble Metalloporphyrins as a Pathway to Functionalized Polycycles

Ivana Šagud🄳 and **Irena Škorić**🄳

Department of Organic Chemistry, Faculty of Chemical Engineering and Technology, University of Zagreb, Marulićev trg 19, 10000 Zagreb, Croatia

Correspondence should be addressed to Irena Škorić; iskoric@fkit.hr

Academic Editor: Maria da Graça P. Neves

Photocatalytic processes are present in natural biochemical pathways as well as in the organic synthetic ones. This minireview will cover the field of photocatalysis that uses both the free-base and specially metallated porphyrins as catalysts. While free-base porphyrins are valuable sensitizers to output singlet oxygen, metalloporphyrins are even more adjustable as photocatalysts because of their coordination capacity, generating a wider range of oxidation reactions. They can be applied in autooxidation reactions, hydroxylations, or direct oxygen transfer producing epoxides. This review will mainly focus on how manganese and some iron porphyrins can be utilized for the functionalization of compounds that have a polycyclic skeleton in their structure. These kinds of compounds are notoriously taxing to obtain and difficult to further functionalize by conventional organic synthetic methods. We have focused on photocatalytic oxygenation reactions in mild conditions with the use of water-soluble porphyrins, as this has been proven to be a good tool for these transformations. In the photocatalytic reactions of some polycyclic heteroaromatic compounds, new polycyclic epoxides, enediones, ketones, alcohols, and/or hydroperoxides are yielded, depending on the catalyst applied. The application of anionic and cationic Mn(III) porphyrins under different reaction parameters results in different reaction pathways generating a vast number of photocatalytic products. Recently, Co and Ni complexes have been also photophysically investigated and confirmed as potential photocatalysts for the functionalization of organic substrates.

1. Introduction

Photocatalytic processes have been demonstrated to be numerous in both natural and artificial surroundings, such as photosynthesis, which is the basis of the food chain on Earth [1], as well as oxidative degradation of manifold damaging organic pollutants [2] and surfactants [3]. These processes are also used in photodynamic therapy (PDT) and oxidation of organic compounds. Living organisms can also profit from the application of these processes where different sensitizers such as porphyrins can, by excitation, lead to in situ production of singlet oxygen and/or superoxide radical anion in the tissue of malignant tumors [4] as oxidative agents. Singlet oxygen can be employed for preparative reasons in organic synthetic chemistry, like for the synthesis of oxygenated derivatives of organic compounds. When a nonmetallated porphyrin is the photoactive species in the reaction, the longer-lived triplet is the key state in the photoinduced reactions. The porphyrin acts as a sensitizer and produces singlet oxygen via triplet quenching by the dissolved ground-state oxygen molecules. While free-base porphyrins are useful sensitizers for the production of singlet oxygen [5–9], metalloporphyrins are much more versatile photocatalysts due to their coordination ability, promoting a wider range of oxidation reactions as was first represented by Hennig et al. [10–12] (Figure 1).

Metalloporphyrins can be applied in autooxidation, hydroxylation, or direct oxygen transfer giving epoxides [10, 11]. Cationic Mn(III) porphyrins are attested to be effectual catalysts for oxygenation of α-pinene (Scheme 1).

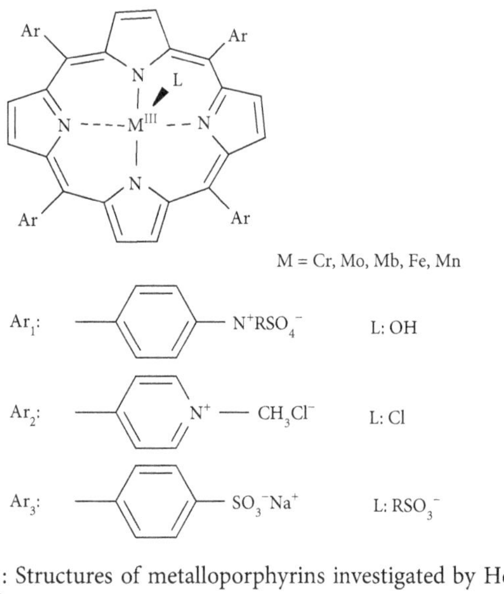

M = Cr, Mo, Mb, Fe, Mn

Ar$_1$: ⟨—⟩—N$^+$RSO$_4^-$ L: OH

Ar$_2$: ⟨—⟩—N$^+$——CH$_3$Cl$^-$ L: Cl

Ar$_3$: ⟨—⟩—SO$_3^-$Na$^+$ L: RSO$_3^-$

FIGURE 1: Structures of metalloporphyrins investigated by Hennig et al. [11].

The selective epoxidation giving compound **2** was detected in aqueous systems at an apropos low substrate/catalyst ratio (S/C = 500), until in aprotic organic solvents, such as benzene or toluene, allylic hydroxylation (**3**–**5**) and keto products (**6,7**) were formed [12]. Using various metalloporphyrins in acetonitrile, photocatalytic epoxidation of cyclooctene was also achieved [13]. Photocatalytic oxygenation of cycloalkenes [12–15] and other unsaturated heteroaromatics [5] was carried out by the application of both metallated and free-base porphyrins [5, 12–15], and it was confirmed that Fe- and Mn-porphyrin complexes give the most effective results as photocatalysts. Resemblances and differences in the photocatalytic oxygenation runways may shed light on the mechanisms of the diverse oxygenation processes, giving an indication for a convenient choice of a catalyst and efficient conversion with high selectivity. The ligand charge, affecting its Lewis basicity, may cause the catalytic activity of the complex through the metal center.

By photocatalytic oxygenation of various alkenes, with dioxygen and (5,10,15,20-tetraarylporphyrinato)iron(III) complexes, allylic oxygenation products and/or epoxides are obtained. The composition of the product mixture is influenced by the nature of the substrate and by the concentrations, but the axial ligands also play a role. Alkenes that have a strained double bond preferentially give epoxides, and allylic oxygenation is observed when unstrained alkenes are used. The proposed reaction mechanisms [16] give the oxoiron(IV) porphyrinate ((P)Fe^{IV}=O) as the catalytically active species. The selectivity of this species is related to the oxygenation of α-pinene with microsomal cytochromes P-450 and P-420 obtained from the yeast strain *Torulopsis apicola* [16]. Oxygenation products observed in both cases give evidence for the occurrence of an oxoiron(IV) heme species in microsomal cytochrome P-450-mediated reactions. The enantio-, regio-, and chemoselectivities of the photooxygenation with the iron(III) porphyrins and molecular oxygen are explained by the abstraction of the allylic hydrogen atom followed by catalyzed autoxidation and direct oxygen-

transfer reactions [16]. Oxoiron(IV) porphyrinate exhibits a broad spectrum of oxygenation reaction pathways as does the microsomal cytochrome P-450. It can be presumed that the (P)Fe^{IV}=O would be an attractive candidate for an alternative and/or a competing analogous iron heme complex in cytochrome P-450-mediated oxygenation reactions.

Besides the water solubility of the metalloporphyrins, their photostability is of great importance and in that way manganese porphyrins are much more stable than the analogous iron complexes [11]. Taking this enhanced photostability of the porphyrin complexes into consideration, the manganese(III) porphyrinates were precisely the ones used in the more recent investigations.

According to previous studies on the oxygenation of cycloalkenes, the mechanism of the photooxygenation reaction is quite complicated, involving at least 3–4 elementary steps [5, 6, 9]. In some of the earlier studies [12, 17–30], as well as in more recent ones [13, 15], it is explained that when Mn(III) porphyrins act as photocatalysts, (P)Mn^{IV}=O and (P)Mn^V=O intermediates play a key function in all of the in situ-produced reactive species in the oxygenation of cycloalkenes (Scheme 2).

The production of (P)Mn^V=O in the primary photoreactions was detected in acetonitrile. (P)Mn^{IV}=O was produced by photoinduced homolysis of the O-Cl or O-N bond of axially coordinated chlorate or nitrate, respectively, in acetonitrile [13, 15], or via a ligand-to-metal charge-transfer process (also a photoinduced homolysis but that of the metal-ligand bond) with chloride or hydroxide axial ligands in aqueous systems [11, 12]. In the latter case, the Mn(II) species was formed in the primary photochemical step (equation 1) followed by the coordination of oxygen (equation 2).

$$(P)Mn^{III}OH + h\nu \rightarrow (P)Mn^{II} + \cdot OH \qquad (1)$$

$$2(P)Mn^{II} + O_2 \rightarrow 2(P)Mn^{IV} = O \qquad (2)$$

Equation 2 is an overall reaction comprising of several steps. When experiments are run in water-acetone as solvent mixtures, hydroxide or water is axially coordinated around the Mn(III) center. Hydroxyl radicals formed in the primary photochemical step most probably react with the organic solvent. The Mn(IV) complexes are readily disproportionate and give highly reactive manganese(V)-oxo species (equation 3) [13, 15].

$$2(P)Mn^{IV} = O + H^+ \rightleftharpoons (P)Mn^V = O + (P)Mn^{III}OH \qquad (3)$$

Disproportionation is much faster than synproportionation in the equilibrium system. A polar solvent promotes the disproportionation process, and in this case, the process comes over with nearly a diffusion-controlled rate constant [31]. The rate constants for epoxidation of olefins are several orders of magnitude higher for Mn(V)-oxo porphyrins than for the convenient Mn(IV) species. It is considered that the (P)Mn^V=O species is the principal oxidant in the photocatalytic oxygenations in the investigated systems.

Starting with the idea that a polycyclic structure can be easily modified and functionalized by the utilization of a

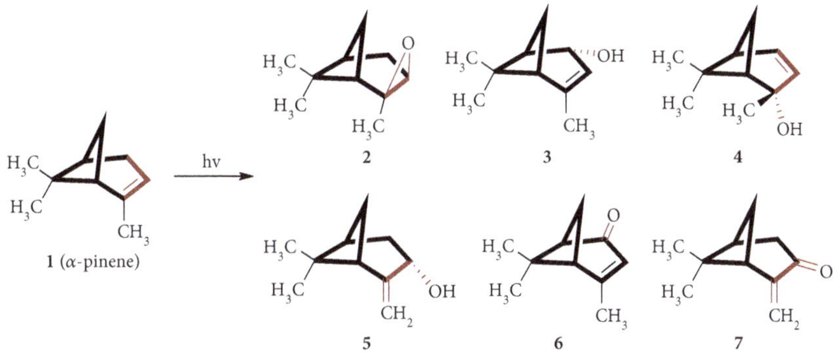

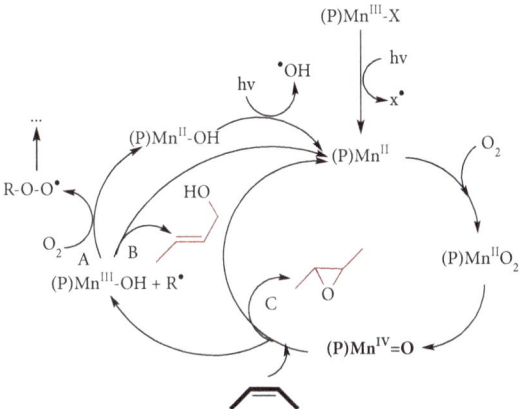

Scheme 1: Product distribution for photocatalytic oxygenation of α-pinene (1) [11, 17].

Scheme 2: Photocatalytic oxygenation of alkenes in the presence of metal porphyrinates [11] (A—autooxidation; B—oxygen rebound mechanism; and C—direct oxygen transfer).

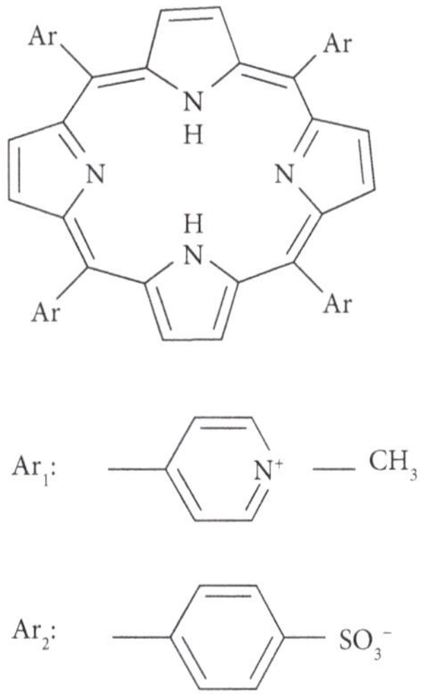

Figure 2: Structures of cationic porphyrin (Ar₁) and anionic porphyrin (Ar₂).

photocatalytic approach, Kikaš et al. and Vuk et al. have made significant progress in this area. They have been proliferative in the field of photocatalytic oxygenation using water-soluble manganese porphyrins applied to bicycloalkenes [32–35]. From their viewpoint, it was very important to consider the structure of the natural terpene, α-pinene (Scheme 1), as a compound having the bicyclo(3.1.1)hexene structure, very similar to the structure of photoproducts obtained by a cycloaddition reaction in very good yields. Recently, these yields have even been improved by the utilization of the flow-photochemical methodology [36]. The manganese(III) complexes of the cationic 5,10,15,20-tetrakis(1-methyl-4-pyridinium)porphyrin (Mn(III)TMPyP⁵⁺) and the anionic 5,10,15,20-tetrakis(4-sulfonatophenyl)porphyrin (Mn(III)TSPP³⁻) along with the anionic free base were used in the experiments (Figure 2).

In the first published paper by these authors, an investigation was performed on the benzobicyclo(3.2.1)octadiene system 8 where the oxygenation runways of this furobicyclic skeleton in a thermal reaction involving mCPBA and photocatalytic processes mediated by nonmetallated and Mn(III) porphyrins are compared. In the thermal reaction of 8 and 9 using mCPBA (Scheme 3), the enedione 11 is obtained via the rearrangement of intermediate epoxide 10 produced primarily from 8.

There are two epoxidation possibilities on the furan derivative 8, but the authors corroborate with the literature and propose that the initial epoxidation comes over at the substituted cyclohexyl side of the furan ring as signified in the left sequence in Scheme 3. This is judicious on the basis of the idea that epoxidation should occur at the double bond holding the more electron-donating functional group. Further oxygenation to 12 does not take place. Compounds 9a-c were also subjected to additional thermal transformations using mCPBA, producing all trans-14a-c (formed via intermediates 13a-c). Further oxygenation to the product 15 was not observed.

Light-initiated oxygenation of 8 was carried out using various porphyrins as photocatalysts (Scheme 4).

It was concluded that the longer-lived triplet state plays a role in the photoinduced reactions if a free-base porphyrin is the photoactive species used. This catalyst operates as a sensitizer thus producing singlet oxygen via triplet quenching. The product is the hydroxybutenolide 20 (Scheme 4), which is not detected in any other oxygenation process. This obviously demonstrates that this is the only case when singlet oxygen is the oxidative agent. Thus, due to previous studies on the oxygenation of cycloalkenes, the mechanism

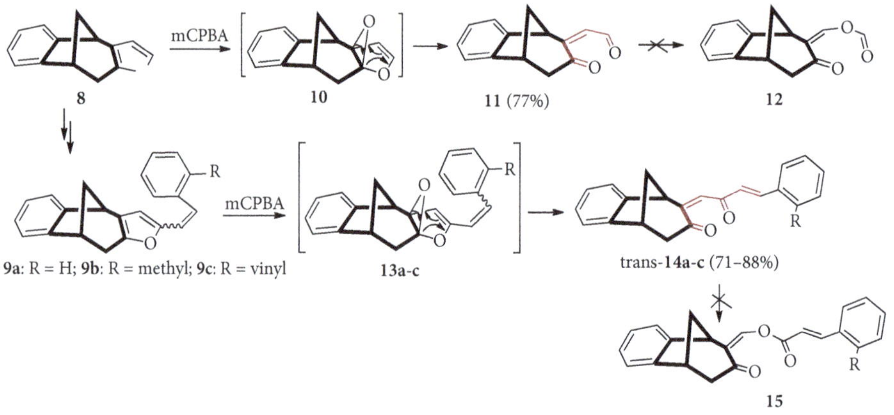

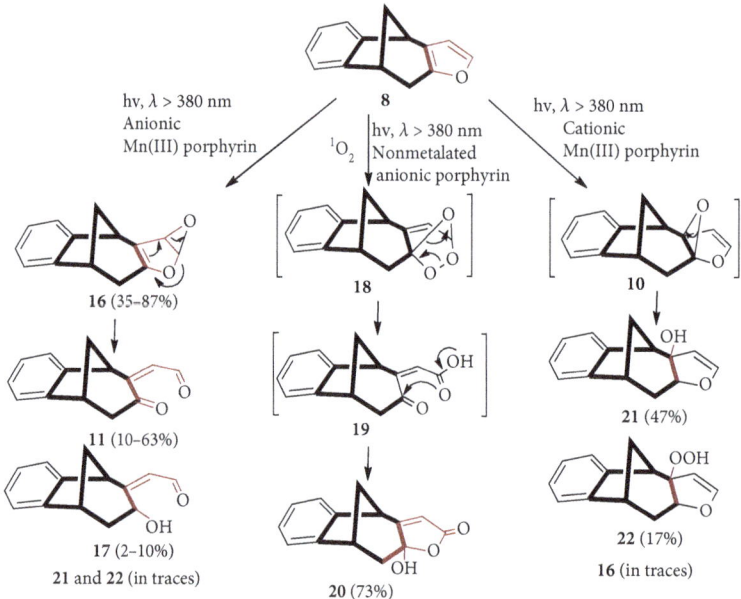

SCHEME 3: Proposed reaction pathways for the thermal reactions of **8** and **9a-c**.

SCHEME 4: Proposed reaction runways for the photocatalytic oxygenation of **8** (pH = 7, air saturation).

of this reaction is sufficiently complicated [5, 6, 9]. In the case when anionic Mn(III)TSPP^{3-} was used, the epoxy-derivative **16** and furan ring-opened **11** and **17** products were observed (Scheme 4). When the Mn(III)TMPyP^{5+} porphyrin is used, hydroxy **21** and hydroperoxy **22** derivatives are the major products. The change in the sign of the charge of the ligand alters the assignment of the products. The lower Lewis basicity of the porphyrin ligand improves oxygenation on C=C bonds, and this was verified by flash photolysis experiments with Mn(III) porphyrins holding substituents of different electron demands [15, 31]. When the authors modified the experimental conditions (pH or oxygen concentration), there were variations in the ratio of the products, but this did not affect the establishment of further novel species. Enhancement of pH to 10 magnified the quantity of the epoxide **16**. They explain that higher pH disrupts further oxidation of this derivative, and this can be linked to the function protons play in the disproportionation producing an Mn(V) species. Bubbling oxygen instead of air

significantly enhanced the amounts of the compounds **16** and **17**. Studies [12, 13, 15] have pointed out that in the case of Mn(III) porphyrins as photocatalysts, (P)Mn^{IV}=O and (P)Mn^{V}=O intermediates play a key function in the oxygenation of cycloalkenes. The production of (P)Mn^{V}=O was observed in acetonitrile as a consequence of heterolytic cleavage of the O-Cl bond in the axially coordinated perchlorate counterion [15, 31]. (P)Mn^{IV}=O was produced by photoinduced homolysis of the O-Cl or O-N bond of axially coordinated chlorate or nitrate, respectively, in acetonitrile [15, 31] or via a ligand-to-metal charge-transfer process with chloride or hydroxide axial ligands in aqueous systems [11, 12]. The mechanism of this complex set of reactions is as described previously by Hennig et al. and already presented earlier in the minireview. As indicated in Scheme 4, the positively charged porphyrin ligand advances the electrophilic attack to the inside double bond of the furan ring. The anionic catalyst favors the outside double bond. Likened to the outside C=C bond, the accessibility of

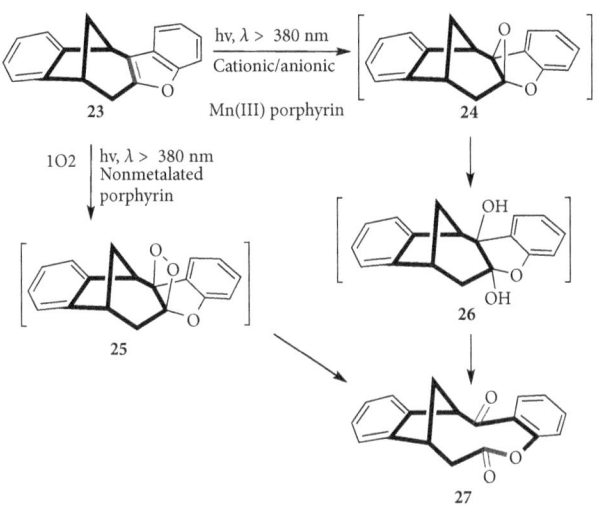

SCHEME 5: Proposed reaction pathways for photocatalytic oxygenation of 23.

the inside double bond for the oxygen atom coordinated to the lumbering macrocyclic skeleton is sterically interfered with its bicyclic environment. From an electronic aspect, this bond is more favored for an electrophilic attack as an outcome of the electron-donating consequence of the nearby hydrocarbon (bicycloalkyl) parts of the molecule. Because of the lower Lewis basicity of the cationic complex, the corresponding Mn(V)-oxo intermediate has a much more electrophilic performance than the anionic one. For the lesser electrophilic anionic complex, steric disturbance is the predominant result, promoting the attack at the more accessible outside bond. To study the effect of the increased steric hindrance of the oxidative attack at the outer bond of the furan ring, photocatalytic oxygenation experiments using both the anionic and cationic manganese(III) porphyrins (Mn(III)TSPP^{3-} and Mn(III)TMPyP^{5+}) and their corresponding free bases (H$_2$TSPP^{4-} and H$_2$TMPyP^{4+}) were carried out with the annulated derivative 23 (Scheme 5) [33].

The major product was the same in every case, no matter which catalyst was used under diverse conditions [33]. Structure determination and characterization unambiguously showed that 10-membered ketolactone 27 was obtained. This result propounds that, beside a strong steric disturbance, a considerable electronic consequence was enforced by the annulation of a benzene ring to the outside of the furan ring.

In their later paper [34], the same authors investigate more polycyclic substrates containing oxygen and sulfur in their structures. When the structure that is studied has a (2,3-b-furo) moiety incorporated into the skeleton, the results with both cationic and anionic Mn(III) porphyrin catalysts differ (Scheme 6) from the previously studied (3,2-b)furo-octadienes [32].

While in the case of 8, using the anionic Mn(III) porphyrin, epoxide and furan ring-opened derivatives were the main products, and photocatalytic oxygenation of 28 led to the formation of hydroxy 29 and hydroperoxy 30 derivatives. In the presence of the cationic Mn(III) porphyrin, only one product was formed (Scheme 6), the hydroxybutenolide derivative 31,

which is similar to that observed in the photocatalytic oxygenation of 8 (Scheme 4, formed when using the free-base catalyst where the photochemically generated singlet oxygen was the oxidative agent) [32]. The results in this study suggest that in the compound 28 the inner double bond is more preferred by the attack of the cationic Mn(III) porphyrin than by the anionic one. The replacement of oxygen by sulfur resulted in changes in photocatalytic reactivity (Schemes 7 and 8).

The reactivity of these thienyl substrates is much lower than that of the corresponding furan ones, and this is attributed to the much higher aromaticity of the thiophene ring. The products formed from 32 suggest that the attack by hydroxyl radicals (equation 1) play a more determining role in this system than the Mn(V)=O species do, and this is in accordance with the catalyst-independent yields.

Continuing this study of photocatalytic oxygenations of various bicyclic organic compounds, derivatives with an isolated/free double bond were investigated [35]. These compounds also contained a phenyl group (unsubstituted or substituted) close to the free double bond, which significantly affected the mechanism of manganese(III) porphyrin-based photocatalytic oxygenation and the products gained (Scheme 9).

A considerable π-stacking interaction between that phenyl ring and the porphyrin catalyst promoted the functionalization of the carbon atom resulting in the formation of the suitable hydroperoxy derivatives of 38a and b. No effect of the porphyrin charge was observed in these cases, and the main oxygenation reaction of the methoxy derivative 38c was efficient only with the cationic complex; this is probably due to its interaction with the electron-rich free double bond. These results further corroborate that both steric and electronic effects govern the mechanisms of the photocatalytic oxygenations of these compounds.

All the presented successful results confirm that the use of the photocatalytic activity of water-soluble Mn(III) porphyrins for the oxygenation of benzobicyclo(3.2.1)octadienes 8, 23, 28, 32, 36, and 38 was justified, especially as they possess a basic core very similar to those previously analyzed and naturally coming over cycloalkenes, which are bioactive and significant substances isolated from nature [37].

Recently, Co and Ni complexes have also been photophysically investigated by Horváth et al. as potential photocatalysts as well as for the functionalization of organic substrates by photocatalytic oxygenation in comparison to Mn(III) porphyrins [38–40]. The obtained results well demonstrated how the size of the metal center determines the structure and, thus, the photoinduced behavior of the porphyrin complexes, along with the substituents on the ligand. Co(III) porphyrin complexes showed similar photophysical characteristics as the depicted Mn(III) porphyrins, while Ni complexes display somewhat different photophysical behavior and function as special sensitizers, which immediately transmit their excitation energy to the electron donor, promoting the direct charge transfer toward the acceptor. All those results well demonstrate that both Co(III) and Ni(II) porphyrin complexes may be applicable for solar energy utilization in the visible range and probably as oxidative

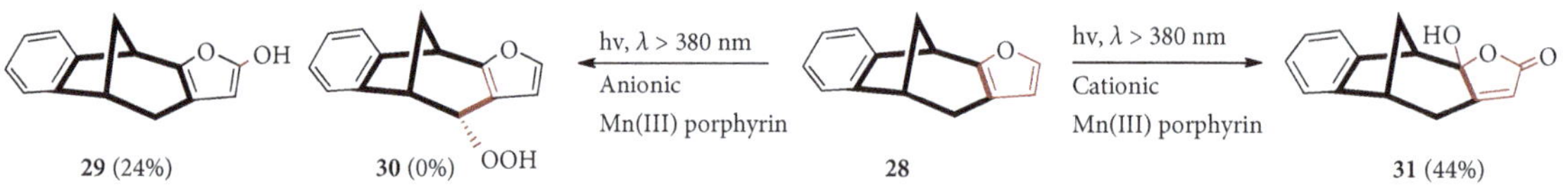

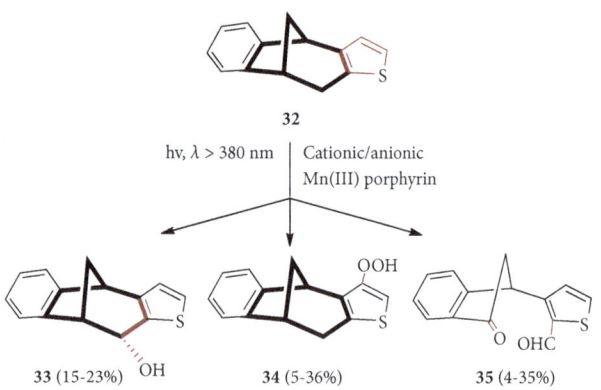

SCHEME 6: Proposed reaction pathways for the photocatalytic oxygenation reactions of **28** (pH = 7, oxygen saturation) [34].

SCHEME 7: Reaction pathway for the photocatalytic oxygenation reactions of **32** (pH = 7, oxygen saturation) [34].

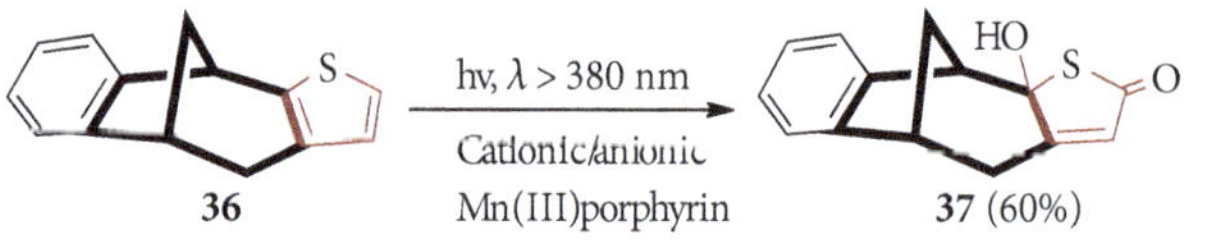

SCHEME 8: Photocatalytic oxygenation of **36** to photoproduct **37** (pH = 7, oxygen saturation) [34].

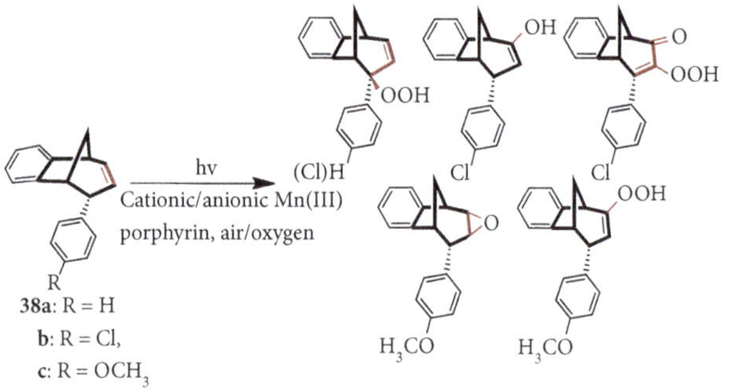

SCHEME 9: Photocatalytic oxygenation of **38** (pH = 7, air/oxygen saturation) [35].

reagents for photocatalytic oxygenation of the described unsubstituted photoproducts to give new functionalized polycycles very similar to the structures of some terpenes from nature.

2. Conclusions

It is shown that free-base and metallated porphyrins are extremely useful in photocatalysis. This minireview has focused on the utilization of these porphyrins for the functionalization of compounds that have a polycyclic skeleton in their structure. These kinds of compounds are notoriously taxing to obtain and difficult to further functionalize using conventional organic synthetic methods, so photocatalytic oxygenation is a good tool for these transformations. In these photocatalytic processes, novel polycyclic epoxides, enediones, ketones, alcohols, and/or hydroperoxides are formed, subordinate on the catalyst used. The application of anionic and cationic Mn(III) porphyrins under different reaction parameters resulted in different reaction pathways thus generating a vast number of photocatalytic products. As a future development in the field, Co and Ni complexes have also been recently photophysically investigated and confirmed as good potential photocatalysts for further functionalization of organic substrates.

Conflicts of Interest

The authors declare that they have no conflicts of interest.

References

[1] L. R. Milgrom, *The Colours of Life: An Introduction to the Chemistry of Porphyrins and Related Compounds*, Oxford University Press, Oxford, 1997.

[2] O. T. Woo, W. K. Chung, K. H. Wong, A. T. Chow, and P. K. Wong, "Photocatalytic oxidation of polycyclic aromatic hydrocarbons: intermediates identification and toxicity testing," *Journal of Hazardous Materials*, vol. 168, no. 2-3, pp. 1192–1199, 2009.

[3] E. Szabó-Bárdos, O. Markovics, O. Horváth, N. Törő, and G. Kiss, "Photocatalytic degradation of benzenesulfonate on colloidal titanium dioxide," *Water Research*, vol. 45, no. 4, pp. 1617–1628, 2011.

[4] S. Weimin, Z. Gen, D. Guifu, Z. Yunxiao, Z. Jin, and T. Jingchao, "Synthesis and in vitro PDT activity of miscellaneous porphyrins with amino acid and uracil," *Bioorganic & Medicinal Chemistry*, vol. 16, no. 10, pp. 5665–5671, 2008.

[5] K. Gollnick and A. Griesbeck, "Singlet oxygen photooxygenation of furans," *Tetrahedron*, vol. 41, no. 11, pp. 2057–2068, 1985.

[6] N. J. Turro, *Modern Molecular Photochemistry*, University Science Books, Sausalito, 1991.

[7] M. R. Iesce, F. Cermola, A. Guitto, R. Scarpati, and M. L. Graziano, "Carbonyl oxide chemistry. 4. Novel observations on the behavior of 1-methoxy-2,3,7-trioxabicyclo(2.2.1)hept-5-ene," *The Journal of Organic Chemistry*, vol. 60, no. 16, pp. 5324–5327, 1995.

[8] C. J. P. Monteiro, M. M. Pereira, M. E. Azenha et al., "A comparative study of water soluble 5,10,15,20-tetrakis(2,6-dichloro-3-sulfophenyl)porphyrin and its metal complexes as efficient sensitizers for photodegradation of phenols," *Photochemical & Photobiological Sciences*, vol. 4, no. 8, pp. 617–624, 2005.

[9] T. Montagnon, M. Tofi, and G. Vassilikogiannakis, "Using singlet oxygen to synthesize polyoxygenated natural products from furans," *Accounts of Chemical Research*, vol. 41, no. 8, pp. 1001–1011, 2008.

[10] L. Weber, R. Hommel, J. Behling, G. Haufe, and H. Hennig, "Photocatalytic oxygenation of hydrocarbons with (tetraarylporphyrinato)iron(III) complexes and molecular oxygen. Comparison with microsomal cytochrome P-450 mediated oxygenation reactions," *Journal of the American Chemical Society*, vol. 116, no. 6, pp. 2400–2408, 1994.

[11] H. Hennig, J. Behling, R. Meusinger, and L. Weber, "Photocatalytic oxygenation of selected cycloalkenes in aqueous solutions induced by water-soluble metal porphyrin complexes," *Chemische Berichte*, vol. 128, no. 3, pp. 229–234, 1995.

[12] H. Hennig, "Homogeneous photo catalysis by transition metal complexes," *Coordination Chemistry Reviews*, vol. 182, no. 1, pp. 101–123, 1999.

[13] M. Hajimohammadi, F. Bahadoran, S. S. H. Davarani, and N. Safari, "Selective photocatalytic epoxidation of cyclooctene by molecular oxygen in the presence of porphyrin sensitizers," *Reaction Kinetics, Mechanisms and Catalysis*, vol. 99, pp. 243–250, 2009.

[14] A. Maldotti, L. Andreotti, A. Molinari, G. Varani, G. Cerichelli, and M. Chiarini, "Photocatalytic properties of iron porphyrins revisited in aqueous micellar environment: oxygenation of alkenes and reductive degradation of carbon tetrachloride," *Green Chemistry*, vol. 3, no. 1, pp. 42–46, 2001.

[15] M. Newcomb, R. Zhang, Z. Pan et al., "Laser flash photolysis production of metal-oxo derivatives and direct kinetic studies of their oxidation reactions," *Catalysis Today*, vol. 117, no. 1-3, pp. 98–104, 2006.

[16] L. Weber and G. Haufe, "Qualitative und quantitative analyse von monoterpenoiden produktgemischen mittels13C-NMR-spektroskopie," *Journal für praktische Chemie*, vol. 330, no. 2, pp. 319–322, 1988.

[17] H. Hennig, D. Rehorek, R. Stich, and L. Weber, "Photocatalysis induced by light-sensitive coordination compounds," *Pure and Applied Chemistry*, vol. 62, no. 8, pp. 1489–1494, 1990.

[18] H. Hennig, L. Weber, R. Stich, M. Grosche, and D. Rehorek, *Homogeneous Photo Complex Catalysis and Organic Synthesis*, vol. 6, CRC Press, Boca Raton, 1992.

[19] H. Hennig, L. Weber, and D. Rehorek, *Photosensitive Metal–Organic Systems: Mechanistic Principles and Applications*, C. Kutal and N. Serpone, Eds., American Chemical Society, 1993.

[20] H. Hennig, R. Billing, and H. Knoll, *Photosensitization and Photocatalysis Using Inorganic and Organometallic Compounds*, K. Kalyanasundaram and M. Gratzel, Eds., Kluwer Academic Press, Dordrecht, 1993.

[21] H. Hennig and R. Billing, "Advantages and disadvantages of photocatalysis induced by light-sensitive coordination compounds," *Coordination Chemistry Reviews*, vol. 125, no. 1-2, pp. 89–100, 1993.

[22] H. Hennig, *Contributions to Development of Coordination Chemistry*, G. Ondrejovic and A. Sirota, Eds., Slovak Technical University Press, Bratislava, 1993.

[23] H. Hennig, S. Knoblauch, and D. Scholz, "Photons, attractive reagents in coordination chemistry: the photocatalytic activation of dioxygen by transition-metal porphyrinates," *ChemInform*, vol. 28, no. 43, 1997.

[24] H. Hennig and D. Luppa, "Photokatalytische Aktivierung von molekularem Sauerstoff mittels Eisen(III)porphyrin-Komplexen," *Journal für praktische Chemie*, vol. 341, pp. 757–767, 1999.

[25] H. Hennig, *Transition Metals for Fine Chemicals and Organic Synthesis*, C. Bolm and M. Beller, Eds., VCH, Weinheim, 1998.

[26] H. Hennig, K. Hofbauer, K. Handke, and R. Stich, "Ungewöhnliche reaktionswege bei der photolyse von diazido(phosphan)-nickel(II)-komplexen: bildung von nickel(o)-komplexen über nitrenintermediate," *Angewandte Chemie*, vol. 109, no. 4, pp. 373–375, 1997.

[27] H. Hennig, R. Billing, and K. Ritter, "Sensibilisierte photolyse von bis(dimethylglyoximato)cobalt(III)-komplexen mit axial koordiniertem azid bzw. Thiophenolat als photochemischen opferliganden," *Journal für Praktische Chemie*, vol. 339, no. 1, pp. 272–276, 1997.

[28] H. Hennig, K. Ritter, A. K. Chibisov et al., "Comparative time-resolved IR and UV spectroscopic study of monophosphine and diphosphine platinum(II) azido complexes," *Inorganica Chimica Acta*, vol. 271, no. 1-2, pp. 160–166, 1998.

[29] H. Henning, K. Hofbauer, K. Handke, and R. Stich, "Unusual reaction pathways in the photolysis of diazido(phosphane)-nickel(II) complexes: nitrenes as intermediates in the formation of nickel(0) complexes," *Angewandte Chemie International Edition in English*, vol. 36, no. 4, pp. 408–410, 1997.

[30] L. Weber, G. Haufe, D. Rehorek, and H. Hennig, "Photocatalytic oxygenation of strained alicyclic alkenes with μ-oxo-bis(tetraphenylporphyrinatoiron(III)) and molecular oxygen," *Journal of the Chemical Society, Chemical Communications*, vol. 502, no. 7, pp. 502–503, 1991.

[31] R. Zhang, J. H. Horner, and M. Newcomb, "Laser flash photolysis generation and kinetic studies of porphyrin–manganese–oxo intermediates. Rate constants for oxidations effected by porphyrin–MnV–oxo species and apparent disproportionation equilibrium constants for porphyrin–MnIV–oxo species," *Journal of the American Chemical Society*, vol. 127, no. 18, pp. 6573–6582, 2005.

[32] I. Kikaš, O. Horváth, and I. Škorić, "Functionalization of the benzobicyclo(3.2.1)octadiene skeleton via photocatalytic and thermal oxygenation of a furan derivative," *Tetrahedron Letters*, vol. 52, no. 47, pp. 6255–6259, 2011.

[33] I. Kikaš, O. Horváth, and I. Škorić, "Functionalization of the benzobicyclo(3.2.1)octadiene skeleton via photocatalytic oxygenation of furan and benzofuran derivatives," *Journal of Molecular Structure*, vol. 1034, pp. 62–68, 2013.

[34] D. Vuk, I. Kikaš, K. Molčanov, O. Horváth, and I. Škorić, "Functionalization of the benzobicyclo(3.2.1)octadiene skeleton via photocatalytic oxygenation of thiophene and furan derivatives: the impact of the type and position of the heteroatom," *Journal of Molecular Structure*, vol. 1063, pp. 83–91, 2014.

[35] D. Vuk, O. Horváth, Ž. Marinić, and I. Škorić, "Functionalization of the benzobicyclo(3.2.1) octadiene skeleton possessing one isolated double bond via photocatalytic oxygenation," *Journal of Molecular Structure*, vol. 1107, pp. 70–76, 2016.

[36] A. Ratković, Ž. Marinić, and I. Škorić, "Flow-photochemical synthesis of the functionalized benzobicyclo(3.2.1)octadiene skeleton," *Journal of Molecular Structure*, vol. 1168, pp. 165–174, 2018.

[37] E. D. Coy, L. E. Cuca, and M. Sefkow, "Macrophyllin-type bicyclo(3.2.1)octanoid neolignans from the leaves of

Pleurothyrium cinereum," *Journal of Natural Products*, vol. 72, no. 7, pp. 1245–1248, 2009.

[38] O. Horváth, Z. Valicsek, M. A. Fodor et al., "Visible light-driven photophysics and photochemistry of water-soluble metalloporphyrins," *Coordination Chemistry Reviews*, vol. 325, pp. 59–66, 2016.

[39] M. M. Major, O. Horváth, M. A. Fodor et al., "Photophysical and photocatalytic behavior of nickel(II) 5,10,15,20-tetrakis(1-methylpyridinium-4-yl)porphyrin," *Inorganic Chemistry Communications*, vol. 73, pp. 1–3, 2016.

[40] M. A. Fodor, O. Horváth, L. Fodor, G. Grampp, and A. Wankmüller, "Photophysical and photocatalytic behavior of cobalt(III) 5,10,15,20-tetrakis(1-methylpyridinium-4-yl)porphyrin," *Inorganic Chemistry Communications*, vol. 50, pp. 110–112, 2014.

2

C-, N-, S-, and F-Doped Anatase TiO$_2$ (101) with Oxygen Vacancies: Photocatalysts Active in the Visible Region

Julio César González-Torres,[1] Enrique Poulain,[1] Víctor Domínguez-Soria,[2] Raúl García-Cruz (ID),[1] and Oscar Olvera-Neria (ID)[1]

[1]*Área de Física Atómica Molecular Aplicada (FAMA), CBI, Universidad Autónoma Metropolitana-Azcapotzalco, Av. San Pablo 180, Col. Reynosa Tamaulipas, 02200 Ciudad de México, Mexico*
[2]*Área de Química Aplicada, CBI, Universidad Autónoma Metropolitana-Azcapotzalco, Av. San Pablo 180, Col. Reynosa Tamaulipas, 02200 Ciudad de México, Mexico*

Correspondence should be addressed to Oscar Olvera-Neria; oon@correo.azc.uam.mx

Academic Editor: Xuxu Wang

Anatase TiO$_2$ presents a large bandgap of 3.2 eV, which inhibits the use of visible light radiation ($\lambda > 387$ nm) for generating charge carriers. We studied the activation of TiO$_2$ (101) anatase with visible light by doping with C, N, S, and F atoms. For this purpose, density functional theory and the Hubbard U approach are used. We identify two ways for activating the TiO$_2$ with visible light. The first mechanism is broadening the valence or conduction band; for example, in the S-doped TiO$_2$ (101) system, the valence band is broadened. A similar process can occur in the conduction band when the undercoordinated Ti atoms are exposed on the TiO$_2$ (101) surface. The second mechanism, and more efficient for activating the anatase, is to generate localized states in the gap: N-doping creates localized empty states in the bandgap. For C-doping, the surface TiO$_2$ (101) presents a "cleaner" gap than the bulk TiO$_2$, resulting in fewer recombination centers. The dopant valence electrons determine the number and position of the localized states in the bandgap. The formation of charge carriers with visible light is highly favored by the oxygen vacancies on TiO$_2$ (101). The catalytic activity of C-doping using visible radiation can be explained by its high absorption intensity generated by oxygen vacancies on the surface. The intensity of the visible absorption spectrum of doped TiO$_2$ (101) follows the order: C > N > F > S dopant.

1. Introduction

In the late '70s, photocatalysis took a turn when researchers discovered the TiO$_2$ ability to degrade stable organic compounds as they were studying the water photoelectrolysis [1–3]. Since then, there has been a great interest to improve the degradation efficiency of organic pollutants using TiO$_2$ [4–7].

Anatase TiO$_2$ has an energy bandgap of 3.2 eV, that is, UV radiation is mandatory to promote electrons from the valence band (VB) to the conduction band (CB). Until now, the primary challenge has been to generate charge carriers using visible light rather than UV radiation. Different methodologies have been used to modify the TiO$_2$ absorption properties, such as doping with transition metals [8–14] or main group elements [15–24], and synthesize composites like ZrO$_2$/TiO$_2$, reducing the recombination of charge carriers by aligning the energy gaps [25, 26]. Charge carriers act as oxidation and reduction centers, which enable the creation of reactive species such as hydroxyl radicals, peroxides, or acid compounds. These species promote the degradation of pollutants [27–29].

Atanelov et al. [15] theoretically studied TiO$_2$ doped with C and N atoms at two different concentrations. They found that dopants decrease the bandgap, which enhances the TiO$_2$ photocatalytic activity. It is crucial that states created by doping process localize close to the valence band maximum (VBM) or the conduction band minimum (CBM), since states spread along the bandgap may act as a recombination center [15, 30–32]. Sakthivel and Kisch [33] found that

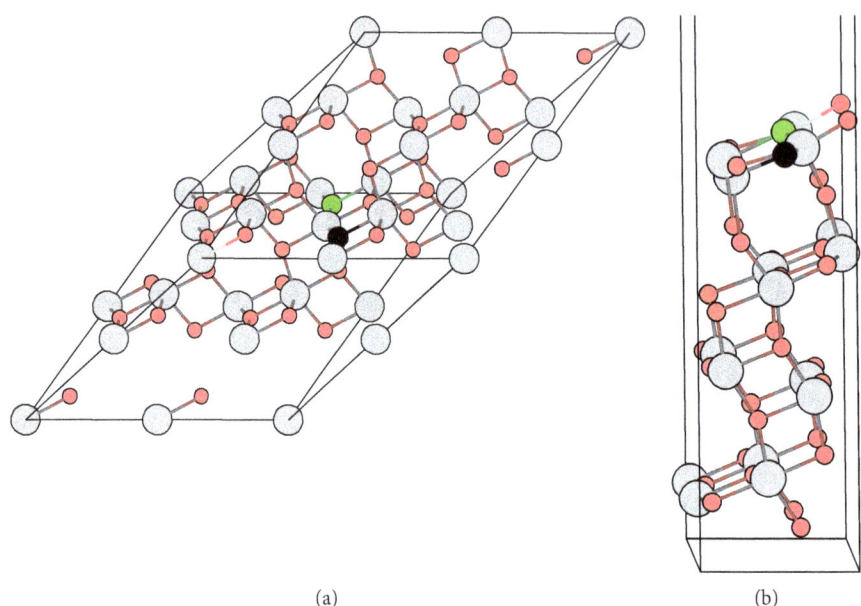

(a)	(b)

FIGURE 1: Doping sites on anatase TiO$_2$. (a) The $3 \times 3 \times 3$ bulk supercell. (b) The $2 \times 1 \times 1$ surface in the (101) direction with four layers. Both models constituted by 48 atoms. One three-coordinated O atom was replaced by C, N, S, or F atom. Red balls are O atoms; gray ones are Ti. Black balls indicate the substitution site, and green balls show the oxygen vacancy.

C-doping TiO$_2$ is five times more efficient than N-doping in the degradation of 4-chlorophenol when artificial light ($\lambda \geq 455$ nm) is used. Tian and Liu [34] showed that high S concentrations as a TiO$_2$ dopant induced a light absorption redshift. Whereas F-doped TiO$_2$ does not reduce the bandgap [16, 35], Yu et al. [36] obtained a gap of 2.90 eV for an anatase and rutile mixture doped with F$^-$ ions, where the interface may reduce the recombination of photogenerated electrons and holes. Czoska et al. [16] reported that Ti^{3+} states separate from the CBM when the local density approximation (LDA) + U parameter method is used, making this procedure suitable to describe the F-doping on the anatase TiO$_2$.

TiO$_2$ usually presents oxygen vacancies; these defects can be controlled experimentally based on nonstoichiometric conditions of synthesis (Ti excess or oxygen deficiency) or by heating the sample in a vacuum at 900 K [37–39]. Zhang et al. [40] showed that there is a synergistic effect of N dopant and oxygen vacancy in TiO$_2$ contributing to the significant enhancement of the visible light photoactivity. The synergy between species N-Ti and the oxygen vacancy also occurs in heterojunctions such as NiO/N-TiO$_2$ catalysts [41], but in general, doped TiO$_2$ and oxygen vacancies have been studied separately. For other dopants, it is not understood how the simultaneous interaction between dopants and vacancies affects the optical response of the TiO$_2$ (101) surface in the visible region. This knowledge will help optimize doped TiO$_2$ for the degradation of organic compounds.

This work aims to make a systematic study on C-, N-, S-, and F-doped TiO$_2$ (101) to obtain the most suitable element to reduce the bandgap or to generate localized states in the gap. Furthermore, the generalized gradient approximation (GGA) and the Hubbard U method are used to explain the intraband states' nature. Notably, this

article focuses on the simultaneous interaction of oxygen vacancies and main group dopants to analyze and evaluate the anatase surface response to the visible radiation.

2. Computational Details

The electronic structure of the anatase TiO$_2$ (101) surface and bulk doped with the C, N, S, and F elements was studied using the VASP5.4.1 program [42, 43] that implements the density functional theory (DFT) [44, 45]. The Perdew-Burke-Ernzerhof (PBE) functional [46, 47] was used to take into account the exchange-correlation energy within the GGA framework [48–51].

Standard DFT calculations underestimate the bandgap because of its difficulty in representing empty states [52]. For recovering of the electron correlation effects—at least in pairs—and the Ti 3d orbitals additional repulsion, we use the Hubbard U parameter [53] as suggested by Morgade and Cabeza for TiO$_2$ [54]. A value of $U = 5.5$ eV was considered only for Ti 3d electrons following the recommendation by Calzado et al. [55].

The projector augmented wave (PAW) approach is used for describing the ion-electron interactions [56, 57]. The plane-wave basis for all elements with an energy cutoff of 400 eV was employed throughout calculations. We used a k-point mesh of $3 \times 3 \times 3$ for the bulk systems and $5 \times 5 \times 1$ for the surface. The total energy convergence criterion was 1×10^{-4} eV.

A $3 \times 3 \times 3$ supercell was built for modeling the bulk. The (101) plane is the most stable for anatase TiO$_2$ [58, 59]. Then, the surface was represented by a $2 \times 1 \times 1$ supercell considering four layers in the (101) direction; both the bulk and surface models contain 48 atoms (see Figure 1). A vacuum

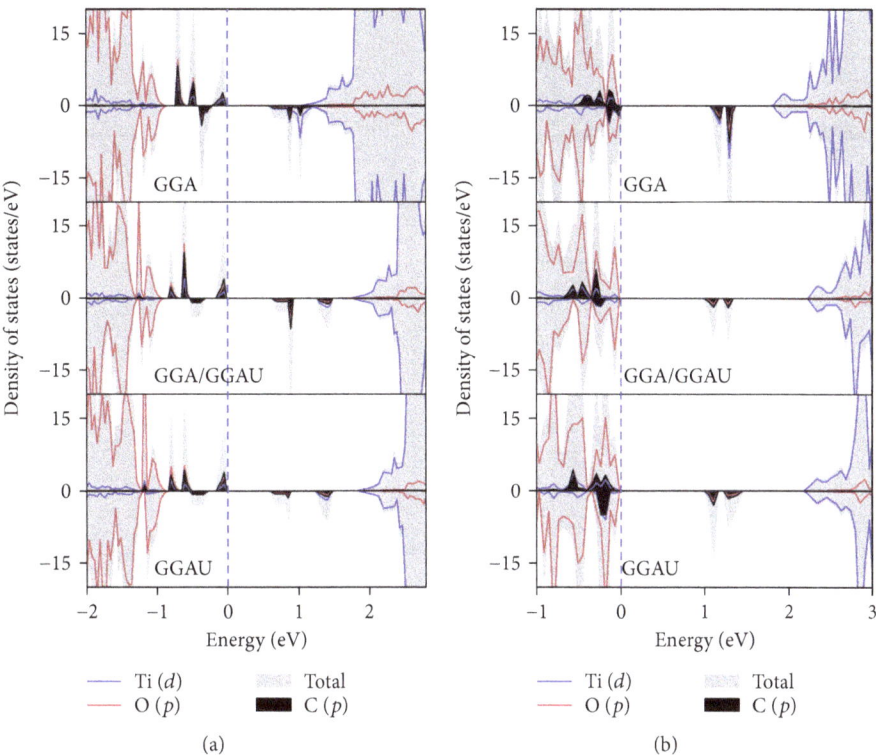

FIGURE 2: The total and partial density of states of the three relaxation methodologies for C-doped anatase TiO$_2$. (a) The bulk supercell. (b) The TiO$_2$ (101) surface. The Fermi level corresponds to the zero energy.

of 16 Å was included to ensure that the two adjacent surfaces do not interact.

For the bulk and surface models, the three-coordinated oxygen atom was substituted, yielding a doping concentration of 2.08% mole. On the surface, the second most exposed oxygen atom was replaced by the dopants (see Figure 1(b)). The bottom layer of the slab was constrained to simulate the bulk environment.

Three methodologies were used to study the electronic structure of doped TiO$_2$. The first, an electronic and geometry relaxation using the PBE functional were performed (GGA); for the second approach, the relaxed geometry obtained by the first method was used to accomplish a single point calculation with the U parameter (GGA/GGAU). Finally, both geometry and electronic relaxation were determined with the U parameter (GGAU). For all calculations, spin polarization was considered.

For studying the simultaneous interaction between impurities and oxygen vacancies, we detach the nearest surface oxygen atom to the dopant on the doped TiO$_2$ (101); in the pristine TiO$_2$ surface, we removed the same oxygen for comparison (see Figure 1(b)). The absorption spectra were obtained with the VASPKIT program [60], using the real and imaginary parts of frequency-dependent complex dielectric function [61], after the electronic ground state has been determined. The systems with oxygen vacancies were studied with the GGAU methodology, which represents the most reliable method of those tested in this work.

3. Results and Discussion

The optimized parameters for the bulk TiO$_2$ are a = 3.8248 Å, c = 9.6909 Å, and z = 0.2068, which agree with experimental values obtained by Horn et al. [62]. For the GGA methodology in the bulk anatase, we obtain an energy bandgap (E_g) of 2.19 eV. For the GGA/GGAU and GGAU methodologies, the E_g broadens to 2.81 and 2.76 eV, respectively, and there are no significant changes in the electronic structure for the bulk system (see Figure S1(a) in the Supplementary Materials). The main effect of the U parameter is to move the Ti $3d$ states toward more positive energies, which increases the bandgap because the CBM is displaced. From now on, every time the three bandgaps are mentioned, we are referring to the values obtained with the GGA, GGA/GGAU, and GGAU approaches.

For the TiO$_2$ (101) surface, the E_g narrows to 1.85, 2.31, and 2.25 eV for each used methodology, respectively. We attribute this narrowness to the bond unsaturation at the material surface, corresponding to undercoordinated Ti species located at the beginning of the CB, which promotes a bandgap reduction. The density of states (DOS) for the surface depicts characteristic localized empty states below the CB, Figure S1(b). These states have a well-defined shape and are distinct from the occupied states created by the Ti^{3+} species.

3.1. C-Doped TiO$_2$. For the bulk anatase, C introduces occupied states above the original VB, so the Fermi level is located

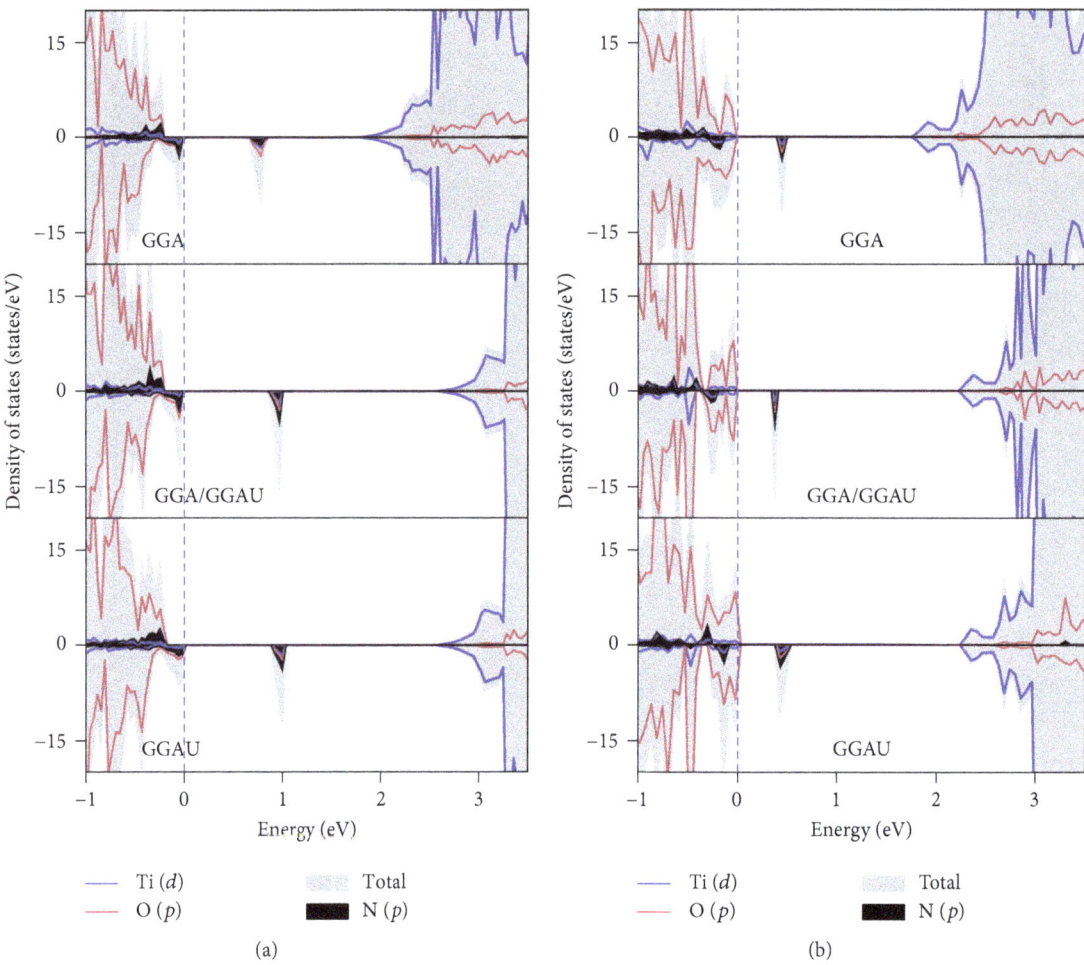

FIGURE 3: The total and partial density of states of the three relaxation methodologies for N-doped anatase TiO$_2$. (a) The bulk supercell. (b) The TiO$_2$ (101) surface.

at the end of these states. Furthermore, C-doping generates unoccupied states that lie in the bandgap region. The calculated bandgaps are 2.44, 2.66, and 2.66 eV (without considering the states introduced by dopants) for the three methodologies (see Figure 2(a) and S2(a)).

The O atoms in the anatase TiO$_2$ have an oxidation state of −2, which implies that O accepts two electrons in its 2p empty orbitals. Carbon has 2 electrons less than O, so its 2p orbitals are not going to be full by receiving 2 electrons like O atoms. This process will leave two unoccupied C 2p orbitals that are not in the same energy level that Ti 3d states therefore will appear below the CB.

When the U correlation is considered, the empty states introduced by C-doping are separated from the Ti 3d orbitals, and the extra repulsion pushes Ti 3d orbitals towards the CBM. This behavior also occurs on the TiO$_2$ (101) surface; the calculated bandgaps are 1.87, 2.36, and 2.30 eV, which are similar to the pristine system, but the C atom generates two empty states in the bandgap (see Figure 2(b) and S2(b)).

C-doped TiO$_2$ (101) presents a magnetic moment of 2 μ_B per supercell, which has also been reported for both rutile and anatase phases [15, 63]. Our calculations confirm that one substitutional doping yields a paramagnetic material

with a magnetic moment of 2 μ_B, but only 0.70 μ_B on the C atom was measured within the Wigner-Seitz radius of 1.0 Å. The experimental value observed by Ye et al. [64] was 0.0236 μ_B per carbon in carbide state, that is, about 30 times less than the theoretical value.

3.2. N-Doped TiO$_2$. The N atom has one electron less than O, meaning that charge transfer by Ti will not fulfill all the N 2p orbitals, leaving one unpaired electron and also an empty spin-orbital. This process generates N 2p localized occupied states at the top of the VB and one empty state in the gap, near the VB (see Figure 3 and S3).

The empty states within the bandgap are not affected by the U correlation, which means that they are not hybridized with Ti 3d empty orbitals, neither the bulk nor the surface. These results agree with the data reported by Lin et al. [65] for anatase and Atanelov et al. [15] for rutile phase. For N-doped TiO$_2$, it is easier to excite electrons from the surface (~0.4 eV) than from the bulk system (~1.0 eV), which are in agreement with previous results [66].

On the surface, there is a reduction of the energy bandgap to 2.30 eV (GGAU), whereas the reported experimental value is 2.5 eV [67]. The substitution of O by N leaves an unpaired

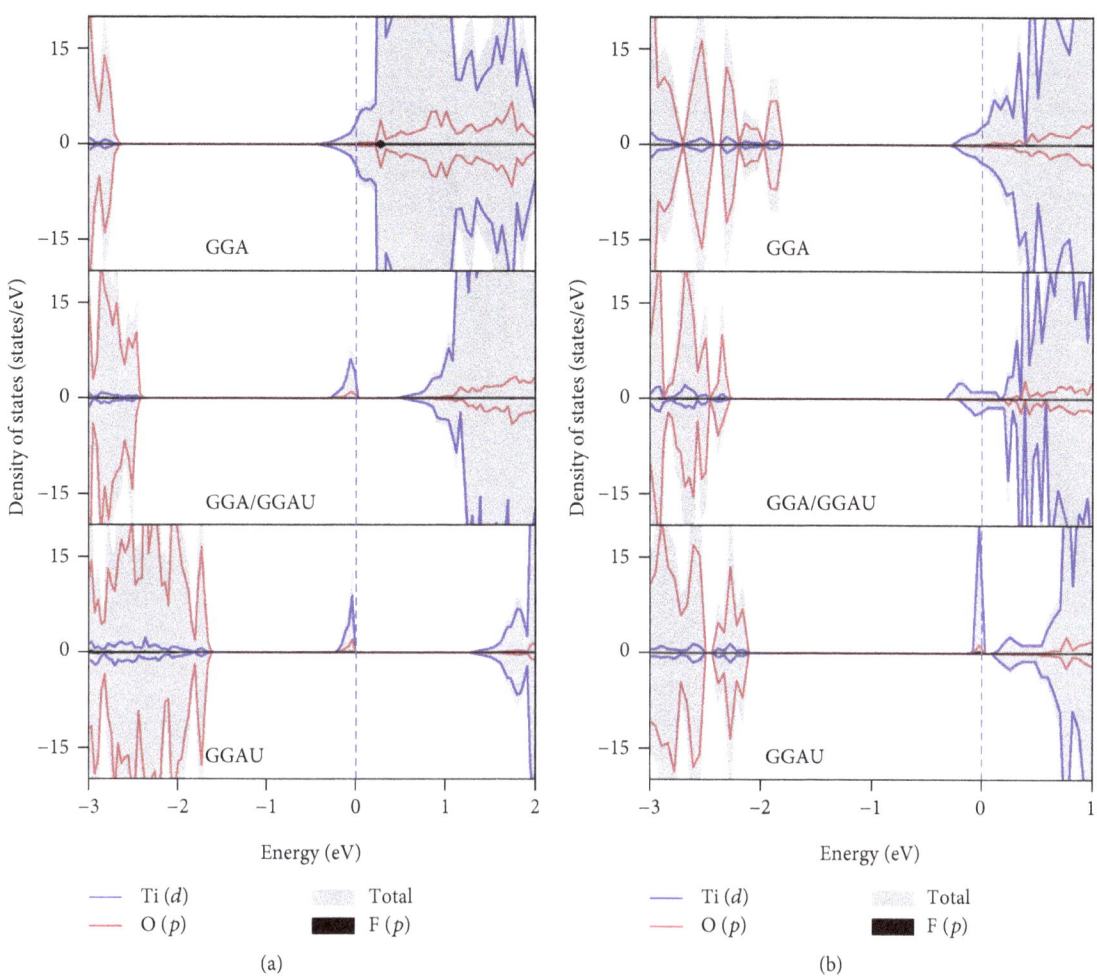

FIGURE 4: The total and partial density of states of the three relaxation methodologies for F-doped anatase TiO$_2$. (a) The bulk supercell. (b) The TiO$_2$ (101) surface.

electron into the system, yielding a magnetic moment of 1 μ_B per supercell, but only 0.503 μ_B on the N atom was measured within the Wigner-Seitz radius of 0.979 Å. An experiment for rutile crystals reported a paramagnetic behavior for the N-doping [68]; our calculations also reproduced this result for the N-doped anatase surface and bulk.

The required energy to obtain charge carries is reduced regarding the pristine TiO$_2$ due to localized states within the bandgap generated by dopants. C- and N-doped TiO$_2$ have localized states within the bandgap, reducing the excitation energy, but it has been reported that there is not a direct correlation between the excitation energy to form charge carriers and the photocatalytic activity [69]. The performance of the catalyst also depends on the recombination time, that is, the lifetime before holes and electrons are destroyed by mutual interaction. The position of the localized states within the bandgap affects the recombination time [32].

3.3. F-Doped TiO$_2$. The F atom has one electron more than oxygen. The substitution of F$^-$ instead of O^{2-} in the lattice leaves one unpaired electron; F can accept only one electron to acquire its closed shell configuration [36]. There are three possible configurations for F-doped TiO$_2$: (i) the unpaired

electron locates in the surrounds of the fluorine; (ii) fluorine transfers the one electron excess to their neighbors, or (iii) the unpaired electron delocalizes within the CB. As Di Valentin et al. [70] have pointed out, the last two options have the same probability of occurrence.

Our results show that the Fermi level is displaced towards the CB, and for the bulk and surface, the GGA method delocalizes the electron in the conduction band. For the bulk system, adding the U correlation parameter to the bulk system localizes occupied states below the CB, separated by 0.6 and 1.31 eV for the GGA/GGAU and GGAU methods (see Figure 4(a) and S4(a)). These occupied states are Ti 3d orbitals, so it confirms the charge transfer and the reduction from Ti^{4+} to Ti^{3+}. The surface system, with the GGA/GGAU scheme, does not bring a clear picture of these Ti^{3+} states; it seems these states are degenerated with the Ti surface states. Nonetheless, in the GGAU relaxation, the Ti^{3+} states are separated from the CBM, so we can observe the different nature of the surface Ti and Ti^{3+} states. The fluorine states, for the surface and bulk systems, are located at the VB bottom (see Figure S4).

Fluorine inhibits the full charge transfer from the Ti atoms, leading one Ti^{3+} atom. These states are depicted on

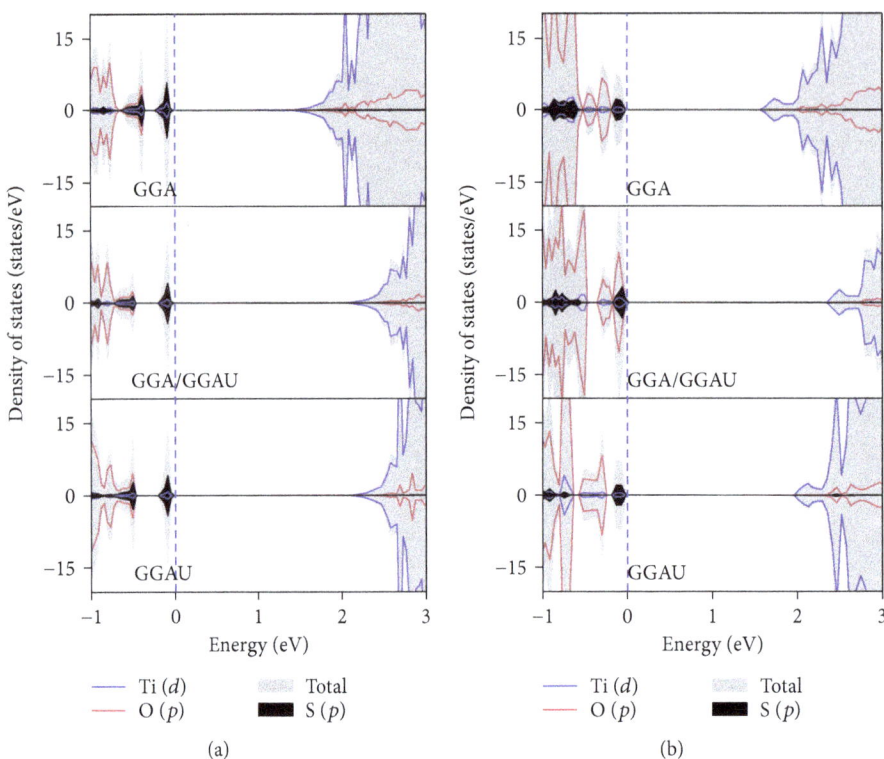

FIGURE 5: The total and partial density of states of the three relaxation methodologies for S-doped anatase TiO_2. (a) The bulk supercell. (b) The TiO_2 (101) surface.

TABLE 1: Summary of bandgaps (E_g) for the bulk and surface models for all doped and pristine TiO_2. Energy units are in eV.

Dopant	Bulk			Surface		
	GGA	GGA/GGAU	GGAU	GGA	GGA/GGAU	GGAU
C	2.44	2.66	2.66	1.87	2.36	2.30
N	2.00	2.80	2.77	1.85	2.29	2.30
F	2.19	2.89	2.89	1.63	2.00	2.30
S	1.80	2.51	2.55	1.69	2.40	2.08
Pure	2.19	2.81	2.76	1.85	2.31	2.25

the DOS for the small dopant concentration using the GGAU scheme, which localizes an occupied state without F $2p$ orbital participation; this means the electron is in the Ti $3d$ orbital.

The F-doping TiO_2 system must be treated with caution because the F-doping creates Ti^{3+} states, that is not adequately described by GGA functionals, as in the case of the oxygen vacancies [71], but hybrid functionals recover the physical picture [16]. Here, it has been shown that the PBE functional gives delocalized solutions with the bulk and surface models, which is consistent with reports where F distributes its unpaired electron towards neighbors Ti atoms [16, 35]. The GGAU scheme recovers the localization of Ti^{3+} with a lower computational cost compared with hybrid functionals.

3.4. S-Doped TiO_2. We have argued that the number of localized states—empty and occupied—and its electron

TABLE 2: Formation energy for anatase TiO_2 doped with the main group elements. Results obtained with the GGAU scheme.

Dopant	Formation energy (eV)	
	Bulk	Surface
C	3.52	3.30
N	3.36	1.98
F	1.85	2.45
S	5.81	3.01

distribution is directly related to the electron configuration of the dopant p orbitals; thus, we finally examined an element which is in the same oxygen group, the sulfur.

Sulfur has 16 electrons and presents a similar electron distribution and oxidation state like oxygen, which leads to an energy bandgap with no empty states as Figure 5 shows

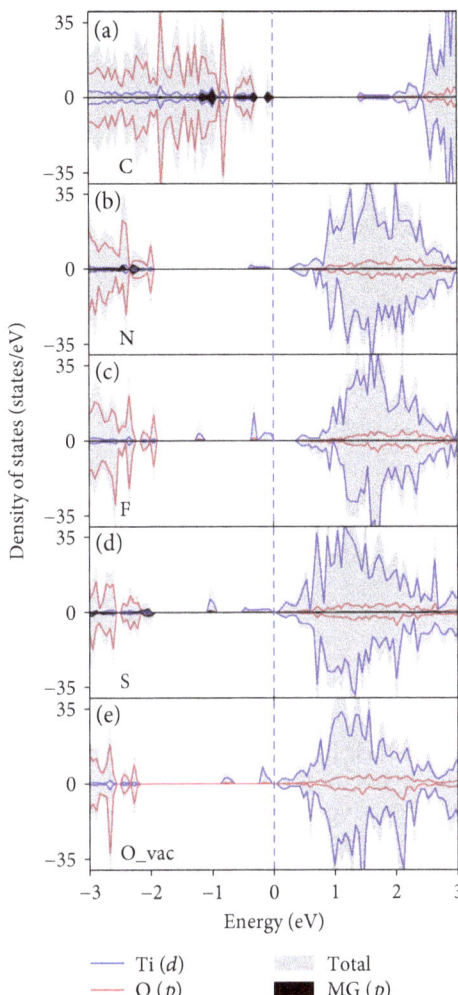

FIGURE 6: The total and partial density of states for pure and C-, N-, S-, F-doped anatase TiO$_2$ (101) with one oxygen vacancy. Results obtained with the GGAU methodology.

for all cases. The E_g reduction (see Table 1) is due to the occupied states that sulfur introduces at the top of the VB; it is important to note that spin-up and spin-down DOS are symmetrical (Figure S5) meaning all orbitals are fully occupied, confirming our configuration model. For the bulk system, these states are localized near the VB edge; this means there is some sulfur occupied states that show some repulsion from the occupied O 2p orbitals. This behavior is reflected in the high formation energy for S-doping (5.81 eV), as showed in Table 2. On the surface, in all cases, these states are closer to the VB than in the bulk. For the GGA/GGAU scheme, O 2p, S 3p, and Ti 3d occupied states are mixed just below the Fermi level. In the GGAU case, the occupied states under Fermi level present S 3p and Ti 3d character only.

The covalent radius of S (1.02 Å) differs considerably from the O atom (0.73 Å), and then the orbital O 2p and S 3p hybridization is inhibited as GGAU scheme shows. The S-doping generates some occupied localized states, mainly S 3p character, below the Fermi level. However, although these states are occupied, they do not overlap with

the VB (see Figure 5(a)). This may explain why experimentally high concentrations of sulfur are needed to enhance the photoactivity in the pollutant degradation [34]. If the sulfur substitution is located far from the surface, the charge carriers do not reach the surface, inhibiting the degradation activity. At high concentration, the probability that dopants reach the surface increases as well as the reactivity. This fact is carried out with large formation energies (5.81 eV, bulk) due to significant structural changes by S covalent radius.

The GGA/GGAU and GGAU methods give, for almost all cases, a good picture for the localized electronic states within the bandgap. However, the GGA/GGAU methodology has to be carefully treated because of the lack of ionic relaxation some states may not be well localized. It has been reported that the Hubbard U parameter depends heavily on the used geometry, so we encourage doing the full optimization GGAU methodology.

The Hubbard correction slightly widens the bandgap for all systems (see Table 1). U correlation is mandatory to describe TiO$_2$ systems if there is a hybridization between the dopant and Ti 3d orbitals. The Hubbard model is also necessary when unpaired electrons occupy Ti 3d states because the extra electron repulsion in the valence electrons localizes states below the CBM.

3.5. Oxygen Vacancies on Doped TiO$_2$ (101). We also studied the effect on the electronic structure of oxygen vacancies on the doped TiO$_2$ (101) surface systems. The formation energy of one oxygen vacancy follows the order: F (5.740) > S (5.734) > C (4.084) > N (3.963 eV). Experimentally, the oxygen vacancies are promoted when the surface is doped with N and C, which is consistent with our theoretical results [72–74].

Removing one oxygen atom from the pure TiO$_2$ surface yields Ti^{3+} states that are below the conduction band [71, 75] (see Figure 6(e)). C-doping shows a different behavior, that is, the two electrons available after removing the neutral oxygen go to the C 2p orbital (empty states generated by C-doping), above the valence band (see Figure 6(a)). Since oxygen and sulfur are isoelectronic, the oxygen vacancy on the S-doped system generates Ti^{3+} states just as with the pristine TiO$_2$ (101) (see Figure 6(d)).

For F-doping, there are three intrabandgap states, and one of them corresponds to the extra electron of fluorine compared to oxygen, which is in agreement with Czoska et al. [16] results. The other two states correspond to the ones generated by the oxygen vacancy. The electrons from the oxygen vacancies create Ti^{3+} species near the Fermi level; the Ti^{3+} state created by the F electron is the nearest to the valence band.

For N-doping, one electron from the oxygen vacancy locates at N 2p orbital; the other creates a Ti^{3+} state near the CB (see Figure 6(b)). C, as N atoms, accepts the extra charge derived from the oxygen vacancy. Zhang et al. [40] also found that the visible light photoactive centers of N-doped TiO$_2$ are the substitutional N species with a diamagnetic [O-Ti^{4+}-N^{3-}-Ti^{4+}-V$_O^-$] cluster containing an oxygen vacancy and an N atom. It has been shown that the charge excess can be trapped by N impurities leading to a

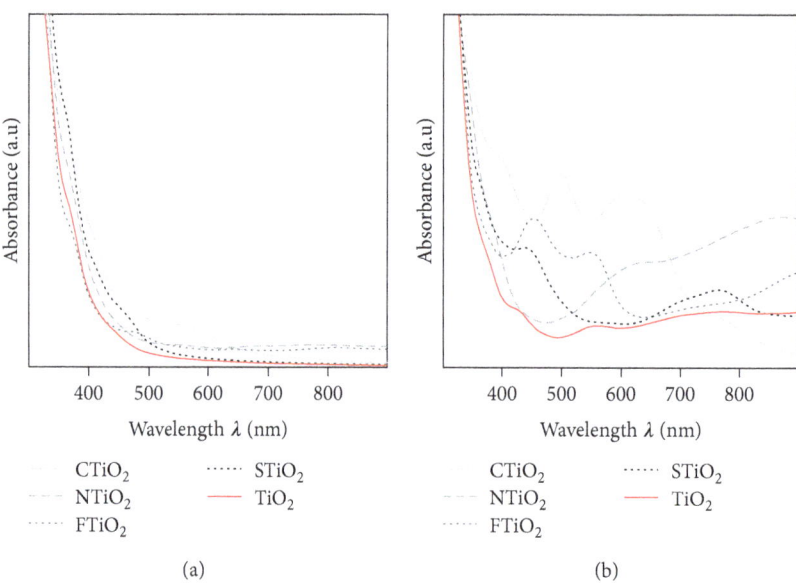

FIGURE 7: The UV-visible spectrum of anatase surface. (a) C-, N-, F-, and S-doped TiO_2 (101). (b) Doped surface with one oxygen vacancy. Results obtained with the GGAU methodology.

catalytic efficiency enhancement [74]. Finazzi et al. [66] reported a similar behavior when substituting two O atoms by N; the unpaired electrons fill the dopant N $2p$ states giving a closed shell configuration. The synergy between the dopants and oxygen vacancies cleans the gap, but for nitrogen aside to clean the gap, these new states are well mixed with the O $2p$ states from the valence band, so it enhanced the charge mobility.

For the C-doped TiO_2 bulk system, Kamisaka et al. [76] concluded that oxygen vacancies diminish the photocatalytic activity by filling the holes generated by the dopant, inhibiting the charge mobility. In our case, for the surface TiO_2 (101) anatase, C $2p$ states lie above the valence band, which is beneficial for the photocatalytic activity. Because the C dopant is localized near the surface, the generated charge carriers will already be at a reactive site, that is, these localized states may enhance the charge carrier lifetime.

3.6. *Optical Response of Doped TiO$_2$ (101) with One Oxygen Vacancy.* The optical absorption of the C-, N-, F-, and S-doped surface TiO_2 (101) was also determined (see Figure 7(a)). To correct the DFT gap underestimation, the optical absorption also includes the scissor operator (with a value of 0.44 eV). This method has been successfully applied to several materials [77–79]. The same scissor operator is used for the doped surface with and without oxygen vacancies.

The UV-visible spectrum (Figure 7(a)) agrees with the experimental data reported by Chen et al. [80], where the C-doping shows the maximum absorbance. In our case, there is a shoulder at $\lambda_{max} = 500$ nm (blue), which corresponds to transitions from valence band O $2p$ states to C $2p$ empty localized in the center of the bandgap (see Figures 3(b) and 6(a)). When the oxygen vacancy is considered, the optical activity in the visible region is significantly increased for

doped TiO_2, which favors the generation of charge carriers using visible light (see Figure 7(b)).

Due to that oxygen vacancies are present in almost all synthesized TiO_2 materials, the simultaneous presence of main group elements and one oxygen vacancy is frequent. According to experimental results [80], when the nitrogen concentration increases, it also rises the oxygen vacancies and Ti^{3+}, leading to the enhancement of photocatalytic activity. From our theoretical results, N-doping presents the lowest formation energy of oxygen vacancies (3.963 eV). For the N-doped system with an oxygen vacancy, there are two shoulders at $\lambda_{max} \sim 550$ and 750 nm, which correspond to transitions from Ti $3d$ occupied states to Ti $3d$ empty states (see Figure 6(b)).

The F atom hinders the formation of oxygen vacancy. These results contrast with Hattori et al. [81] results. They stated that F ion promotes the formation of oxygen vacancies. The partial DOS (Figure 6(c)) shows that the Fermi level is displaced about 0.2 eV from the conduction band when an oxygen vacancy is present, which does not reduce the TiO_2 chemical potential. The F-doping without oxygen vacancy presents a similar optical activity than TiO_2 with oxygen vacancies, which agrees with experimental evidence that F-TiO_2 can hardly displace the light absorption towards the visible region. In spite of this, it has been reported that F-doped TiO_2 has catalytic activity under visible light. It is speculated that this activity may be due to other species like F_2 or HF adsorbed on the surface [36, 82].

For all the studied cases, oxygen vacancies on the TiO_2 (101) surface displace the light spectrum towards the visible region. A severe increase in the optical absorption intensity is obtained following the order C > N > F > S. The reason that F and S do not present a high photocatalytic efficiency is that they promote Ti^{3+} states; the charge carriers have not the oxidation potential required to form the $OH^\bullet$ species. The high absorbance of C-doping in the visible light (Figure 7(b)) as well as the localized state above the bandgap explains why it is five times more efficient than N-doping in the degradation of 4-chlorophenol when it is used as an artificial light ($\lambda \geq 455$ nm) [33].

4. Conclusions

The electronic structure of the bulk and surface (101) models of anatase TiO_2 doped with the main group elements C, N, F, and S, with and without oxygen vacancies, was studied using DFT and the Hubbard U approach.

The GGA approach alone underestimated the bandgap for all cases, with a systematic failure of the method. The best and simplest method for describing the doped TiO_2 electronic structure is the ionic relaxation of the TiO_2 models with the GGA + U scheme, which obtains localized solutions at the same computational cost of the GGA approach.

Two ways for reducing the TiO_2 bandgap were identified. The first mechanism is broadening the VB or CB. The presence of occupied states near the VBM reduces the bandgap, and then less energy is required to excite electrons from VB to CB. The S-doping TiO_2 (101) is an example of that mechanism. A similar process can occur, but in the CB when the TiO_2 (101) surface is formed, the exposed Ti atoms present undercoordination, giving rise to empty states very close the CBM.

The second mechanism for reducing the bandgap is to generate localized states in the gap. However, this process has the inconvenience that these states may act as recombination centers. For example, the C- and N-doping form localized empty states in the gap. For the C-doping, the surface TiO_2 (101) presents a "cleaner" gap than the bulk TiO_2, which is favorable to eliminate recombination centers. Finally, the F-doping represents a similar case to oxygen vacancies, where the fluorine atom promotes the Ti^{3+} species formation, which appears below the CBM.

The number of valence electrons of the dopant element determines the number of states and its position within the gap when the O atom is replaced.

Both doping and oxygen vacancies on TiO_2 (101) displace the light spectrum towards the visible region. A great increase in the optical absorption intensity is obtained with the following order C > N > F > S. Therefore, the high catalytic activity of C-doping using visible radiation is consistent with its high absorption intensity generated by oxygen vacancies on the surface.

Conflicts of Interest

The authors declare that they have no conflicts of interest.

Acknowledgments

Oscar Olvera-Neria thanks the CONACYT-México for the financial support for the project CB-2011-01/166246. The authors are indebted to the UAM Iztapalapa for the computing time in the Yoltla supercomputer. Julio César González-Torres is grateful to the CONACYT for the studentship granted to pursue his doctoral studies.

Supplementary Materials

Figure S1: the total and partial density of states of the three relaxation methodologies for anatase TiO_2. (a) The bulk supercell. (b) The TiO_2 (101) surface. The Fermi level corresponds to the zero energy. Figure S2: the total and partial density of states of the three relaxation methodologies for C-doped anatase TiO_2. (a) The bulk supercell. (b) The TiO_2 (101) surface. Figure S3: the total and partial density of states of the three relaxation methodologies for N-doped anatase TiO_2. (a) The bulk supercell. (b) The TiO_2 (101) surface. Figure S4: the total and partial density of states of the three relaxation methodologies for F-doped anatase TiO_2. (a) The bulk supercell. (b) The TiO_2 (101). Figure S5: the total and partial density of states of the three relaxation methodologies for S-doped anatase TiO_2. (a) The bulk supercell. (b) The TiO_2 (101) surface. (Supplementary Materials)

References

[1] M. Fujihira, Y. Satoh, and T. Osa, "Heterogeneous photocatalytic oxidation of aromatic compounds on TiO_2," Nature, vol. 293, no. 5829, pp. 206–208, 1981.

[2] A. Fujishima and K. Honda, "Electrochemical photolysis of water at a semiconductor electrode," Nature, vol. 238, no. 5358, pp. 37-38, 1972.

[3] T. Inoue, A. Fujishima, S. Konishi, and K. Honda, "Photoelectrocatalytic reduction of carbon dioxide in aqueous suspensions of semiconductor powders," Nature, vol. 277, no. 5698, pp. 637-638, 1979.

[4] S. Hamzezadeh-Nakhjavani, O. Tavakoli, S. P. Akhlaghi, Z. Salehi, P. Esmailnejad-Ahranjani, and A. Arpanaei, "Efficient photocatalytic degradation of organic pollutants by magnetically recoverable nitrogen-doped TiO_2 nanocomposite photocatalysts under visible light irradiation," Environmental Science and Pollution Research, vol. 22, no. 23, pp. 18859–18873, 2015.

[5] M. Romero-Sáez, L. Y. Jaramillo, R. Saravanan et al., "Notable photocatalytic activity of TiO_2-polyethylene nanocomposites for visible light degradation of organic pollutants," Express Polymer Letters, vol. 11, no. 11, pp. 899–909, 2017.

[6] T. Sreethawong, S. Ngamsinlapasathian, and S. Yoshikawa, "Positive role of incorporating P-25 TiO_2 to mesoporous-assembled TiO_2 thin films for improving photocatalytic dye degradation efficiency," Journal of Colloid and Interface Science, vol. 430, pp. 184–192, 2014.

[7] R. Sidaraviciute, E. Krugly, L. Dabasinskaite, E. Valatka, and D. Martuzevicius, "Surface-deposited nanofibrous TiO_2 for photocatalytic degradation of organic pollutants," Journal of Sol-Gel Science and Technology, vol. 84, no. 2, pp. 306–315, 2017.

[8] P. Haowei, L. Jingbo, L. Shu-Shen, and X. Jian-Bai, "First-principles study of the electronic structures and magnetic properties of 3d transition metal-doped anatase TiO_2," *Journal of Physics: Condensed Matter*, vol. 20, no. 12, article 125207, 2008.

[9] K. Song, X. Han, and G. Shao, "Electronic properties of rutile TiO_2 doped with 4d transition metals: first-principles study," *Journal of Alloys and Compounds*, vol. 551, pp. 118–124, 2013.

[10] S. Yang and H. Lee, "Determining the catalytic activity of transition metal-doped TiO_2 nanoparticles using surface spectroscopic analysis," *Nanoscale Research Letters*, vol. 12, no. 1, p. 582, 2017.

[11] P. Ribao, M. J. Rivero, and I. Ortiz, "TiO_2 structures doped with noble metals and/or graphene oxide to improve the photocatalytic degradation of dichloroacetic acid," *Environmental Science and Pollution Research*, vol. 24, no. 14, pp. 12628–12637, 2017.

[12] S. N. R. Inturi, T. Boningari, M. Suidan, and P. G. Smirniotis, "Visible-light-induced photodegradation of gas phase acetonitrile using aerosol-made transition metal (V, Cr, Fe, Co, Mn, Mo, Ni, Cu, Y, Ce, and Zr) doped TiO_2," *Applied Catalysis B: Environmental*, vol. 144, pp. 333–342, 2014.

[13] D. Toprek, V. Koteski, J. Belošević-Čavor, V. Ivanovski, and A. Umićević, "Ab initio study of electronic and optical properties of *Fe* doped anatase *TiO_2* (101) surface," vol. 1120, pp. 17–23, 2017.

[14] Y. Lin, Z. Jiang, C. Zhu et al., "The electronic structure, optical absorption and photocatalytic water splitting of (Fe + Ni)-codoped TiO_2: a DFT+U study," *International Journal of Hydrogen Energy*, vol. 42, no. 8, pp. 4966–4976, 2017.

[15] J. Atanelov, C. Gruber, and P. Mohn, "The electronic and magnetic structure of *p*-element (C,N) doped rutile-TiO_2; a hybrid DFT study," *Computational Materials Science*, vol. 98, pp. 42–50, 2015.

[16] A. M. Czoska, S. Livraghi, M. Chiesa et al., "The nature of defects in fluorine-doped TiO_2," *Journal of Physical Chemistry C*, vol. 112, no. 24, pp. 8951–8956, 2008.

[17] H. Gao, J. Zhou, D. Dai, and Y. Qu, "Photocatalytic activity and electronic structure analysis of N-doped anatase TiO_2: a combined experimental and theoretical study," *Chemical Engineering & Technology*, vol. 32, no. 6, pp. 867–872, 2009.

[18] K. Yang, Y. Dai, and B. Huang, "Study of the nitrogen concentration influence on N-doped TiO_2 anatase from first-principles calculations," *The Journal of Physical Chemistry C*, vol. 111, no. 32, pp. 12086–12090, 2007.

[19] H. Hou, F. Gao, M. Shang et al., "Enhanced visible-light responsive photocatalytic activity of N-doped TiO_2 thoroughly mesoporous nanofibers," *Journal of Materials Science: Materials in Electronics*, vol. 28, no. 4, pp. 3796–3805, 2017.

[20] S. K. Kassahun, Z. Kiflie, D. W. Shin, S. S. Park, W. Y. Jung, and Y. R. Chung, "Facile low temperature immobilization of N-doped TiO_2 prepared by sol–gel method," *Journal of Sol-Gel Science and Technology*, vol. 83, no. 3, pp. 698–707, 2017.

[21] J. Zhao, W. Li, X. Li, and X. Zhang, "Low temperature synthesis of water dispersible F-doped TiO_2 nanorods with enhanced photocatalytic activity," *RSC Advances*, vol. 7, no. 35, pp. 21547–21555, 2017.

[22] X. Zhang, Z. Li, Z. Zhan, and L. Di, "Preparation of F-doped TiO_2 photocatalysts by gas–liquid plasma at atmospheric pressure," *Topics in Catalysis*, vol. 60, no. 12-14, pp. 980–986, 2017.

[23] R. Ata, O. Sacco, V. Vaiano, L. Rizzo, G. Y. Tore, and D. Sannino, "Visible light active N-doped TiO_2 immobilized on polystyrene as efficient system for wastewater treatment," *Journal of Photochemistry and Photobiology A: Chemistry*, vol. 348, pp. 255–262, 2017.

[24] Y. H. Lin, H. T. Hsueh, C. W. Chang, and H. Chu, "The visible light-driven photodegradation of dimethyl sulfide on S-doped TiO_2: characterization, kinetics, and reaction pathways," *Applied Catalysis B: Environmental*, vol. 199, pp. 1–10, 2016.

[25] D. Guerrero-Araque, D. Ramírez-Ortega, P. Acevedo-Peña, F. Tzompantzi, H. A. Calderón, and R. Gómez, "Interfacial charge-transfer process across ZrO_2-TiO_2 heterojunction and its impact on photocatalytic activity," *Journal of Photochemistry and Photobiology A: Chemistry*, vol. 335, pp. 276–286, 2017.

[26] I. Permadani, D. A. Phasa, A. W. Pratiwi, and F. Rahmawati, "The composite of ZrO_2-TiO_2 produced from local zircon sand used as a photocatalyst for the degradation of methylene blue in a single batik dye wastewater," *Bulletin of Chemical Reaction Engineering & Catalysis*, vol. 11, no. 2, pp. 133–139, 2016.

[27] K. Nakata and A. Fujishima, "TiO_2 photocatalysis: design and applications," *Journal of Photochemistry and Photobiology C: Photochemistry Reviews*, vol. 13, no. 3, pp. 169–189, 2012.

[28] K. Hashimoto, H. Irie, and A. Fujishima, "TiO_2 photocatalysis: a historical overview and future prospects," *Japanese Journal of Applied Physics*, vol. 44, no. 12, article 8269, 2005.

[29] M. A. Henderson, "A surface science perspective on TiO_2 photocatalysis," *Surface Science Reports*, vol. 66, no. 6-7, pp. 185–297, 2011.

[30] W. Shockley and W. T. Read, "Statistics of the recombinations of holes and electrons," *Physical Review*, vol. 87, no. 5, pp. 835–842, 1952.

[31] R. N. Hall, "Electron-hole recombination in germanium," *Physical Review*, vol. 87, no. 2, pp. 387–387, 1952.

[32] R. Asahi, T. Morikawa, T. Ohwaki, K. Aoki, and Y. Taga, "Visible-light photocatalysis in nitrogen-doped titanium oxides," *Science*, vol. 293, no. 5528, pp. 269–271, 2001.

[33] S. Sakthivel and H. Kisch, "Daylight photocatalysis by carbon-modified titanium dioxide," *Angewandte Chemie International Edition*, vol. 42, no. 40, pp. 4908–4911, 2003.

[34] F. H. Tian and C. B. Liu, "DFT description on electronic structure and optical absorption properties of anionic S-doped anatase TiO_2," *The Journal of Physical Chemistry B*, vol. 110, no. 36, pp. 17866–17871, 2006.

[35] T. Yamaki, T. Umebayashi, T. Sumita et al., "Fluorine-doping in titanium dioxide by ion implantation technique," *Nuclear Instruments and Methods in Physics Research Section B: Beam Interactions with Materials and Atoms*, vol. 206, pp. 254–258, 2003.

[36] J. C. Yu, J. Yu, W. Ho, Z. Jiang, and L. Zhang, "Effects of F^- doping on the photocatalytic activity and microstructures of nanocrystalline TiO_2 powders," *Chemistry of Materials*, vol. 14, no. 9, pp. 3808–3816, 2002.

[37] U. Diebold, "The surface science of titanium dioxide," *Surface Science Reports*, vol. 48, no. 5-8, pp. 53–229, 2003.

[38] K. Iijima, M. Goto, S. Enomoto et al., "Influence of oxygen vacancies on optical properties of anatase TiO_2 thin films," *Journal of Luminescence*, vol. 128, no. 5-6, pp. 911–913, 2008.

[39] R. Chandana, P. Mohanty, A. C. Pandey, and N. C. Mishra, "Oxygen vacancy induced structural phase transformation in

TiO$_2$ nanoparticles," *Journal of Physics D: Applied Physics*, vol. 42, no. 20, article 205101, 2009.

[40] Z. Zhang, X. Wang, J. Long, Q. Gu, Z. Ding, and X. Fu, "Nitrogen-doped titanium dioxide visible light photocatalyst: spectroscopic identification of photoactive centers," *Journal of Catalysis*, vol. 276, no. 2, pp. 201–214, 2010.

[41] L. Chen, Q. Gu, L. Hou et al., "Molecular p-n heterojunction-enhanced visible-light hydrogen evolution over a N-doped TiO$_2$ photocatalyst," *Catalysis Science & Technology*, vol. 7, no. 10, pp. 2039–2049, 2017.

[42] G. Kresse and J. Furthmüller, "Efficient iterative schemes for *ab initio* total-energy calculations using a plane-wave basis set," *Physical Review B*, vol. 54, no. 16, pp. 11169–11186, 1996.

[43] G. Kresse and J. Furthmüller, "Efficiency of ab-initio total energy calculations for metals and semiconductors using a plane-wave basis set," *Computational Materials Science*, vol. 6, no. 1, pp. 15–50, 1996.

[44] P. Hohenberg and W. Kohn, "Inhomogeneous electron gas," *Physical Review*, vol. 136, no. 3B, pp. B864–B871, 1964.

[45] W. Kohn and L. J. Sham, "Self-consistent equations including exchange and correlation effects," *Physical Review*, vol. 140, no. 4A, pp. A1133–A1138, 1965.

[46] J. P. Perdew, K. Burke, and M. Ernzerhof, "Generalized gradient approximation made simple," *Physical Review Letters*, vol. 77, no. 18, pp. 3865–3868, 1996.

[47] J. P. Perdew, K. Burke, and M. Ernzerhof, "Generalized gradient approximation made simple [Phys. Rev. Lett. 77, 3865 (1996)]," *Physical Review Letters*, vol. 78, no. 7, pp. 1396–1396, 1997.

[48] A. D. Becke, "Density-functional exchange-energy approximation with correct asymptotic behavior," *Physical Review A*, vol. 38, no. 6, pp. 3098–3100, 1988.

[49] A. D. Becke, "Density functional calculations of molecular bond energies," *Journal of Chemical Physics*, vol. 84, no. 8, pp. 4524–4529, 1986.

[50] J. P. Perdew, "Density-functional approximation for the correlation energy of the inhomogeneous electron gas," *Physical Review B*, vol. 33, no. 12, pp. 8822–8824, 1986.

[51] J. P. Perdew and W. Yue, "Accurate and simple density functional for the electronic exchange energy: generalized gradient approximation," *Physical Review B*, vol. 33, no. 12, pp. 8800–8802, 1986.

[52] J. P. Perdew, "Density functional theory and the band gap problem," *International Journal of Quantum Chemistry*, vol. 28, no. S19, pp. 497–523, 1985.

[53] S. L. Dudarev, G. A. Botton, S. Y. Savrasov, C. J. Humphreys, and A. P. Sutton, "Electron-energy-loss spectra and the structural stability of nickel oxide: an LSDA+U study," *Physical Review B*, vol. 57, no. 3, pp. 1505–1509, 1998.

[54] C. I. N. Morgade and G. F. Cabeza, "Synergetic interplay between metal (Pt) and nonmetal (C) species in codoped TiO$_2$: a DFT+U study," *Computational Materials Science*, vol. 111, pp. 513–524, 2016.

[55] C. J. Calzado, N. C. Hernández, and J. F. Sanz, "Effect of on-site Coulomb repulsion term U on the band-gap states of the reduced rutile (110) TiO$_2$ surface," *Physical Review B*, vol. 77, no. 4, article 045118, 2008.

[56] P. E. Blöchl, "Projector augmented-wave method," *Physical Review B*, vol. 50, no. 24, pp. 17953–17979, 1994.

[57] G. Kresse and D. Joubert, "From ultrasoft pseudopotentials to the projector augmented-wave method," *Physical Review B*, vol. 59, no. 3, pp. 1758–1775, 1999.

[58] M. Lazzeri, A. Vittadini, and A. Selloni, "Structure and energetics of stoichiometric TiO$_2$ anatase surfaces," *Physical Review B*, vol. 63, no. 15, article 155409, 2001.

[59] M. Lazzeri, A. Vittadini, and A. Selloni, "Erratum: Structure and energetics of stoichiometric TiO$_2$ anatase surfaces [Phys. Rev. B 63, 155409 (2001)]," *Physical Review B*, vol. 65, no. 11, article 119901, 2002.

[60] V. Wang, "VASPKIT, postprocessing tool for VASP code," 2014, http://vaspkit.sourceforge.net.

[61] M. Gajdoš, K. Hummer, G. Kresse, J. Furthmüller, and F. Bechstedt, "Linear optical properties in the projector-augmented wave methodology," *Physical Review B*, vol. 73, no. 4, article 045112, 2006.

[62] M. Horn, C. F. Schwerdtfeger, and E. P. Meagher, "Refinement of the structure of anatase at several temperatures*," *Zeitschrift für Kristallographie*, vol. 136, no. 3-4, pp. 273–281, 1972.

[63] K. Yang, Y. Dai, B. Huang, and M.-H. Whangbo, "On the possibility of ferromagnetism in carbon-doped anatase TiO$_2$," *Applied Physics Letters*, vol. 93, no. 13, p. 132507, 2008.

[64] X. J. Ye, W. Zhong, M. H. Xu, X. S. Qi, C. T. Au, and Y. W. Du, "The magnetic property of carbon-doped TiO$_2$," *Physics Letters A*, vol. 373, no. 40, pp. 3684–3687, 2009.

[65] Z. Lin, A. Orlov, R. M. Lambert, and M. C. Payne, "New insights into the origin of visible light photocatalytic activity of nitrogen-doped and oxygen-deficient anatase TiO$_2$," *The Journal of Physical Chemistry B*, vol. 109, no. 44, pp. 20948–20952, 2005.

[66] E. Finazzi, C. Di Valentin, A. Selloni, and G. Pacchioni, "First principles study of nitrogen doping at the anatase TiO$_2$(101) surface," *The Journal of Physical Chemistry C*, vol. 111, no. 26, pp. 9275–9282, 2007.

[67] N. C. Khang, N. Van Khanh, N. H. Anh, D. T. Nga, and N. Van Minh, "The origin of visible light photocatalytic activity of N-doped and weak ferromagnetism of Fe-doped TiO$_2$ anatase," *Advances in Natural Sciences: Nanoscience and Nanotechnology*, vol. 2, no. 1, article 015008, 2011.

[68] L. Chun-Ming, X. Xia, Z. Yan, J. Yong, and Z. Xiao-Tao, "Magnetism of a nitrogen-implanted TiO$_2$ single crystal," *Chinese Physics Letters*, vol. 28, no. 12, article 127201, 2011.

[69] D. Dvoranová, V. Brezová, M. Mazúr, and M. A. Malati, "Investigations of metal-doped titanium dioxide photocatalysts," *Applied Catalysis B: Environmental*, vol. 37, no. 2, pp. 91–105, 2002.

[70] C. Di Valentin, E. Finazzi, G. Pacchioni et al., "Density functional theory and electron paramagnetic resonance study on the effect of N–F codoping of TiO$_2$," *Chemistry of Materials*, vol. 20, no. 11, pp. 3706–3714, 2008.

[71] N. S. Portillo-Vélez, O. Olvera-Neria, I. Hernández-Pérez, and A. Rubio-Ponce, "Localized electronic states induced by oxygen vacancies on anatase TiO2 (101) surface," *Surface Science*, vol. 616, pp. 115–119, 2013.

[72] M. Batzill, E. H. Morales, and U. Diebold, "Influence of nitrogen doping on the defect formation and surface properties of TiO$_2$ rutile and anatase," *Physical Review Letters*, vol. 96, no. 2, article 026103, 2006.

[73] C. Di Valentin, G. Pacchioni, and A. Selloni, "Theory of carbon doping of titanium dioxide," *Chemistry of Materials*, vol. 17, no. 26, pp. 6656–6665, 2005.

[74] C. Di Valentin, G. Pacchioni, A. Selloni, S. Livraghi, and E. Giamello, "Characterization of paramagnetic species in N-doped TiO_2 powders by EPR spectroscopy and DFT calculations," *The Journal of Physical Chemistry B*, vol. 109, no. 23, pp. 11414–11419, 2005.

[75] G. Pacchioni, "Oxygen vacancy: the invisible agent on oxide surfaces," *Chemphyschem*, vol. 4, no. 10, pp. 1041–1047, 2003.

[76] H. Kamisaka, T. Adachi, and K. Yamashita, "Theoretical study of the structure and optical properties of carbon-doped rutile and anatase titanium oxides," *The Journal of Chemical Physics*, vol. 123, no. 8, article 084704, 2005.

[77] M. Guo and J. Du, "First-principles study of electronic structures and optical properties of Cu, Ag, and Au-doped anatase TiO_2," *Physica B: Condensed Matter*, vol. 407, no. 6, pp. 1003–1007, 2012.

[78] J. Li, S.-H. Wei, S.-S. Li, and J.-B. Xia, "Design of shallow acceptors in ZO: first-principles band-structure calculations," *Physical Review B*, vol. 74, no. 8, article 081201, 2006.

[79] H. Weng, J. Dong, T. Fukumura, M. Kawasaki, and Y. Kawazoe, "First principles investigation of the magnetic circular dichroism spectra of co-doped anatase and rutile TiO_2," *Physical Review B*, vol. 73, no. 12, article 121201, 2006.

[80] D. Chen, Z. Jiang, J. Geng, Q. Wang, and D. Yang, "Carbon and nitrogen co-doped TiO_2 with enhanced visible-light photocatalytic activity," *Industrial & Engineering Chemistry Research*, vol. 46, no. 9, pp. 2741–2746, 2007.

[81] A. Hattori, K. Shimoda, H. Tada, and S. Ito, "Photoreactivity of sol–gel TiO_2 films formed on soda-lime glass substrates: effect of SiO_2 underlayer containing fluorine," *Langmuir*, vol. 15, no. 16, pp. 5422–5425, 1999.

[82] H. Sun, S. Wang, H. M. Ang, M. O. Tadé, and Q. Li, "Halogen element modified titanium dioxide for visible light photocatalysis," *Chemical Engineering Journal*, vol. 162, no. 2, pp. 437–447, 2010.

Photocatalytic Degradation and Hydrogen Production of TiO$_2$/Carbon Fiber Composite Using Bast as a Carbon Fiber Source

Guomei Tang ⓘ,^{1,2,3} Asim Abas,^{1} and Shanpeng Wang^1

^{1}*School of Physical Science and Technology, Lanzhou University, Lanzhou 730000, China*
^{2}*School of Mathematics and Computer Science Institute, Northwest Minzu University, Lanzhou, Gansu 730030, China*
^{3}*State Key Laboratory of Advanced Processing and Recycling of Non-Ferrous Metals, Lanzhou University of Technology, Lanzhou 730050, China*

Correspondence should be addressed to Guomei Tang; tanggm13@lzu.edu.cn

Academic Editor: Xuxu Wang

TiO$_2$/carbon fiber composite is achieved by loading TiO$_2$ nanoparticles on biomass carbon fiber, which originates from the carbonized natural bast. The carbonized process and the loading amount of TiO$_2$ are researched in detail. It is found that the carbonized bast fiber shows robust adsorption characteristics for TiO$_2$ nanoparticles in aqueous dispersion, and TiO$_2$ nanoparticles with ~15 wt.% in total weight are uniformly loaded onto the fiber surface. The photocatalytic properties of TiO$_2$/carbon fiber composite are evaluated by photocatalytic degradation of rhodamine B and water splitting for hydrogen production. The results indicate that 90% RhB molecules could be attacked in 60 min under UV light irradiation, and the hydrogen production rate of water splitting is up to 338.51 μmol/h. The highlight is that TiO$_2$/carbon fiber composite is easy to be recycled due to the incorporation of macroscopical biomass carbon fiber.

1. Introduction

Since Fujishima and Honda reported the groundbreaking research work on splitting water to produce hydrogen on TiO$_2$ electrode [1], the photocatalysis had aroused great interest. TiO$_2$ nanomaterials possess commendable performance in terms of its nontoxic, cost effectiveness, strong oxidizing activity, and long-term stability against photocorrosion, which have a very broad application prospect in many fields, such as air pollution, waste water treatment, hydrogen production, and sterilization [2–7]. The application of TiO$_2$ nanomaterials is greatly limited due to the poor visible light response and high charge recombination rate, originating from the wide band gap (3.2 eV for anatase and 3.0 eV for rutile) and the relatively high electrical resistance. TiO$_2$ is combined with carbon species such as carbon nanotubes, graphene, carbon fiber, and graphite-like carbon to form TiO$_2$/carbon composites [8–12] due to the good adsorbability and strong charge transport ability [13, 14].

Carbon fiber is a kind of one-dimensional carbon material with excellent properties, such as high tensile strength, low weight, high chemical resistance, high temperature tolerance, and excellent electrical conductivity, making them very popular in many fields. However, commercialized carbon fiber is relatively expensive due to the raw materials and fabrication process, which is not conducive to the photocatalyst application in the wide range. Biomass fiber carbonization is a feasible route to obtain low-cost carbon fiber, which can meet some applications with no high requirement on mechanical behavior, such as photocatalyst and solar cells.

In the liquid phase system, TiO$_2$ nanomaterials are difficultly recycled after reaction; the poison and aggregate of TiO$_2$ powder are also disadvantageous factors [15]. They all seriously restrict the practical application of photocatalysis. Therefore, it is an effective way to solve the recovery of photocatalyst by combining TiO$_2$ nanomaterials with carbon fiber.

Natural bast fiber was used in textile in the ancient times. Its main component is cellulose, which takes up about 75% content of the fiber. The bast becomes a kind of high-

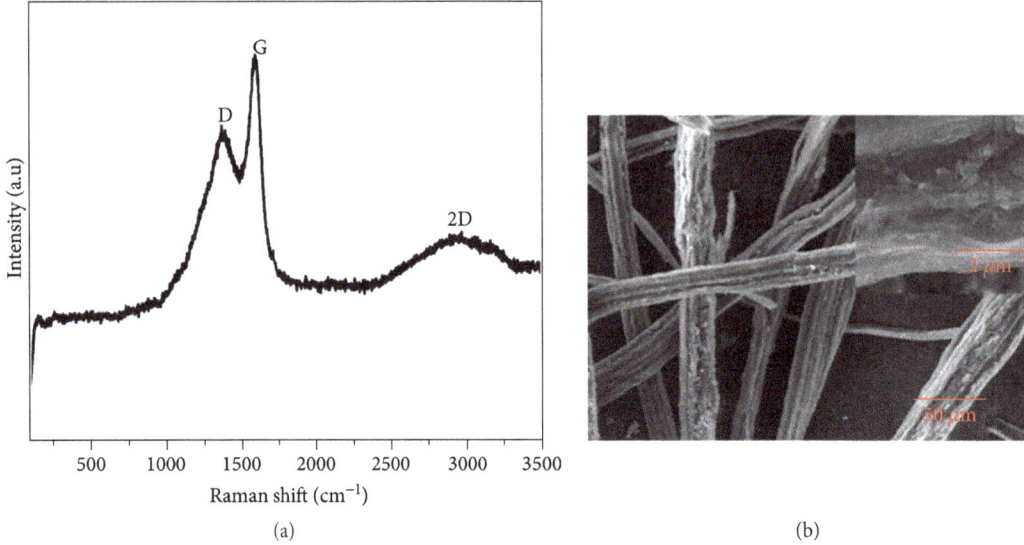

FIGURE 1: The Raman spectrum (a) and SEM image (b) of biomass carbon fiber.

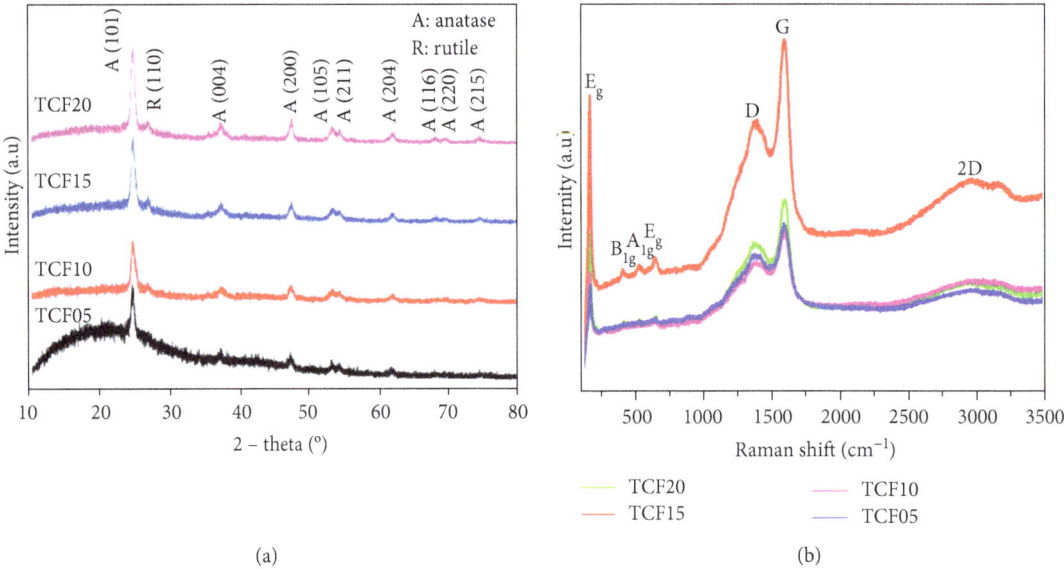

FIGURE 2: XRD patterns (a) and Raman spectra (b) of TiO_2/carbon fiber micronanocomposites.

strength, low elongation fiber, so it is an ideal raw material of biomass carbon fiber. However, the electrical conductivity, high temperature resistance, and the chemical and physical adsorptions of bast fiber are disadvantageous. Biomass fiber carbonization is a suitable way to improve electrical conduction and structural properties, including increasing porous structure and fiber roughness, which are conducive to enhance the adhesion and load amount of TiO_2 nanomaterials on the surface of biomass carbon fiber. Because bast fiber is widely accessible and low cost, the biomass carbon fiber using bast fiber as a raw material has great advantages in practical application of catalyst carrier.

Here, the bast was carbonized to get biomass carbon fiber. Then, the carbon fiber was put into TiO_2 nanoparticle dispersion with different concentrations to obtain optimal

TiO_2/carbon fiber composite. Using the recycled photocatalyst, the photocatalytic degradation of rhodamine B (RhB) and water splitting for hydrogen production were performed, 90% RhB molecules were attacked in 60 min under UV light irradiation, and the hydrogen production rate was up to 338.51 μmol/h. A theoretical model was proposed to explain the photocatalytic mechanism of the composites.

2. Experimental Section

2.1. Preparation of the Sample. TiO_2/carbon fiber composites were prepared by the following process. A certain amount of bast fiber was located into the tube furnace. In the hydrogen ambience with 100 sccm, the fiber was heated to 300°C at a rate of 10°C/min and then maintained for 1.5 h. The biomass

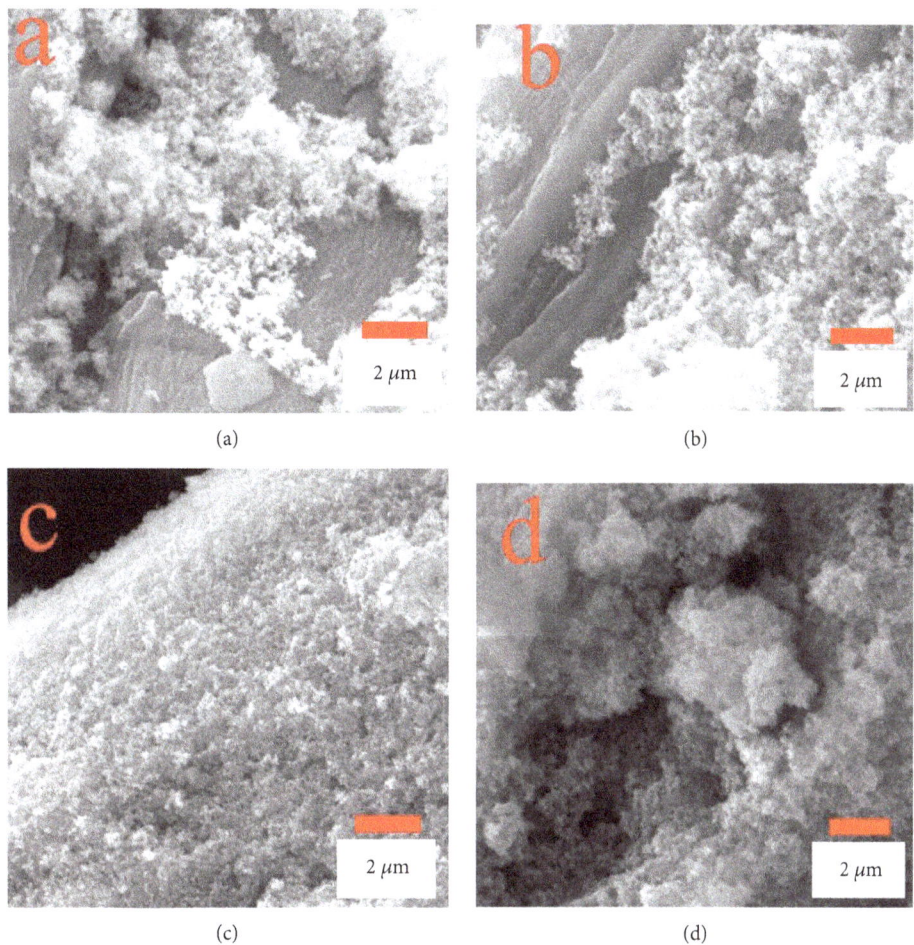

FIGURE 3: (a–d) SEM images of TCF05, TCF10, TCF15, and TCF20.

carbon fiber was obtained after cooling down. Different amounts of TiO_2 nanoparticles (P25) were dispersed into 50 mL deionized water with strong ultrasonic to attain homogenous dispersion (5, 10, 15, and 20 g/L, resp.). After immersing the biomass carbon fiber into TiO_2 dispersion for several minutes, the composites were annealed at 450°C under H_2 ambience. TiO_2/carbon fiber composites with different concentrations of TiO_2 dispersions, labelled as TCF05, TCF10, TCF15, and TCF20, could be achieved by repeating the above-depicted procedures to increase the load amount. The load amount of TiO_2 nanoparticles was estimated to be about 15 wt.% in total mass.

2.2. Characterization.
The morphology of the samples was characterized by field emission scanning electron microscopy (FESEM, Hitachi S4800). Raman spectroscopy (Horiba Jobin Yvon LabRAM HR800) and grazing-angle X-ray diffraction (Rigaku D/MAX-2400) were employed to characterize the crystal structure of the samples. UV-visible absorption spectrum was recorded using Hitachi U-3900H spectrophotometer to evaluate the photocatalytic degradation performance. Water splitting for hydrogen production was applied to evaluate the photocatalysis on photocatalytic hydrogen

production system (Labsolar-IIIAG, Perfectlight Technology Co. Ltd., China).

2.3. Photocatalytic Activity Test

2.3.1. Photocatalytic Degradation of RhB Molecules.
The photocatalytic activities were evaluated by photodegradation of RhB molecules. The procedures had been reported in the previous works [16–18]. 0.2 g TiO_2/carbon fiber composite was dispersed into 100 ml RhB solution (10 mg/L). The suspensions were magnetically stirred in dark for 30 minutes to reach the adsorption-desorption equilibrium. The degradation was carried out under high pressure xenon lamp (300 W) with a cutoff filter ($\lambda \geq 400$ nm). During the irradiation, 4.5 ml dispersion was collected and centrifuged to remove the catalysts with a time interval. The filtrates were analyzed by recording the changes of the absorption bands (around 554 nm) of RhB using Hitachi U-3900H spectrophotometer. After the photocatalyst reaction, TiO_2/carbon fiber composite was rinsed by ethanol and deionized water and dried for recycling.

2.3.2. Photocatalytic Hydrogen Production from Water Splitting.
0.15 g TiO_2/carbon fiber composite was added into

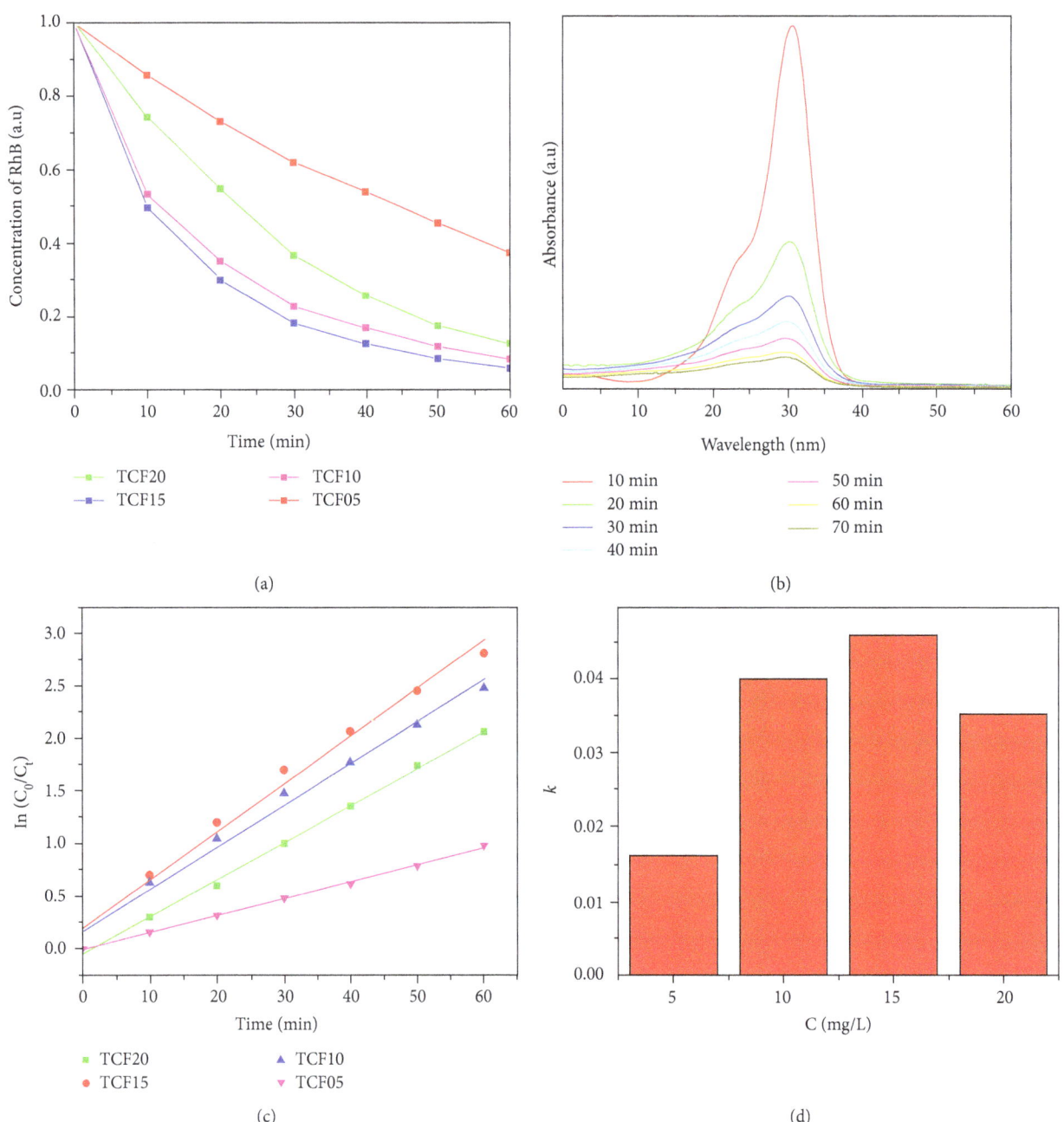

FIGURE 4: (a) The photocatalytic degradation of RhB with different TiO$_2$/carbon fiber micronanocomposites under UV light irradiation. (b) Absorption spectral changes of RhB solution under UV light irradiation in the presence of TCF15. (c) The degradation kinetic data of RhB under UV light irradiation at different sampling time (d). Degradation rate K value of different samples.

the solution mixed by 90 ml water and 10 ml methanol. 2.5 uL chloroplatinic acid solution (10 mg/L), as the assistant catalyst, was added into the solution. Under the 300 W xenon lamp (with cutoff filter) irradiation, the hydrogen amount was analyzed by gas chromatography (GC 7900).

3. Results and Discussion

Figures 1(a) and 1(b) are Raman and SEM images of biomass carbon fiber. Figure 1(a) shows that D peak and G peak of the biomass fiber were located at around 1360 cm^{-1} and

1580 cm^{-1}, which are consistent with the Raman shift peaks of carbon [19]. This indicates that natural bast fiber has been transformed into inorganic carbon fiber after 300°C carbonization process in hydrogen ambience. Figure 1(b) shows that the surface of carbon fiber is much rough. There is a lot of hollow structure on the surface of biomass carbon fiber (the inset), which is in favor of enhancing the adhesion and loading amount of TiO$_2$ nanoparticles on carbon fiber. In addition, during the measurement process of SEM images, we find that the electrical conduction of the biomass carbon fiber is obviously good,

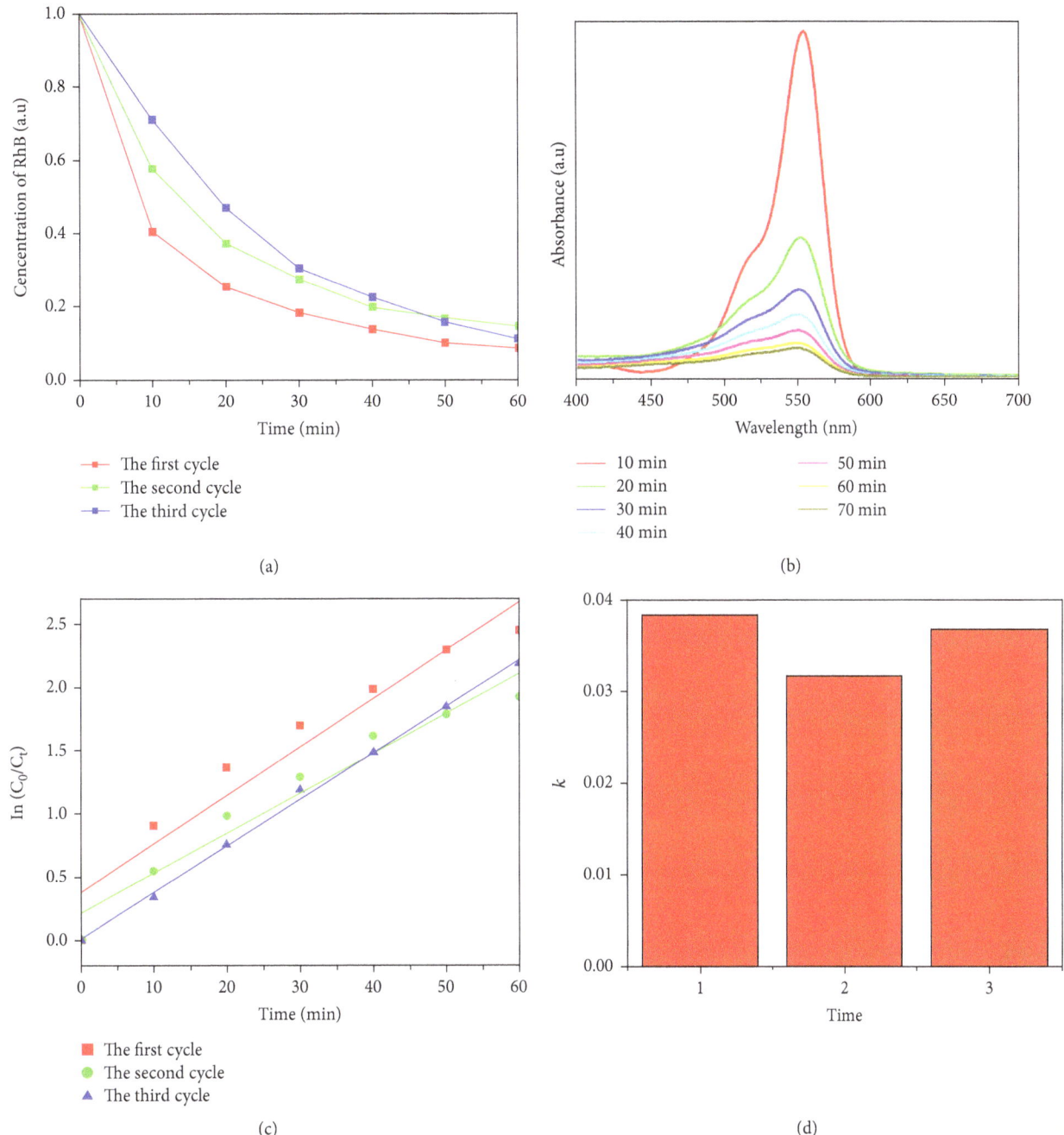

FIGURE 5: (a) The cycle test of TCF15 for photocatalytic degradation of RhB under UV light irradiation. (b) Absorption spectral changes of RhB solution under UV light irradiation in the presence of TCF15. (c) The degradation kinetic data of RhB under UV light irradiation at different cycle times. (d) Degradation rate K value data calculated from (c).

which facilitates the photocatalytic activity of TiO_2/carbon fiber composite.

Figure 2 is XRD patterns and Raman spectra of TCF05, TCF10, TCF15, and TCF20 samples. In Figure 2(a), the peaks sited at approximately 25.21, 37.81, 47.81, 53.91, 55.11, 62.81, 68.81, 70.51, 75.31, and 27.48° are corresponding to the crystal plane (101), (004), (200), (105), (211), (204), (116), (220), and (215) of anatase TiO_2 and (110) of rutile TiO_2, respectively. The broad diffraction peak located at around 20° is related to amorphous carbon, and

the peak gradually decreases due to the increase of TiO_2 load amount from TCF05 to TCF20. These results indicate that the loaded photocatalyst on the biomass carbon fiber is a mixture of anatase and rutile TiO_2. Figure 2(b) shows the Raman spectra of TiO_2/carbon fiber composites. Besides the peaks of carbon fiber, the Raman shift peaks located at 144, 400, 518, and 640 cm^{-1} are related to the Eg, B1g, A1g, and Eg Raman vibrations of anatase TiO_2. It illustrates that there are a large number of TiO_2 nanoparticles on the biomass carbon fiber, which is consistent with XRD results.

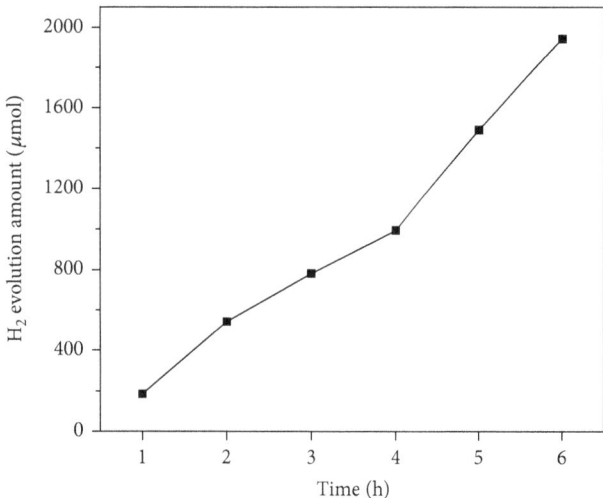

FIGURE 6: The hydrogen production of the TCF15 under UV light irradiation.

We can believe that TCF15 might be better in the four different TiO$_2$/carbon fiber composites based on the XRD and Raman spectra.

Figure 3 shows the SEM images of TCF05, TCF10, TCF15, and TCF20. For the TCF05 (Figure 3(a)) and TCF10 (Figure 3(b)), TiO$_2$ nanoparticles are locally adhered on the surface of biomass carbon fiber. After increasing TiO$_2$ dispersion concentration, the load amount of TiO$_2$ nanoparticles has been greatly improved for the TCF15 (Figure 3(c)). However, the exceeded dispersion concentration leads to the aggregation of TiO$_2$ nanoparticles on the biomass carbon fiber (Figure 3(d)). Due to the weak adhesion between TiO$_2$ nanoparticles, the aggregated nanoparticles will break off during the following photocatalysis experiment. Therefore, it is obvious that the optimal TiO$_2$/carbon fiber composite is TCF15 due to the suitable load amount and uniformity of TiO$_2$ nanoparticles.

Figure 4 shows the photocatalytic degradation of RhB molecules operated by TiO$_2$/carbon fiber composite under UV light irradiation. It can be seen that the TCF15 has the optimal photocatalytic efficiency from Figure 4(a). Photocatalytic reaction kinetic process should conform to the Langmuir-Hinshelwood model [20]:

$$-\frac{dC}{dt} = v = kH = kK[C](1 + K[C]). \qquad (1)$$

Here, C is the concentration of RhB and K is a constant. When the initial concentration of reactants is low, the formula can make the appropriate mathematical transform:

$$Ln\left(\frac{C_D}{C}\right) = Kt. \qquad (2)$$

Here, C_0 represents the initial concentration of RhB and C is the concentration of RhB after irradiation t min. This formula is a reaction kinetic equation. It is found that TiO$_2$ photocatalytic degradation reactions abide by the first-order reaction kinetic equation [21, 22]. Based on the kinetic data

in Figures 4(c) and 4(d), we can get the optimum load amount of TiO$_2$ nanoparticles on the TiO$_2$/carbon fiber composites by comparing the degradation rate K value. It is obviously seen that TCF15 is the optimal composites in our research range. TiO$_2$ nanoparticles are uniformly loaded on the TCF15 surface, and an ideal photocatalytic efficiency can be expected. The higher concentration will lead to the aggregation of TiO$_2$ nanoparticles and the exfoliation in the photocatalyst experiment process.

To characterize the stabilization of the photocatalysis properties, TCF15 is done in the cycle test for photocatalytic degradation of RhB molecules (Figure 5). It can be seen that the photocatalyst efficiency of TCF15 has no obvious decrease after three cycles. The degradation of RhB molecules was more than 90% within 60 min. Therefore, the composite maintains the high photocatalytic degradation performance. It indicates that the TiO$_2$/carbon fiber composite has the reusability without the decrease of photocatalysis efficiency.

Besides the photocatalytic degradation of organic molecules, water splitting for hydrogen production is another excellent property of photocatalyst. The TiO$_2$/carbon fiber composite can also be used to produce hydrogen energy. Figure 6 shows the photocatalytic hydrogen production under UV light irradiation with TCF15. It is found that the amount of production hydrogen with TCF15 photocatalyst linearly increases with the increase of irradiation time. The average hydrogen production rate reaches to 338.51 μmol/h. Therefore, we believe that the TiO$_2$/carbon fiber composite not only has superior photocatalytic degradation reusability, but also exhibits excellent photocatalytic hydrogen production performance.

The above-depicted results indicate that TiO$_2$/carbon fiber composite has the superior photocatalytic degradation of RhB molecules and water splitting for hydrogen production. To check the reliability of the composite, XRD and SEM are carried out on the TCF15 sample after performing the photocatalysis experiments. In Figure 7(a), the XRD peaks of TiO$_2$ are still very obvious, which indicate that TiO$_2$ nanoparticles existed after photocatalytic reaction. SEM image (Figure 7(b)) also proves that the TiO$_2$ nanoparticles are still uniformly loaded on the surface of carbon fibers, which indicates that the exfoliation of TiO$_2$ nanoparticle is not obvious.

Based on the above experimental results, a model is proposed to explain the photocatalytic mechanism of TiO$_2$/biomass carbon fiber composite. Figure 7(c) shows the schematic diagram of the photocatalysis. The biomass carbon fiber surface is very rough, and there is a large number of hollow structures with robust physical adsorption characteristics. After loading TiO$_2$ nanoparticles, the roughness of the fibers is further increased, which is beneficial to the absorption of RhB molecules. As seen from Figure 2(a), TiO$_2$ nanoparticles are made of anatase and rutile phases with 80 : 20 ratio, which formed the heterojunctions (the schematic diagram in Figure 7(d)) [23]. Under the UV light irradiation, the electrons are excited from the valance band (E_v) into the conduction band (E_c) in TiO$_2$, leaving the holes in the valance band. Due to the separation effect of TiO$_2$ heterojunctions for the photogenerated electron-hole pairs, the

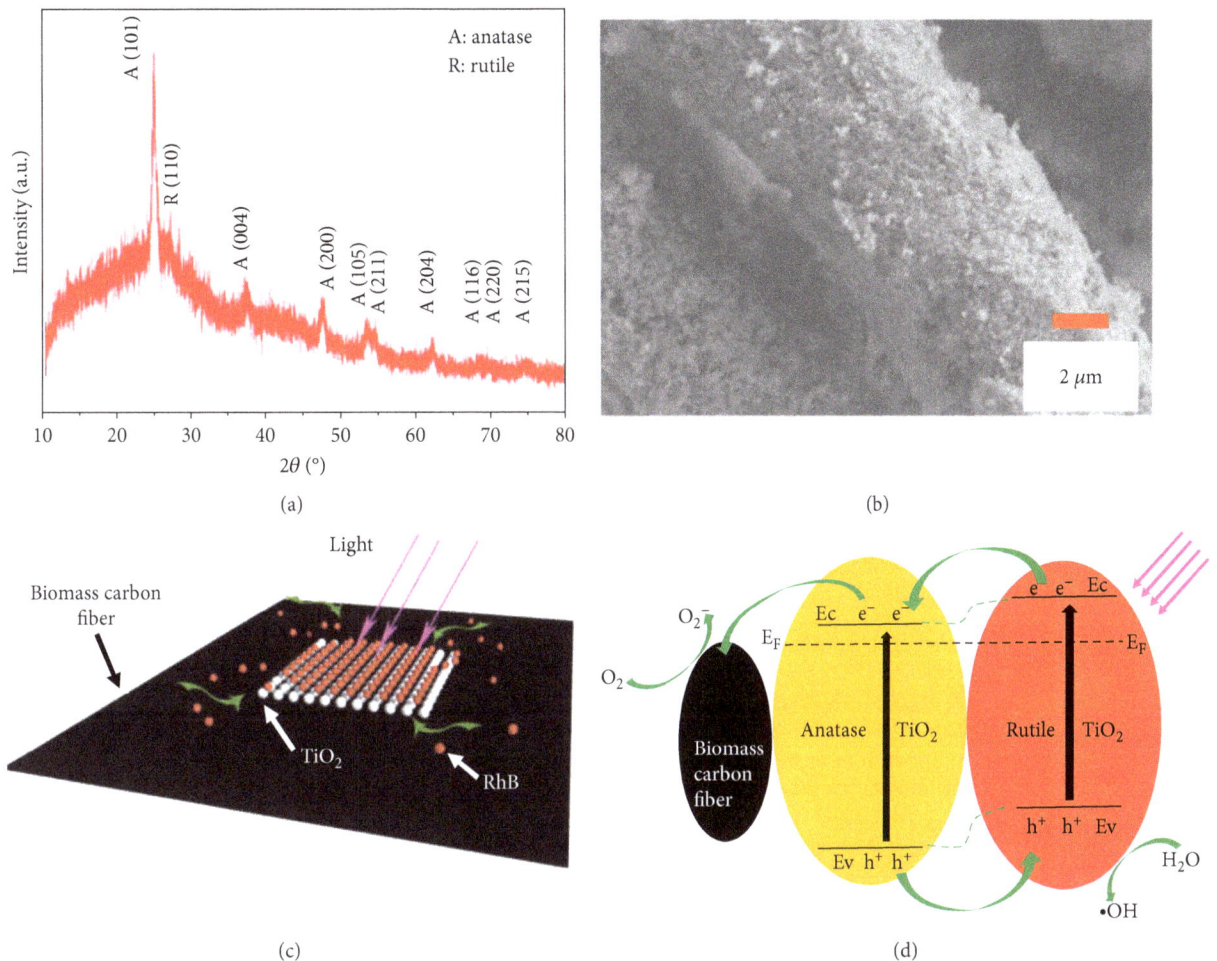

FIGURE 7: XRD pattern (a) and SEM (b) image of TCF15 after the photocatalytic degradation cycle. (c, d) Photocatalytic reaction mechanism diagram.

photogenerated electrons would transfer from the excited rutile TiO_2 phase to the E_c band of anatase TiO_2 phase. At the same time, the photogenerated holes would collect in the E_v band of rutile TiO_2 phase. It is noted that TiO_2 heterojunctions are loaded on the surface of carbon fibers. The carbon fibers have the ability of electron capture since they show high electron storage capacity and electrical conductivity [24]. Due to lower Fermi level of carbon fibers than that of anatase TiO_2 phase, electrons transferred from anatase TiO_2 phase to carbon fibers can be prompted. The above-depicted two processes can minimize the recombination chance of the photogenerated electron-hole pairs, which improves the efficiency of the photocatalytic activity. After being separated, the photogenerated holes combine with water molecules to form hydroxyl radicals. Meanwhile, superoxide radicals can be formed by combining the photogenerated electrons with oxygen molecules. These radicals possess the potential to oxidize the adsorbed RhB molecules on the photocatalyst surfaces [25]. Photocatalytic degradation process of RhB molecules is dominated by the oxidation of chemical reaction under the UV light irradiation. In addition, the strong adsorption ability of carbon fibers increases the contact opportunities of RhB molecules with the composites, which

are also beneficial to the enhancement of photocatalytic properties. Photocatalytic hydrogen production is a dominant reduction reaction of water splitting. Methanol is used to consume the photogenerated hole carriers. The electrons reduce water molecules to produce hydrogen. Therefore, through the design of micronanostructure, we have achieved the recycled, high-efficient TiO_2/biomass carbon fiber composite photocatalyst.

4. Conclusions

As the raw material, natural bast fiber is carbonized to get biomass carbon fiber. TiO_2/carbon fiber composites are achieved by combining biomass carbon fiber and TiO_2 nanoparticles with anatase and rutile mixed phase. Based on the superior adsorbability and electrical conduction of the biomass carbon fiber, the composites show the excellent catalytic activities on photocatalytic degradation and water splitting for hydrogen production. The rate of photocatalytic hydrogen production can reach 338.51 μmol/h. The very favorable feature for TiO_2/carbon fiber composite is easy to recycle as the photocatalyst.

Conflicts of Interest

The authors declare that they have no conflicts of interest.

Acknowledgments

This work was supported by a grant from the Fundamental Research Funds for the Central Universities (Grant no. 31920170036) and the State Key Laboratory of Advanced Processing and Recycling of Non-Ferrous Metals, Lanzhou University of Technology (SKLAB02014003).

References

[1] A. Fujishima and K. Honda, "Electrochemical photolysis of water at a semiconductor electrode," *Nature*, vol. 238, no. 5358, pp. 37-38, 1972.

[2] M. Choi, J. Lim, M. Baek, W. Choi, W. Kim, and K. Yong, "Investigating the unrevealed photocatalytic activity and stability of nanostructured brookite TiO_2 film as an environmental photocatalyst," *ACS Applied Materials & Interfaces*, vol. 9, no. 19, pp. 16252–16260, 2017.

[3] J. Yao and C. Wang, "Decolorization of methylene blue with TiO_2 sol via UV irradiation photocatalytic degradation," *International Journal of Photoenergy*, vol. 2010, Article ID 643182, 6 pages, 2010.

[4] G. Yi, B. Xing, J. Jia et al., "Fabrication and characteristics of macroporous TiO_2 photocatalyst," *International Journal of Photoenergy*, vol. 2014, Article ID 783531, 7 pages, 2014.

[5] Q. Sun, Y. Lu, J. Tu, D. Yang, J. Cao, and J. Li, "Bulky macroporous TiO_2 photocatalyst with cellular structure via facile wood-template method," *International Journal of Photoenergy*, vol. 2013, Article ID 649540, 6 pages, 2013.

[6] A. Nishimura, N. Ishida, D. Tatematsu et al., "Effect of Fe loading condition and reductants on CO2 reduction performance with Fe/TiO_2 photocatalyst," *International Journal of Photoenergy*, vol. 2017, Article ID 1625274, 11 pages, 2017.

[7] Z. Fang, Y. Wang, D. Xu, Y. Tan, and X. Liu, "Blue luminescent center in ZnO films deposited on silicon substrates," *Optical Materials*, vol. 26, no. 3, pp. 239–242, 2004.

[8] C. Zhu, B. Lu, Q. Su, E. Xie, and W. Lan, "A simple method for the preparation of hollow ZnO nanospheres for use as a high performance photocatalyst," *Nanoscale*, vol. 4, no. 10, pp. 3060–3064, 2012.

[9] L. Wang, L. Shen, Y. Li, L. Zhu, J. Shen, and L. Wang, "Enhancement of photocatalytic activity on TiO_2-nitrogen-doped carbon nanotubes nanocomposites," *International Journal of Photoenergy*, vol. 2013, Article ID 824130, 7 pages, 2013.

[10] L. C. Sim, K. H. Leong, S. Ibrahim, and P. Saravanan, "Graphene oxide and Ag engulfed TiO_2 nanotube arrays for enhanced electron mobility and visible-light-driven photocatalytic performance," *Journal of Materials Chemistry A*, vol. 2, no. 15, pp. 5315–5322, 2014.

[11] L. Ji, Y. Zhang, S. Miao, M. Gong, and X. Liu, "In situ synthesis of carbon doped TiO_2 nanotubes with an enhanced photocatalyticperformance under UV and visible light," *Carbon*, vol. 125, pp. 544–550, 2017.

[12] C. Chen, S. Cao, H. Long et al., "Highly efficient photocatalytic performance of graphene oxide/TiO_2-Bi_2O_3 hybrid coating for organic dyes and NO gas," *Journal of Materials Science: Materials in Electronics*, vol. 26, no. 6, pp. 3385–3391, 2015.

[13] Y. Cui, Q. Wei, H. Park, and C. M. Lieber, "Nanowire nanosensors for highly sensitive and selective detection of biological and chemical species," *Science*, vol. 293, no. 5533, pp. 1289–1292, 2001.

[14] A. K. Geim and K. S. Novoselov, "The rise of graphene," *Nature Materials*, vol. 6, no. 3, pp. 183–191, 2007.

[15] J. Yu, G. Wang, B. Cheng, and M. Zhou, "Effects of hydrothermal temperature and time on the photocatalytic activity and microstructures of bimodal mesoporous TiO_2 powders," *Applied Catalysis B: Environmental*, vol. 69, no. 3-4, pp. 171–180, 2007.

[16] P. Jing, W. Lan, Q. Su, M. Yu, and E. Xie, "Visible-light photocatalytic activity of novel $NiTiO_3$ nanowires with rosary-like shape," *Science of Advanced Materials*, vol. 6, no. 3, pp. 434–440, 2014.

[17] P. Jing, W. Lan, Q. Su, and E. Xie, "High photocatalytic activity of V-doped SrTiO3 porous nanofibers produced from a combined electrospinning and thermal diffusion process," *Beilstein Journal of Nanotechnology*, vol. 6, pp. 1281–1286, 2015.

[18] C. Zhu, Y. Li, Q. Su et al., "Electrospinning direct preparation of SnO_2/Fe_2O_3 heterojunction nanotubes as an efficient visible-light photocatalyst," *Journal of Alloys and Compounds*, vol. 575, pp. 333–338, 2013.

[19] A. E. Lewandowska, C. Soutis, L. Savage, and S. J. Eichhorn, "Carbon fibres with ordered graphitic-like aggregate structures from a regenerated cellulose fibre precursor," *Composites Science and Technology*, vol. 116, pp. 50–57, 2015.

[20] J. M. Herrmann, "Photocatalysis fundamentals revisited to avoid several misconceptions," *Applied Catalysis B: Environmental*, vol. 99, no. 3-4, pp. 461–468, 2010.

[21] D. Monllor-Satoca, R. Gomez, M. Gonzalez-Hidalgo, and P. Salvador, "The "direct-indirect" model: an alternative kinetic approach in heterogeneous photocatalysis based on the degree of interaction of dissolved pollutant species with the semiconductor surface," *Catalysis Today*, vol. 129, no. 1-2, pp. 247–255, 2007.

[22] G. Xue, H. Liu, Q. Chen, C. Hills, M. Tyrer, and F. Innocent, "Synergy between surface adsorption and photocatalysis during degradation of humic acid on TiO_2/activated carbon composites," *Journal of Hazardous Materials*, vol. 186, no. 1, pp. 765–772, 2011.

[23] B. Lu, C. Zhu, Z. Zhang, W. Lan, and E. Xie, "Preparation of highly porous TiO_2 nanotubes and their catalytic applications," *Journal of Materials Chemistry*, vol. 22, no. 4, pp. 1375–1379, 2012.

[24] J. Shi, J. Chen, G. Li, T. An, and H. Yamashita, "Fabrication of Au/TiO_2 nanowires@carbon fiber paper ternary composite for visible-light photocatalytic degradation of gaseous styrene," *Catalysis Today*, vol. 281, no. 3, pp. 621–629, 2017.

[25] T. An, L. Sun, G. Li, Y. Gao, and G. Ying, "Photocatalytic degradation and detoxification of o-chloroaniline in the gas phase: mechanistic consideration and mutagenicity assessment of its decomposed gaseous intermediate mixture," *Applied Catalysis B: Environmental*, vol. 102, no. 1-2, pp. 140–146, 2011.

4

A Comparative Analysis of 2-(Thiocyanomethylthio)-Benzothiazole Degradation Using Electro-Fenton and Anodic Oxidation on a Boron-Doped Diamond Electrode

Armando Vázquez,[1] Lucía Alvarado,[2] Isabel Lázaro ⓘ,[1] Roel Cruz,[1] José Luis Nava,[3] and Israel Rodríguez-Torres ⓘ[1]

[1]Instituto de Metalurgia, Facultad de Ingeniería, Universidad Autónoma de San Luis Potosí, Av. Sierra Leona 550, 78210 San Luis Potosí, SLP, Mexico
[2]Departamento de Ingeniería en Minas, Metalurgia y Geología, Universidad de Guanajuato, Ex. Hacienda de San Matías s/n Fracc. San Javier, 36025 Guanajuato, GTO, Mexico
[3]Departamento de Ingeniería Geomática e Hidráulica, Universidad de Guanajuato, Av. Juárez 77, 36000 Guanajuato, GTO, Mexico

Correspondence should be addressed to Israel Rodríguez-Torres; learsi@uaslp.mx

Academic Editor: Reyna Natividad-Rangel

2-(Thiocyanomethylthio)-benzothiazole (TCMTB) is used as fungicide in the paper, tannery, paint, and coatings industries, and its study is important as it is considered toxic to aquatic life. In this study, a comparison of direct anodic oxidation (AO) using a boron-doped diamond electrode (BDD) and electro-Fenton (EF) processes for TCMTB degradation in acidic chloride and sulfate media using a FM01-LC reactor was performed. The results of the electrolysis processes studied in the FM01-LC reactor showed a higher degradation of TCMTB with the anodic oxidation process than with the electro-Fenton process, reaching 81% degradation for the former process versus 47% degradation for the latter process. This difference was attributed to the decrease in H_2O_2 during the EF process, due to parallel oxidation of chlorides. The degradation rate and current efficiency increased as a function of volumetric flow rate, indicating that convection promotes anodic oxidation and electro-Fenton processes. The results showed that both AO and EF processes could be useful strategies for TCMTB toxicity reduction in wastewaters.

1. Introduction

The paper industry has been identified as a major source of pollutants to aquatic environments due to the large volume of wastewater generated per ton of paper produced. The effluents generated during the paper production process cause damage to the receiving waters as they contain high levels of total organic carbon (TOC) and exhibit a chemical oxygen demand (COD) above the permissible limits [1]. Studies on the treatments of wastewater from the paper industry have reported a content of at least 300 different compounds, including certain compounds of biocide nature [2], such as 2-(thiocyanomethylthio)-benzothiazole (TCMTB), which is often used as a biocide in the wood [3] and tannery industries [4]. TCMTB is listed as hazardous by the Environmental

Protection Agency of the United States, and it is considered highly toxic to freshwater fish, freshwater invertebrates, estuarine/marine fish, and estuarine/marine invertebrates [5]. Hence, it is important to develop and apply techniques for the degradation of TCMTB.

Reemtsma et al. [6] achieved an incomplete TCMTB degradation (75%) in an anaerobic and aerobic wastewater treatment pilot plant, yielding mercaptobenzothiazole (MTB), benzothiazole (BT), and hydroxybenzotriazole (OHBT) as the degradation products, which however are harmful compounds that cause dermatitis [7], cell apoptosis [8], and respiratory tract irritation [9], respectively. De Wever et al. [10, 11] carried out studies to investigate the biodegradation of these compounds and established that MTB is a recalcitrant compound. Recalcitrant compounds or persistent organic pollutants (POPs) are characterized by a high stability against sunlight irradiation and a high resistance to either microbial attack (biological processes) or temperature.

Advanced oxidation processes (AOPs) are effective methods that have been developed for POP treatment. These methods primarily involve hydrogen peroxide, ozone, UV-near visible light in the presence of TiO_2, the Fenton reagent, sonolysis, and the sulfate radical-based AOP [12]. Advanced electrooxidation processes (AEOPs) have been proposed as alternative methods for the removal of organics; these processes employ electrochemical cells in which oxidants are produced in situ on the electrode surface [13].

In recent years, the development of AEOPs has significantly increased, particularly oriented to processes of POP degradation. AEOPs are based on the generation of strong oxidizing species, such as the hydroxyl radical ($^{\cdot}$OH), which can alter the chemical structure of the contaminants [14]. The hydroxyl radical is considered the most important free radical in chemistry due to its strong oxidizing nature ($E^{\circ} = 2.8$ V), which is exceeded only by fluorine ($E^{\circ} = 3.05$ V). The oxidizing power of $^{\cdot}$OH destroys most organic pollutants until total mineralization is achieved, that is, conversion to CO_2, water, and inorganic ions.

In this context, one method of generating the hydroxyl radical is through the Fenton reaction. The Fenton process mechanism is initiated by the formation of the homogeneous hydroxyl radical in accordance with the classical Fenton reaction in acidic medium, as follows [14]:

$$Fe^{2+} + H_2O_2 + H^+ \rightarrow Fe^{3+} + H_2O + {^{\cdot}}OH \qquad (1)$$

This technique becomes an attractive choice because only a small catalytic amount of Fe^{2+} is required during the entire process due to the continuous regeneration of the ion from the Fenton-like reaction [15].

Another interesting aspect of $^{\cdot}$OH production from H_2O_2 is that it can be generated by electrochemical reduction from O_2 in aqueous solution under acidic conditions, as follows [14, 16]:

$$O_2 + 2H^+ + 2e^- \rightarrow H_2O_2 \qquad (2)$$

Electro-Fenton (EF) technology is based on the continuous electrogeneration of H_2O_2 on a suitable cathode (generally, carbon-based), which is fed with either O_2 or air [14], and the addition of an iron catalyst to produce the oxidant hydroxyl radical at the bulk via the Fenton reaction, according to (1).

Another electrochemical process that could generate $^{\cdot}$OH radicals is the anodic oxidation of water [17–19]:

$$H_2O + BDD \rightarrow (^{\cdot}OH)BDD + H^+ + e^- \qquad (3)$$

Hydroxyl radical formation is favored on boron-doped diamond (BDD) thin film anodes; this reaction is based on the use of high O_2 overvoltage anodes favoring heterogeneous hydroxyl radical production, BDD($^{\cdot}$OH). It has been reported that organic compounds can be destroyed by anodic oxidation using BDD electrodes, resulting in their complete mineralization [20].

This study compares TCMTB degradation via the electro-Fenton (EF) process and the anodic oxidation process in a filter press reactor FM01-LC. For the EF process, a reticulated vitreous carbon was used as the cathode along with a dimensionally stable anode (DSA) made of Ti with a cover of IrO_2/Ta_2O_5; for the AO process, a BDD electrode was used as anode, and a stainless steel electrode was used as cathode.

2. Materials and Methods

2.1. Reagents and Physicochemical Analysis. All of the chemicals used were of analytical grade, and deionized water (18 MΩ cm) was employed for the preparation of solutions. The organic compound TCMTB was purchased from Insumos Agrícolas Company (industrial grade, 30% purity). The pH measurements were obtained using a Thermo Orion pH meter 420A. The H_2O_2 concentration was determined using the $Ti(SO_4)_2$ colorimetric method and analyzed by UV-vis spectrophotometry at $\lambda = 410$ nm[21]. The performance of the process was evaluated following TCMTB UV-vis absorbance at $\lambda = 290$ nm [3] in a Shimadzu UV/VIS/NIR spectrophotometer UV-3600 with a scanning stage step of 250 nm. Total organic carbon (TOC) was measured with a Shimadzu total organic carbon analyzer 5000A, and chemical oxygen demand (COD) tests were run according to standard protocols [22].

2.2. Microelectrolysis

2.2.1. Experimental Devices. A three-electrode system was used for voltammetric experiments using a 100 mL Pyrex electrochemical cell. The potential was applied using a Princeton Applied Research potentiostat-galvanostat VersaSTAT, and Versa software was used to record the data. For the AO study, a BDD rotating disc electrode (RDE) with a surface area of 0.03141 cm^2 and a DSA (Ti with a cover of IrO_2/Ta_2O_5) with a surface area of 0.096 cm^2 were used as working electrodes. For the cathodic production of hydrogen peroxide, RDE made of vitreous carbon and stainless steel 304 were used, both with a surface area of 0.196 cm^2. The BDD was cleaned using an anodic polarization treatment (1 M $HClO_4$) for 30 min at 10 mA cm^{-2} [23].

A graphite rod was used as the counter electrode in both sets of experiments. Potential measurements were obtained versus a saturated mercurous sulfate reference electrode (SSE) with a potential of 0.6415 V. All the potential measurements shown in this study were referred to the standard hydrogen electrode (SHE). To ensure reproducibility, all of the experiments were performed in triplicate.

2.2.2. Voltammetric Studies. To determine the best current density and electrode potential domain to be applied to favor both the electro-Fenton process (cathodic H_2O_2) and the anodic oxidation (anodic $^{\cdot}$OH), a microelectrolysis study was performed. For this study, two types of solutions were used: (a) a blank solution (0.02 M NaCl and 0.03 M Na_2SO_4) and (b) a synthetic solution (0.02 M NaCl, 0.03 M Na_2SO_4, and 0.07 M TCMTB). The concentration of the blank solution was fixed to an ionic strength similar to that registered in wastewater from the paper industry [24]. The TCMTB concentration was set to achieve a $570 \, mg \, L^{-1}$ TOC (similar to the concentration found in paper industry effluents), which is equivalent for a turbidity of approximately 383 NTU (nephelometric turbidity unit).

Prior to starting the EF experiments, each solution was aerated for 60 min to insure O_2 saturation and acidified with 1 M H_2SO_4 to reach a pH of 3.

A series of anodic and cathodic potential pulses were applied on static electrodes for 30 s from the open-circuit potential (OCP) to positive potentials for the anodic process and to negative potentials for the cathodic process. From the current transients obtained, j-E curves were constructed using current density data sampled at different times for each potential applied.

2.3. Macroelectrolysis

2.3.1. Experimental Devices. Macroelectrolysis experiments were performed in an FM01-LC electrochemical reactor [25, 26]. Figure 1(a) shows an expanded view, including the turbulence promoter type D [27]. The flow distributor thickness was 0.6 cm; a stainless steel plate ($64 \, cm^2$ exposed area) and reticulated vitreous carbon (RVC) of $16 \times 4 \, cm$ and 0.4 cm of thickness (10 pores per inch (ppi), porosity of 0.99, and specific surface area of $4.92 \, cm^{-1}$) were used as cathodes, and BDD and DSA plates ($64 \, cm^2$) were used as anodes. The volume of electrolyte to fill the reactor was $28.2 \, cm^3$. A mercury/mercurous sulfate reference electrode Hg/Hg_2SO_4 was connected to the electrochemical reactor to measure the electrode potential. More details of the FM01-LC are described in detail in [25].

An undivided mode configuration with a single electrolyte compartment and electrolyte flow circuit for the FM01-LC cell is shown in Figure 1(b). The electrolyte was contained in a 2.5 L acrylic reservoir; a Marathon Electric™ 1/3 HP centrifugal coupled pump 335AD-MD was used, and flow rates were measured by a Cole-Parmer variable area plastic flow meter F44500. The electrolyte flow circuit was constructed using 0.5-inch internal diameter PVC tubing and valves as well as three-way connectors constructed of the same material. The experiments were conducted using a Sorensen high-power DC power supplies. In experiments to evaluate H_2O_2 production, the electrochemical reactor was fitted with a RVC as cathode and a stainless steel electrode as anode. For the EF process, the reactor was equipped with the RVC as cathode and DSA as anode; for the AO method, a stainless steel electrode as cathode and a BDD as anode were used.

2.3.2. Electrochemical Degradation of TCMTB in a FM01-LC Filter Press-Type Electrochemical Cell. All of the experiments in the FM01-LC cell were performed at three different volumetric flows (Q_v): 5.67, 9.46, and $13.24 \, L \, min^{-1}$. For each experiment, the final TOC values were measured, and the integral current efficiency was calculated using [25]

$$\phi = \frac{4FV\left[TOC_{(0)} - TOC_{(t)}\right]}{IT}, \quad (4)$$

where F is the Faraday constant with a value of $96{,}485 \, C \, mol^{-1}$, V is the solution volume (L), I is the current

applied (A), and t is the time of electrolysis (s), which for these experiments was 180 minutes.

For electro-Fenton experiments, H_2O_2 generation was achieved using RVC as the cathode, DSA as the anode, and a blank solution as the electrolyte applying a constant current density. This value was determined by varying the current density until reaching the electrode potential determined by the microelectrolysis experiments, which was within $-1.15 \leq E \leq -0.95$ V/SHE. The RVC cathode was supported on a stainless steel plate (current feeder) using conductive carbon paint glue (SPI supplies™).

Prior to starting the EF experiments, each solution was aerated for 60 min to be saturated with O_2 and acidified with 1 M H_2SO_4 to reach a pH of 3. As an initial step, the concentration of H_2O_2 generated in the blank solution was monitored. For the case of H_2O_2 production in the synthetic solution (in presence of TCMTB), it was not possible to measure its concentration using the colorimetric method because the organic compound caused interference. To determine the current density to be applied to the synthetic solution, the same methodology was used; subsequently, the Fenton reaction was promoted by the addition of 0.5 mM Fe^{2+}, which is a similar concentration to that reported by other studies Peralta et al. [15] and Pérez et al. [28], where it was shown mineralization percentages of around 50%.

For the anodic oxidation tests, a BDD was used as the anode and a stainless steel plate was used as the cathode, applying a constant current density that enabled control of the potential at the anode to obtain ($\cdot$OH)BDD.

3. Results and Discussion

3.1. Microelectrolysis Studies of the Electro-Fenton Process. Typical sampled current density (j-E) curves constructed from current density transients (not shown) are illustrated in Figure 2. The j-E curves were obtained at different constant potential pulses and sampling times from 1 to 30 s, using a vitreous carbon electrode for both the blank solution (continuous lines) and the synthetic solution (semi-continuous and dashed lines).

The curves for the blank solution exhibit three electrochemical processes: Ia electrochemical generation of H_2O_2 from -0.95 to -1.15 V/SHE, (2); IIa electrochemical generation of water from -1.35 to -1.5 V/SHE, (5); and IIIa water reduction below -1.65 V/SHE, (6).

$$O_2 + 4H^+ + 4e^- \rightarrow 2H_2O \quad (5)$$

$$2H_2O + 2e^- \rightarrow H_2 + 2OH^- \quad (6)$$

When the organic compound was added (semi-continuous and dashed lines), a new process Ib is observed at potentials from -0.33 to -0.66 V/SHE, which is related to TMCTB but it does not seem to depend on concentration. Then, a similar Ia process is observed at the same potential region observed for the blank solution (-0.95 to -1.15 V); however, TCMTB seems to favor the electrochemical production of H_2O_2 given that the current density is enhanced as the concentration of this organic compound is increased (from 0.035 to 0.07 M). At potentials around -1.36 V/SHE, the

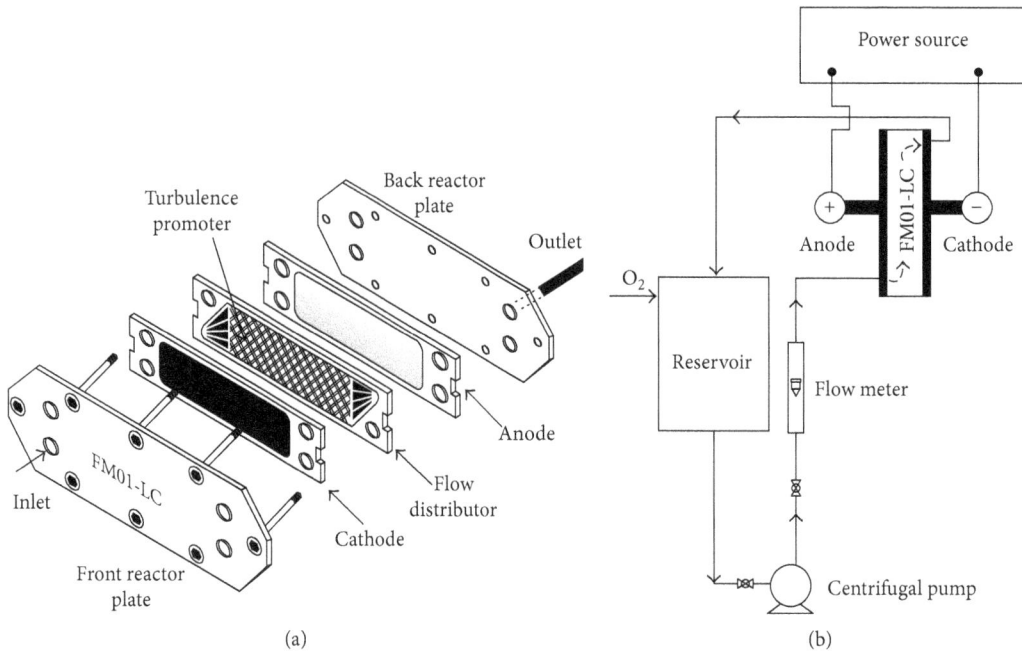

(a)   (b)

FIGURE 1: (a) Expanded view of the FM01-LC cell in the undivided mode. (b) Electrical and flow circuits for the electrochemical flow experiments.

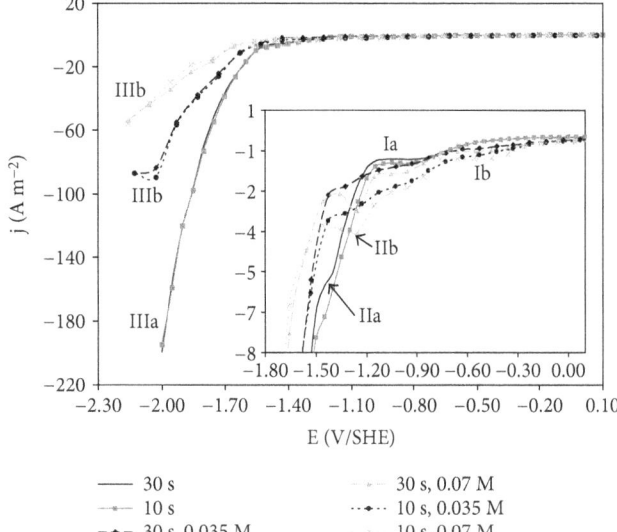

—— 30 s	--·-- 30 s, 0.07 M
--•-- 10 s	···•··· 10 s, 0.035 M
--•-- 30 s, 0.035 M	········ 10 s, 0.07 M

FIGURE 2: Typical j-E curves obtained by sampling current density at different times from current transients obtained at potential pulses between $-2\,V < E > 0.1\,V$ using a vitreous carbon RDE in a blank solution (NaCl 0.02 M and Na_2SO_4 0.03 M, pH 3; continuous lines) and in a synthetic solution (0.02 M NaCl and 0.03 M Na_2SO_4, pH 3, 0.07 M TCMTB—semi-continuous lines; 0.035 M TCMTB—dashed lines). The inset shows an enhanced view. $A_{RDE} = 0.196\,cm^2$.

observed peak (IIb) corresponds to a two-electron reduction process by molecule of TCMTB as reported [4].

Likewise, the reduction of water is shifted to more negative potentials (IIIb), as TCMTB concentration is increased. It is observed that for the two types of solutions, with and without TCMTB, the current density for all of the processes

decreases as the sampling time increases, indicating a mass transport limitation. Moreover, in Figure 2, current density plateaus appeared for the process Ia, which decreased relative to the sampling time, indicating that this process (electrochemical production of H_2O_2) is limited by diffusion.

The anodic processes on the DSA electrode in the blank solution and the synthetic solution were evaluated (Figure 3), and the obtained curves showed that in both solutions, the oxidation becomes important at potentials above 1.4 V/SHE. The low currents obtained in the presence of TCMTB, with respect to blank solution, suggest that in addition to the oxidation processes, part of the energy is also being used for the possible oxidation of the compound. Thus, the oxidation process in the synthetic solution might include the oxidation of hydrogen peroxide, chloride to produce active chlorine, TCMTB with anodically electrogenerated species, and the OER (oxygen evolution reaction); therefore, a careful potential control during the process becomes important [29, 30].

3.2. Microelectrolysis Studies of Anodic Oxidation on the BDD Electrode. A voltammetric study of the cathodic reactions on stainless steel shows only the reduction of the medium from −1.5 V/SHE for both solutions (Figure 4). No sign of TCMTB reduction was observed. A comparison of the results obtained for the synthetic solution on vitreous carbon (Figure 2) and on stainless steel (Figure 4) revealed that the reduction of water in the first case occurred at more negative potentials. This result indicates that the use of stainless steel as a cathode for this system could decrease the cell potential.

Figure 5 shows the curves of current density versus anodic potential pulse on the BDD in different solutions: (a) 1 M $HClO_4$, (b) blank solution, and (c) synthetic solution. In addition, Figure 5 shows a Tafel plot for the $HClO_4$ media, revealing that the current density increases as a function of

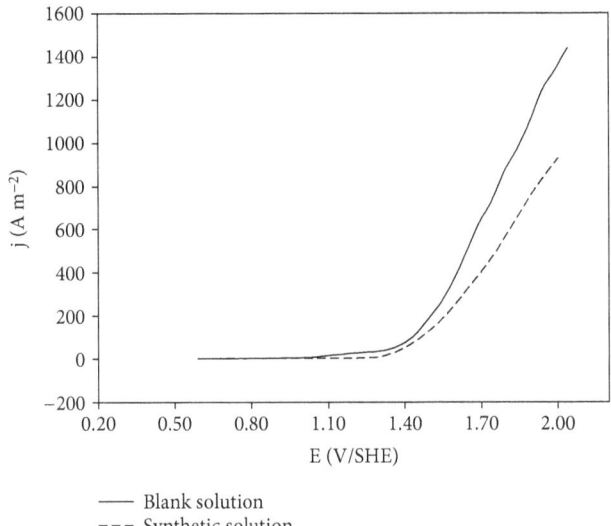

FIGURE 3: Typical j-E curves obtained by sampling current density at different times from typical current transients obtained at potential pulses between 0.5 V < E > 2 V using a DSA electrode in a blank solution (NaCl 0.02 M and Na$_2$SO$_4$ 0.03 M, pH 3; continuous lines) and in a synthetic solution (0.07 M TCMTB, 0.02 M NaCl and 0.03 M Na$_2$SO$_4$, pH 3; dashed lines). $A_{RDE} = 0.09621$ cm^2.

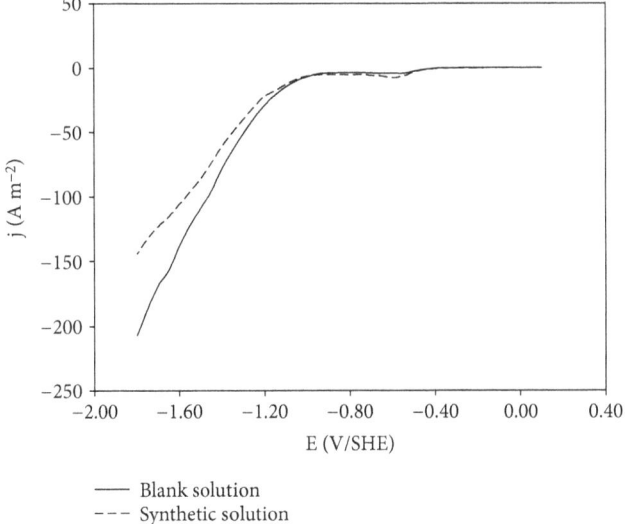

FIGURE 4: Typical j-E curves obtained by sampling current density at different times from typical current transients obtained at potential pulses between −1.8 V < E > 0.1 V using a stainless steel electrode in a blank solution (NaCl 0.02 M and Na$_2$SO$_4$ 0.03 M, pH 3; continuous lines) and in a synthetic solution (0.07 M TCMTB, 0.02 M NaCl, and 0.03 M Na$_2$SO$_4$, pH 3; dashed lines). $A_{RDE} = 0.196$ cm^2.

the imposed potential, which shows that detection of the faradaic current began above 2.3 V, which is consistent with the response of the BDD because this type of electrode requires higher overpotentials to oxidize water [25].

The Tafel slope from Figure 5(a) (insert) was evaluated over a potential range of $2.3 \leq E \leq 2.75$ V/SHE, and a value of 290 mV decade^{-1} was observed, which was similar to that reported by Michaud et al. [17], who determined a value of

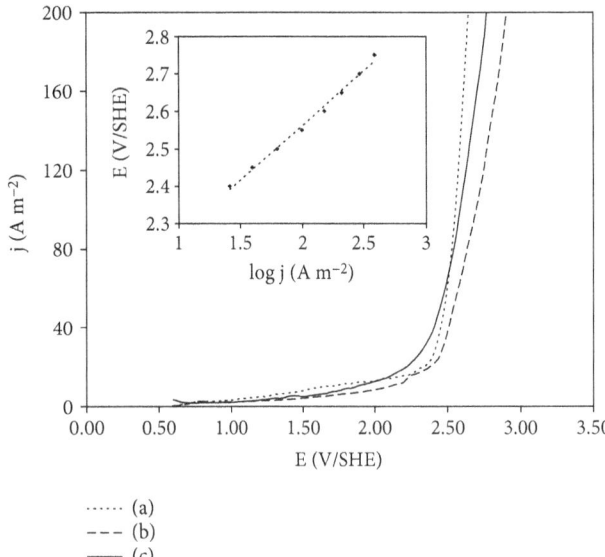

FIGURE 5: Typical j-E curves obtained by sampling current density at different times from typical current transients obtained at potential pulses between 0.6 V < E > 3 V using a BDD RDE for (a) 1 M HClO$_4$, (b) 0.02 M NaCl and 0.03 M Na$_2$SO$_4$ at pH 3, and (c) 0.07 M TCMTB in 0.02 M NaCl and 0.03 M Na$_2$SO$_4$ at pH 3. The inset shows the Tafel plot for j-E curves in (a). $A_{RDE} = 0.031416$ cm^2.

250 mV decade^{-1} over a potential range of $2.4 \leq E \leq 2.9$ V/SHE in which the water oxidation was a one-electron process and the hydroxyl radical formation occurred according to (3). An analysis of the Tafel slopes over the potential range of $2.3 \leq E \leq 2.6$ V for the curves in Figures 5(b) and 5(c) (not shown) exhibited similar values in all cases. In addition, a certain amount of TCMTB adsorption on the BDD surface could be present in Figure 5(c), as indicated by the decrease in the current density. However, TCMTB would be oxidized through ('OH)BDD over the interval of $2.3 \leq E \leq 2.75$ V/SHE in addition to the active chlorine produced on the BDD surface, which will be discussed later.

3.3. Comparison between the Electrochemical Degradation of TCMTB by Electro-Fenton and by Anodic Oxidation on a BDD. A 2.4 mA cm^{-2} cathodic current density was applied to the RVC in the FM01-LC reactor to maintain the electrode potential over the range of $-0.95 \leq E \leq -1.15$ V/SHE and to promote the electrochemical generation of H$_2$O$_2$ on the RVC in the blank solution. As it is shown in Figure 6, the H$_2$O$_2$ concentration increases linearly with the volumetric flow rate over the first 30 min, and then it decreases for a short time; this behavior is repeated. This result is attributed to the instability of the H$_2$O$_2$ molecule, which is accompanied by its exothermic decomposition to oxygen and water, as follows [31]:

$$2H_2O_2 \rightarrow O_2 + 2H_2O \tag{7}$$

At room temperature (ca. 23°C), the rate of decomposition is slow; however, as the temperature increases, the rate of decomposition also increases [32]; in this study, the batch mode employed provoked a temperature solution of 60°C,

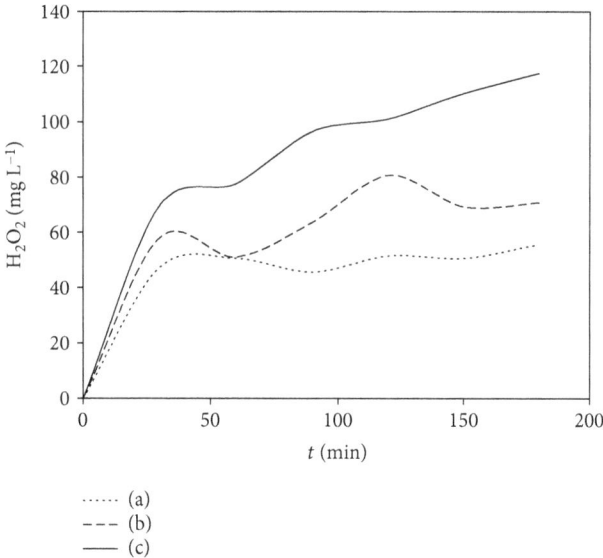

TABLE 1: Current efficiencies for electrochemical generation of H_2O_2 at different volumetric flows ($j = 2.4\,\text{mA cm}^{-2}$).

Q_v (L min^{-1})	Current efficiency (%)
5.67	24
9.46	32
13.24	51

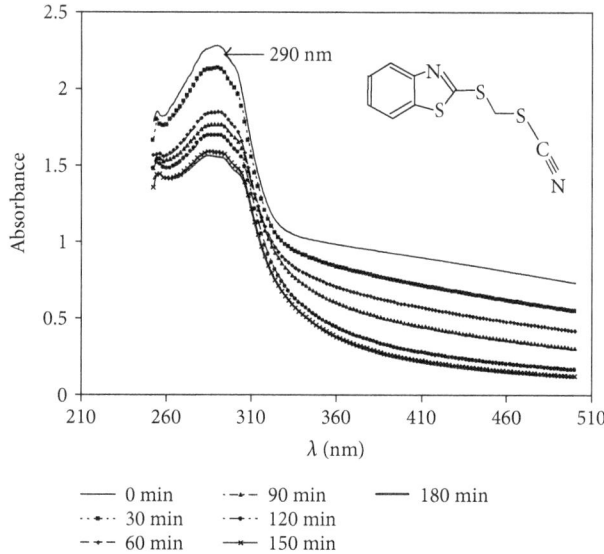

FIGURE 6: Influence of the volumetric flow rate on electrochemical generation of H_2O_2 on the RVC cathode fitted in the FM01-LC reactor. Electrolyte: 0.02 M NaCl and 0.03 M Na_2SO_4 at pH 3. $A_{RVC} = 125.95\,\text{cm}^2$, $A_{stainless\ steel} = 64\,\text{cm}^2$. $2.4\,\text{mA cm}^{-2}$. (a) 5.67 L min^{-1}, (b) 9.46 L min^{-1}, and (c) 13.24 L min^{-1}.

FIGURE 7: Change in the UV-vis spectra of TCMTB as a function of electrolysis time by the electro-Fenton process in the FM01-LC reactor equipped with an RVC cathode and DSA. Electrolyte: 0.02 M NaCl, 0.03 M Na_2SO_4, and 0.5 mM Fe^{2+} at pH 3. $Q_v = 5.67$ L min^{-1}, $2.8\,\text{mA cm}^{-2}$, $A_{RVC} = 125.95\,\text{cm}^2$, $A_{DSA} = 64\,\text{cm}^2$.

because of fluid recirculation. In addition, hydrogen peroxide could react with the chloride ions in solution [30], resulting in the formation of chlorine gas and water (8), which would explain the 8% decrease in the chloride concentration in all of the experiments using the blank solution.

$$H_2O_2 + 2Cl^- + 2H^+ \rightarrow 2H_2O + Cl_2 \qquad (8)$$

Similar to other studies [33], the results obtained in this work showed that there is a mass transport limitation during cathodic H_2O_2 production, which decreases its concentration. However, as it has been reported even with the lowest H_2O_2 concentration (45 mg L^{-1}, $Q_v = 5.6$ L min^{-1}) achieved, it is possible to get an adequate Fenton reaction performance [31].

The current efficiency obtained for the electrochemical generation of H_2O_2 in RVC is shown in Table 1, demonstrating that the current efficiency increases as the volumetric flow increases. This result is attributed to the decrease in the resistance to mass transfer because the convection process is favored.

For the electro-Fenton studies, a current density of $2.8\,\text{mA cm}^{-2}$ was maintained ($-0.65 \leq E \leq -0.8$ V) using the synthetic solution (0.07 M TCMTB, 0.03 M Na_2SO_4, and 0.02 M NaCl at pH 3) and using 0.5 mM Fe^{2+}. The presence of Fe^{2+} allows the generation of the homogeneous ·OH radical. According to Figure 7, TCMTB degradation is possible given that a decrease in absorbance is observed from 2.2 to 1.5 ($\lambda = 290$ nm) at $Q_v = 5.67$ L min^{-1} in 180 min.

For the anodic oxidation, the BDD anodic potential was maintained between $2.3 \leq E \leq 2.6$ V (a potential range over which BDD(·OH) is generated) by setting the current density to $6.8\,\text{mA cm}^{-2}$. All of the electrolysis experiments were

performed in an undivided FM01-LC reactor, and the electrode potential was controlled over the range in which the desirable electrochemical processes are favored. Figure 8 shows that the turbidity (absorbance) decreases with time in all of the experiments ($\lambda = 290$ nm). For the electro-Fenton process, it is observed that by increasing the feed rate, a higher rate of TCMTB degradation is achieved. As previously discussed, an increase in the flow rate increases the H_2O_2 concentration; thus, the amount of ·OH radicals also increases in the bulk solution, favoring the degradation of the TCMTB molecules. A similar phenomenon can be observed in the anodic oxidation process in which an increase in the flow rate also increases the rate of TCMTB degradation. This phenomenon is explained in terms of the effect of increasing the volume flow on the transport of the organic compound from the bulk solution to the electrode surface where radicals are physisorbed BDD(·OH). A comparison of the EF and AO processes shows that higher percentages of degradation are achieved by anodic oxidation (81%) because the BDD(·OH) radical is constantly produced on the BDD surface. In EF, the production of homogeneous radicals is limited by the amount of H_2O_2 available for the process because H_2O_2 reacts not only with Fe^{2+} but also with chloride ions, in addition to undergoing its natural decomposition.

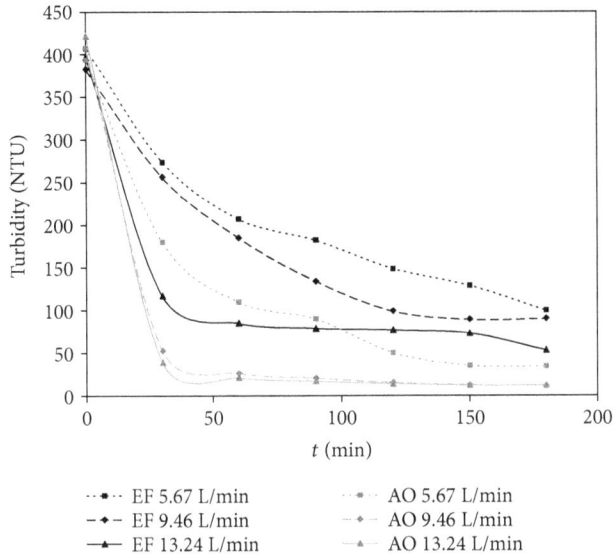

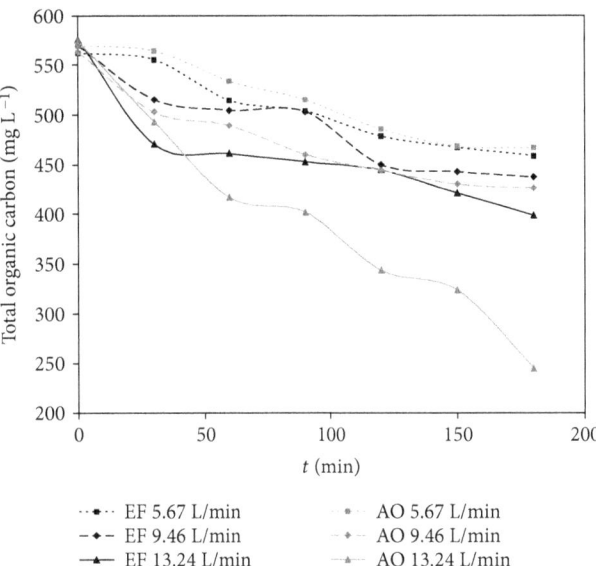

FIGURE 8: Decrease in turbidity as a function of electrolysis time measured at 290 nm during the EF process in the FM01-LC reactor equipped with an RVC cathode and DSA (at 3 mA cm^{-2}, 0.5 mM Fe^{2+}) and the AO process in the FM01-LC reactor equipped with an stainless steel cathode and BDD anode at 6.8 mA cm^{-2}. $A_{RVC} = 125.95$ cm^2, $A_{DSA} = 64$ cm^2, $A_{stainless\ steel} = 64$ cm^2, and $A_{BDD} = 64$ cm^2. Electrolyte: 0.07 M TCMTB in 0.02 M NaCl and 0.03 M Na$_2$SO$_4$.

FIGURE 9: Influence of the volumetric flow rate on TOC removal during the EF and AO processes in the FM01-LC reactor. EF: 2.8 mA cm^{-2}, 0.5 mM Fe^{2+}, $A_{DSA} = 64$ cm^2, $A_{RVC} = 125.95$ cm^2. AO: 6.8 mA cm^{-2}, $A_{BDD} = 64$ cm^2, $A_{stainless\ steel} = 64$ cm^2. Electrolyte: 0.07 M TCMTB in 0.02 M NaCl and 0.03 M Na$_2$SO$_4$.

Although the absorbance for the turbidity values decreases, the TOC values do not decrease at the same rate (Figure 9), indicating that TCMTB is degraded into simpler organic compounds, as has been reported by several authors [10]. It is important to mention that chromatographic studies might help elucidating the differences between the decrease in absorbance and TOC; however, these studies were beyond the scope of this paper. Based on the results obtained here, it is clear that for the anodic oxidation on BDD using a flow rate of 13.24 L min^{-1}, it was possible to achieve up to 57% TCMTB mineralization during the 180 minutes of electrolysis. This achievement was aided by the constant electrogeneration of BDD($^{\cdot}$OH) radicals and the favored transport of TCMTB molecules to the electrode interface.

However, chloride ions affect the percentage of mineralization because they react in a complex mechanism to produce adsorbed chlorine (although weakly sorbed at the electrode surface) and dissolved chlorine, which reacts with water and yields hypochlorous acid, as follows [30]:

$$2Cl^- \rightarrow Cl_{2,ads} + 2e^-$$
$$Cl_{2,ads} \rightarrow Cl_{2,aq} \qquad (9)$$
$$Cl_{2,aq} + H_2O \rightarrow HClO + H^+ + Cl^-$$

These reactions could explain why the concentration of chloride ions decreases significantly in the AO experiments, as shown in Table 2. The formed hypochlorous acid could react with hydroxyl radicals to form chlorine dioxide and

chlorate [30], thus diminishing the concentration of hydroxyl radicals and preventing higher percentages of mineralization.

$$3^{\cdot}OH + HClO \rightarrow ClO_2 + 2H_2O$$
$$4^{\cdot}OH + HClO \rightarrow ClO_3^- + H^+ + 2H_2O \qquad (10)$$

According to Polcaro et al. [30], for chloride ion concentrations on the BDD anode surface, similar to those in our study, approximately 40% of the current would be used in the formation of chlorine, possibly causing the low current efficiencies observed, which are similar other reports in the literature [23]. However, the removal of chlorides was higher when the volumetric flow was increased, resulting in a major mass transport in the reactor. Moreover, the percentage of mineralization also increased due to the higher mass transport. For the case of sulfates, the concentration did not change in this study.

Table 2 shows that the current efficiency and the mineralization of TCMTB are favored by the increase of Q_v. This increase of current efficiency due to convection flow agrees well with the results reported by Panizza et al. [34]. In addition, the formation of active chlorine species can also aid the TCMTB degradation. A higher Q_v value particularly favors anodic oxidation; thus, better results are obtained than those with the electro-Fenton process. However, the consumption of energy for AO is twice that of EF.

4. Conclusions

Macroelectrolysis studies showed that anodic oxidation produces better percentages of degradation than the electro-Fenton process. This result was most likely due to the

TABLE 2: Summary of the results obtained during the different experiments.

Experiment	% removal absorbance	% removal TOC	% removal COD	% removal Cl^-	Integral current efficiency, ϕ (%)	Energy consumption $(kWh\,m^{-3})$
EF $(5.67\,L\,min^{-1})$	40	18	42	8	11	0.113
EF $(9.46\,L\,min^{-1})$	40	23	45	12	13	0.123
EF $(13.24\,L\,min^{-1})$	47	31	72	12	18	0.116
AO $(5.67\,L\,min^{-1})$	71	18	32	42	9	0.248
AO $(9.46\,L\,min^{-1})$	81	24	55	46	12	0.225
AO $(13.24\,L\,min^{-1})$	81	57	78	52	56	0.225

decreased H_2O_2 concentration caused by different reactions that could occur in the solution in the EF process.

Microelectrolysis studies indicated that the degradation and partial mineralization of TCMTB by anodic oxidation were achieved via hydroxyl radicals formed by the oxidation of water in the BDD electrode under galvanostatic conditions.

Electrolysis in the undivided FM01-LC reactor at different volumetric flows at a current density of $6.8\,mA\,cm^{-2}$ revealed that the oxidation rate and current efficiency increased as a function of Q_v. This result demonstrates that convection flow favors the influx of TCMTB to the BDD($^{\cdot}$OH) surface, increasing its degradation.

The electrochemical transformation of TCMTB by the electro-Fenton and anodic oxidation processes could be a useful strategy for toxicity reduction.

Nomenclature

BDD:	Boron-doped diamond
COD:	Chemical oxygen demand $(mol\,L^{-1})$
DSA:	Dimensionless stable anode
F:	Faraday constant $(96,485\,C\,mol^{-1})$
I:	Current applied during electrolysis (A)
j:	Current density $(A\,cm^{-2})$
Q_v:	Volumetric flow $(L\,min^{-1})$
RVC:	Reticulated vitreous carbon
TCMTB:	2-(Thiocyanomethylthio)-benzothiazole
$TOC_{(0)}$:	Initial total organic carbon $(mol\,L^{-1})$
$TOC_{(t)}$:	Total organic carbon at time t $(mol\,L^{-1})$
t:	Time of electrolysis (s)
V:	Solution volume (L)
ϕ:	Current efficiency (%).

Conflicts of Interest

The authors declare that they have no conflicts of interest.

Acknowledgments

The authors are grateful for the SEP-CONACyT grant from the National Council of Science and Technology of Mexico (CONACyT) through the Project no. 240522. Armando Vázquez would like to thank the CONACyT for the doctoral scholarship 217508. The authors also thank the financial support for the publication from SEP-PRODEP. The authors are indebted to Nubia V. Arteaga for laboratory support.

References

[1] D. Pokhrel and T. Viraraghavan, "Treatment of pulp and paper mill wastewater – a review," *Science of The Total Environment*, vol. 333, no. 1-3, pp. 37–58, 2004.

[2] A. Latorre, A. Rigol, S. Lacorte, and D. Barceló, "Organic compounds in paper mill wastewaters," in *The Handbook of Environmental Chemistry*, D. Barceló and A. G. Kostianoy, Eds., vol. 2 of Part O, Springer-Verlag, Berlin Heidelberg, 2005.

[3] K. Tumirah, S. Salamah, A. Rozita, U. Salmiah, and M. A. M. Nasir, "Determination of 2-thiocyanomethylthio benzothiazole (TCMTB) in treated wood and wood preservative using ultraviolet–visible spectrophotometer," *Wood Science and Technology*, vol. 46, no. 6, pp. 1021–1031, 2012.

[4] E. Meneses, M. Arguelho, and J. Alves, "Electroreduction of the antifouling agent TCMTB and its electroanalytical determination in tannery wastewaters," *Talanta*, vol. 67, no. 4, pp. 682–685, 2005.

[5] United States Environmental Protection Agency, "Reregistration eligibility decision for 2-(thiocyanonethylthio) benzothiazole (TCMTB), prevention, pesticides and toxic substances (7510P), EPA739-R-05-003," 2006, October 2017, http://www3.epa.gov/pesticides/chem_search/reg_actions/reregistration/red_PC-035603_1-Aug-06.pdf.

[6] T. Reemtsma, O. Fienh, G. Kalnowski, and M. Jekel, "Microbial transformations and biological effects of fungicide-derived benzothiazoles determined in industrial wastewater," *Environmental Science & Technology*, vol. 29, no. 2, pp. 478–485, 1995.

[7] A. K. Adams and E. M. Warshaw, "Allergic contact dermatitis from mercapto compounds," *Dermatitis*, vol. 17, no. 2, pp. 56–70, 2006.

[8] M. Rajabi, "2-(3,5-Dihydroxyphenyl)-6-hydroxybenzothiazole arrests cell growth and cell cycle and induces apoptosis in breast cancer cell lines," *DNA and Cell Biology*, vol. 31, no. 3, pp. 388–391, 2012.

[9] G. Ginsberg, B. Toal, and T. Kurland, "Benzothiazole toxicity assessment in support of synthetic turf field human health risk assessment," *Journal of Toxicology and Environmental Health, Part A*, vol. 74, no. 17, pp. 1175–1183, 2011.

[10] H. De Wever and H. Verachtert, "Biodegradation and toxicity of benzothiazoles," *Water Research*, vol. 31, no. 11, pp. 2673–2684, 1997.

[11] H. De Wever, H. Verachtert, and P. Besse, "Microbial transformations of 2-substituted benzothiazoles," *Applied Microbiology and Biotechnology*, vol. 57, no. 5-6, pp. 620–625, 2001.

[12] A. Al-Kadasi, A. Idris, K. Saed, and C. T. Guan, "Treatment of textile wastewater by advanced oxidation processes – a review," *Global NEST Journal*, vol. 6, pp. 222–230, 2004,

October 2015, http://journal.gnest.org/sites/default/files/Journal%20Papers/Al-kdasi-222-230.pdf.

[13] H. Särkka, A. Bhatnagar, and M. Sillanppa, "Recent developments of electro-oxidation in water treatment – a review," *Journal of Electroanalytical Chemistry*, vol. 754, pp. 46–56, 2015.

[14] E. Brillas, I. Sirés, and M. Oturan, "Electro-Fenton process and related electrochemical technologies based on Fenton's reaction chemistry," *Chemical Reviews*, vol. 109, no. 12, pp. 6570–6631, 2009.

[15] J. M. Peralta-Hernández, Y. Meas-Vong, F. J. Rodríguez, T. W. Chapman, M. I. Maldonado, and L. A. Godínez, "In situ electrochemical and photo-electrochemical generation of the Fenton reagent: a potentially important new water treatment technology," *Water Research*, vol. 40, no. 9, pp. 1754–1762, 2006.

[16] E. L. Gyenge and C. W. Oloman, "Influence of surfactants on the electro-reduction of oxygen to hydrogen peroxide in acid and alkaline electrolytes," *Journal of Applied Electrochemistry*, vol. 31, no. 2, pp. 233–243, 2001.

[17] P. A. Michaud, M. Panizza, L. Outtara, T. Diaco, G. Foti, and C. H. Comninellis, "Electrochemical oxidation of water on synthetic boron-doped diamond thin film anodes," *Journal of Applied Electrochemistry*, vol. 33, no. 2, pp. 151–154, 2003.

[18] C. A. Martínez-Huitle and E. Brillas, "Decontamination of wastewaters containing synthetic organic dyes by electrochemical methods: a general review," *Applied Catalysis B: Environmental*, vol. 87, no. 3-4, pp. 105–145, 2009.

[19] C. Barrera-Díaz, P. Cañizares, F. J. Fernández, R. Natividad, and M. A. Rodrigo, "Electrochemical advanced oxidation processes: an overview of the current applications to actual industrial effluents," *Journal of the Mexican Chemical Society*, vol. 58, pp. 256–275, 2014.

[20] J. A. Garrido, E. Brillas, P. L. Cabot, F. Centellas, C. Arias, and R. M. Rodríguez, "Mineralization of drugs in aqueous medium by advanced oxidation processes," *Portugaliae Electrochimica Acta*, vol. 25, no. 1, pp. 19–41, 2007.

[21] G. Eisenberg, "Colorimetric determination of hydrogen peroxide," *Industrial and Engineering Chemistry, Analytical Edition*, vol. 15, no. 5, pp. 327-328, 1943.

[22] A. D. Eaton, L. S. Clesceri, E. W. Rice, and A. E. Greenberg, *Standard Methods for the Examination of Water and Wastewater*, APHA, AWWA, & WEF, Washington, USA, 21st edition, 2005.

[23] J. L. Nava, I. Sirés, and E. Brillas, "Electrochemical incineration of indigo. A comparative study between 2D (plate) and 3D (mesh) BDD anodes fitted into a filter-press reactor," *Environmental Science and Pollution Research*, vol. 21, no. 14, pp. 8485–8492, 2014.

[24] A. Vázquez, J. L. Nava, R. Cruz, I. Lázaro, and I. Rodríguez, "The importance of current distribution and cell hydrodynamic analysis for the design of electrocoagulation reactors," *Journal of Chemical Technology and Biotechnology*, vol. 89, no. 2, pp. 220–229, 2014.

[25] J. L. Nava, F. Núñez, and I. González, "Electrochemical incineration of p-cresol and o-cresol in the filter-press-type FM01-LC electrochemical cell using BDD electrodes in sulfate media at pH 0," *Electrochimica Acta*, vol. 52, no. 9, pp. 3229–3235, 2007.

[26] E. Butrón, M. E. Juárez, M. Solis, M. Teutli, I. González, and J. L. Nava, "Electrochemical incineration of indigo textile dye in filter-press-type FM01-LC electrochemical cell using BDD electrodes," *Electrochimica Acta*, vol. 52, no. 24, pp. 6888–6894, 2007.

[27] M. Griffiths, C. Ponce de León, and F. Walsh, "Mass transport in the rectangular channel of a filter-press electrolyzer (the FM01-LC reactor)," *AICHE Journal*, vol. 51, no. 2, pp. 682–687, 2005.

[28] T. Pérez, S. Garcia-Segura, A. El-Ghenymy, J. L. Nava, and E. Brillas, "Solar photoelectro-Fenton degradation of the antibiotic metronidazole using a flow plant with a Pt/air-diffusion cell and a CPC photoreactor," *Electrochimica Acta*, vol. 165, pp. 173–181, 2015.

[29] E. Petrucci, D. Montanaro, and L. Di Palma, "A feasibility study of hydrogen peroxide electrogeneration in seawater for environmental remediation," *Chemical Engineering Transactions*, vol. 28, pp. 91–96, 2012.

[30] A. M. Polcaro, A. Vacca, M. Mascia, S. Palmas, and J. Rodiguez Ruiz, "Electrochemical treatment of waters with BDD anodes: kinetics of the reactions involving chlorides," *Journal of Applied Electrochemistry*, vol. 39, no. 11, pp. 2083–2092, 2009.

[31] A. Alvarez-Gallegos and D. Pletcher, "The removal of low level organics via hydrogen peroxide formed in a reticulated vitreous carbon cathode cell, Part 1. The electrosynthesis of hydrogen peroxide in aqueous acidic solutions," *Electrochimica Acta*, vol. 44, no. 5, pp. 853–861, 1998.

[32] Z. Qiang, J. H. Chang, and C. P. Huang, "Electrochemical generation of hydrogen peroxide from dissolved oxygen in acidic solutions," *Water Research*, vol. 36, no. 1, pp. 85–94, 2002.

[33] K. Cruz-González, O. Torres-López, A. García-León et al., "Determination of optimum operating parameters for Acid Yellow 36 decolorization by electro-Fenton process using BDD cathode," *Chemical Engineering Journal*, vol. 160, no. 1, pp. 199–206, 2010.

[34] M. Panizza, M. Delucchi, and G. Cerisola, "Electrochemical degradation of anionic surfactants," *Journal of Applied Electrochemistry*, vol. 35, no. 4, pp. 357–361, 2005.

5

New Approach of the Oxidant Peroxo Method (OPM) Route to Obtain Ti(OH)$_4$ Nanoparticles with High Photocatalytic Activity under Visible Radiation

André E. Nogueira [1,2] **Lucas S. Ribeiro,**[3] **Luiz F. Gorup,**[4] **Gelson T. S. T. Silva,**[2] **Fernando F. B. Silva,**[2] **Caue Ribeiro** [2] **and Emerson R. Camargo**[3]

[1]*Brazilian Nanotechnology National Laboratory (LNNano), Brazilian Center for Research in Energy and Materials (CNPEM), Zip Code 13083-970 Campinas, São Paulo, Brazil*
[2]*National Laboratory of Nanotechnology for Agrobusiness (LNNA), EMBRAPA-Brazilian Agricultural Research Corporation, Rua XV de Novembro, 1452, São Carlos, SP 13560-970, Brazil*
[3]*Interdisciplinary Laboratory of Electrochemistry and Ceramics (LIEC), Department of Chemistry, Federal University of São Carlos, Rod. Washington Luis km 235, CP 676 São Carlos, SP 13565-905, Brazil*
[4]*FACET-Department of Chemistry, Federal University of Grande Dourados, Dourados, Mato Grosso do Sul 79804-970, Brazil*

Correspondence should be addressed to André E. Nogueira; andreesteves86@hotmail.com

Academic Editor: Juan M. Coronado

Environmental problems related to the generation of wastewater contaminated with organic compounds and the emissions of pollutants from fuel burning have become major global problems. Thus, there is a need for the development of alternative and economically viable technologies for the remediation of the affected ecosystems. Therefore, this work describes the preparation and characterization of a Ti(OH)$_4$ catalyst with the modified surface for application in the photodegradation of organic compounds (methylene blue (MB) dye and the drug amiloride (AML)) and in the artificial photosynthesis process. Characterization results reveal that peroxo groups on the surface of the catalyst had a great influence on the optical properties of the Ti(OH)$_4$ and consequently in its photocatalytic property. This catalyst showed a high photocatalytic activity for the degradation of organic pollutants under visible radiation, reaching approximately 98% removal of both the dye and the drug in 150 min of reaction. In addition, the catalyst presented a great potential for the reduction of CO$_2$ under ultraviolet (UV) radiation when compared to P25, which is a classic catalyst used in photocatalytic processes. The highest photocatalytic activity can be attributed to the strong visible light absorption, due to the narrow band gap, and the effective separation of photogenerated electron-hole pairs caused by the peroxo groups on the Ti(OH)$_4$ surface.

1. Introduction

Photoactivated processes are receiving enormous attention due to important applications as solar cells, CO$_2$ photoreduction, water splitting for hydrogen production, for the degradation of organic compounds, and others [1, 2]. During these processes, semiconductors are irradiated with UV or visible radiation to excite electrons from the valence band (VB) to the conduction band (CB) creating electron/hole pairs. Most of the pairs recombine, but the remaining, via oxidation-reduction reactions, promote the oxidation of water, producing hydroxyl radicals (OH$^.$) from the positive holes in the VB and superoxide anions (O$_2^-$) from the electrons in the CB. In addition, the electrons in the CB can reduce the carbon dioxide (CO$_2$) generating hydrocarbons with a greater added value. These powerful oxidant radicals react with organic and toxic compounds, converting them to water, carbon dioxide, and other simple substances [3, 4].

In the photocatalytic process, two events can occur simultaneously: one involves the oxidation of adsorbed water in the photogenerated holes in the VB, generating hydroxyl radicals (OH⁻) that have a high oxidation power of organic compounds and the second is related to the reduction of an acceptor of electrons (usually dissolved O_2 or CO_2) that were excited to CB.

The photogenerated electrons in the CB react with CO_2, reducing them to carbon monoxide, methane, ethanol, etc. [5].

$$CO_2 + 2e^- + 2H^+ \longrightarrow CO + H_2O, \quad E^o = -0.53\,V$$

$$CO_2 + 6e^- + 6H^+ \longrightarrow CH_3OH + H_2O, \quad E^o = -0.38\,V$$

$$CO_2 + 8e^- + 8H^+ \longrightarrow CH_4 + 2H_2O, \quad E^o = -0.24\,V$$

$$(1)$$

This catalytic CO_2 photoreduction process, also called artificial photosynthesis, is a branch of heterogeneous photocatalysis.

The most widely used semiconductor in photocatalytic processes is TiO_2; however, due to its wide band gap (3.2 eV for the anatase phase), it can be activated only by near-UV light ($k < 385$ nm), which represents a small fraction of solar light (about 3–4%). The inactivity under visible light strongly limits the practical application of the TiO_2 photocatalyst [6]. In recent years, much research has been focused on improving the photocatalytic property of TiO_2, trying to minimize the two main drawbacks associated to the use of TiO_2: (i) high electron and hole recombination rates and (ii) low efficiency under visible irradiation [7].

In order to utilize visible light from solar energy and enhance the photocatalytic reactions, efforts have been focused on exploring novel methods to modify semiconductors such as the process of doping, sensitization, and formation of heterostructures [8]. Photosensitization can be achieved by a photosensitizer that absorbs light energy transforming it into chemical energy, which is transferred to substrates under favorable conditions. The photosensitizers may be adsorbed on the semiconductor surface by an electrostatic or chemical interaction and, upon excitation, inject electrons into the conducting band of the semiconductor. Based on the reported studies in the literature, organic dyes and coordination metal complexes are very effective sensitizers [9–11].

In this work, the OPM route was used with a new approach to obtain nanoparticles of $Ti(OH)_4$ with peroxo groups on the surface, in which its photocatalytic activity was evaluated both in the degradation process of the methylene blue dye and the amiloride drug, and in the process of CO_2 photoreduction, with particular emphasis on the effect of the groups on the optical properties and the photocatalytic efficiency of the catalyst.

2. Materials and Methods

2.1. Synthesis. The synthesis of the catalysts was based on the oxidant peroxo method (OPM) developed by Ribeiro et al. and Camargo et al., in which 250 mg of metallic titanium (98% Aldrich, USA) was added to 100 mL of a 3 : 2 H_2O_2/NH_3 (Synth, Brazil) aqueous solution, which was left in an ice-water cooling bath until complete dissolution of metal, resulting in a transparent yellow solution of the soluble peroxytitanate complex [12–14]. The peroxytitanate complex solution was then heated to 80°C under stirring until the formation of a yellow gel that was dried at 60°C for 24 h to form the $Ti(OH)_4$ with the surface modified with the peroxo group. The $Ti(OH)_4$ was calcined at different temperatures between 300 and 500°C for 1 h at a heating rate of 10°C min⁻¹ in closed alumina boats. The different samples calcined 300, 400, and 500°C are referred as TiO_2-300°C, TiO_2-400°C, and TiO_2-500°C, respectively.

2.2. Characterization. The XPS spectra were performed using an XPS VG Microtech ESCA 3000 (MgKα and AlKα radiations) at an operating pressure of $3 \cdot 10^{-10}$ mbar. The binding energies were corrected for the charging effect by assuming a constant binding energy for the adventitious O1s peak. The LabRAM microspectrometer (HORIBA JobinYvon) was used to obtain the Raman spectra of the catalysts. The excitation wavelength was a 514.5 nm line of a 5.9 mW He-Ne laser as the excitation source is through an Olympus TM BX41 microscope. Thermogravimetric analysis of the catalyst was carried out using a TGA Q500 thermogravimetric analyzer (TA Instruments, New Castle, DE) in air flow and recorded from room temperature to 600°C, at a constant heating rate of 10°C/min. The $Ti(OH)_4$ was characterized by differential scanning calorimetry (DSC 404C controlled by TASC 424/3A, Netzsch, Germany) between 25 and 550°C using an aluminum crucible and a constant heating/cooling rate of 10°C/min with a flux of 0.50 cm³/min. A Rigaku D/MAX 200 diffractometer was used to determine crystal structure of the catalysts, with CuKα radiation and scanned for 2θ values from 5 to 80°. Infrared absorption spectra were recorded by a Fourier transform infrared spectrometer (FTIR) (Bruker EQUIXOX 55, Ettlingen, Germany). Spectra were recorded for KBr disks containing a powdered sample and obtained in the 400–1000 cm⁻¹ range, with 64 scans and 4 cm⁻¹ resolution. The surface morphology of the catalysts was observed by field emission scanning electron microscopy (FEGSEM, ZEISS model-SUPRA 35). The band gap of the catalysts was determined according to the method proposed by Wood and Tauc [15] and Tolvaj et al. [16]. The spectra were recorded using the UV-visible diffuse reflectance spectroscopy (DRS) at room temperature between 200 and 800 nm using a Varian model Cary 5G in the diffuse reflectance mode (R). The specific surface areas of the catalysts were measured with a Micromeritics ASAP-2020 instrument and calculated by the Brunauer-Emmett-Teller (BET) method. The catalysts were pretreated (degasification) by heating at 70°C under vacuum until reaching a degassing pressure of less than 20 mmHg.

2.3. Photodegradation. The photocatalytic activity of the catalyst was evaluated using twenty milliliters of a 10 mg L⁻¹ solution of MB dye or AML drug, mixed with 10 mg of the catalyst, and irradiated with visible (six 15 W Osram lamps, with maximum intensity at 440 nm) and UV (six 15 W Phillips TUV UVC lamps, with maximum intensity at 254 nm) light inside a thermostated photoreactor at 25 ± 3°C. To investigate

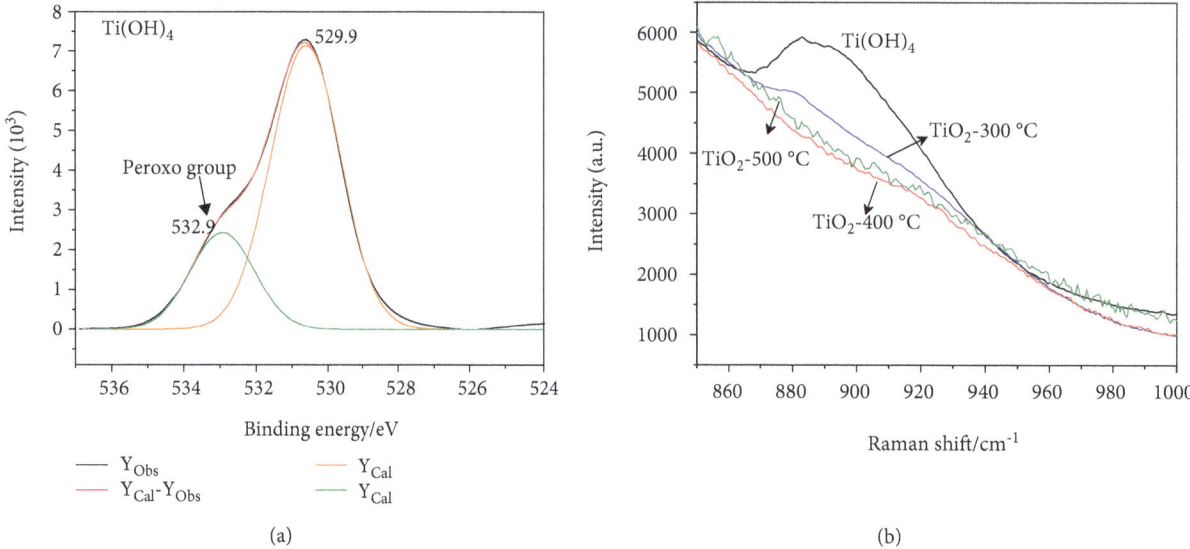

FIGURE 1: (a) XPS profile of Ti(OH)$_4$ and (b) Raman spectra of Ti(OH)$_4$ and TiO$_2$ powders calcined at different temperatures for 1 h.

the influence of radiation on the degradation process of organic compounds in the photocatalysis process, the same experiment was carried out without the presence of the catalyst. For the catalyst adsorption capacity test, we used the same conditions except the presence of radiation. Specified amount of reaction mixture was collected at regular time intervals and analyzed in a UV-Vis spectrophotometer (Shimadzu UV-1601 PC) in the absorbance mode monitor.

2.4. CO$_2$ Photoreduction. The CO$_2$ photoreduction tests were performed using 300 mg of the catalyst suspended in 300 mL of water in a 500 mL steel cylindrical reactor, covered with borosilicate glass. Ultrapure CO$_2$ was bubbled through the reactor for at least 10 min to ensure that all dissolved oxygen was eliminated. The illumination system included a UVC lamp (PHILIPS 11 W) with a wavelength of 253.7 nm in the center of the reactor, and the intensity of the incident light was 9.1 mW/cm^2.

The reaction progress was monitored by collecting and analyzing the sample at regular intervals. Gaseous products were determined with the help of GC-TCD and GC-FID (Varian, CP-3800) using a packed column (HayeSep N (0.5 m × 1.8″)) at a flow rate of 30 mL min^{-1} for H$_2$, 300 mL min^{-1} for air, and 30 mL min^{-1} for N$_2$, injector temperature of 150°C, TCD detector temperature of 200°C, and FID detector temperature of 150°C. For determining the gaseous products, a 2 μL sample was injected in the GC and then the yield was correlated by injecting a standard gaseous mixture. Blank reactions were carried out to ensure that the CH$_4$ originated were photoreduction products of CO$_2$. In the blank reaction, no catalyst was added and all other conditions were maintained the same.

3. Results and Discussion

3.1. Characterization. The synthesis of Ti(OH)$_4$ was carried out following the same principle of peroxo complex hydrolysis by the formation of water through the oxidation of H$_2$O$_2$

as reported by Ribeiro et al. and Camargo et al. [12–14]. However, in this work, instead of the oxidation process of H$_2$O$_2$ occurring through a redox reaction, the solution of peroxo-titanium complex was heated, leading to the decomposition of H$_2$O$_2$ and the formation of O$_2$ and H$_2$O, causing hydrolysis of peroxo complex and the consequent formation of the precipitate [17–20].

The surface of the Ti(OH)$_4$ was characterized by X-ray photoelectron spectroscopy (XPS), in which the XPS spectrum in the O1s region showed two peaks (Figure 1(a)). The main peak at 529.9 eV is related to the oxygen anions, O^{2-}, bound to the metal cations in the lattice, and the second peak at 532.9 eV could be attributed to the formation of peroxo groups on the Ti(OH)$_4$ [20]. Raman spectra of Ti(OH)$_4$ and calcined materials are reported in Figure 1(b). The Ti(OH)$_4$ spectrum is dominated by a broad band between 880 and 940 cm^{-1}, related to the O-O stretching vibration, which is typical for peroxo groups on the surface of titanium oxide as reported by Zou et al. [21]. It was observed that calcination temperatures above 300°C cause a reduction in the amount of peroxo groups on the catalyst surface and in higher temperatures, the presence of this band no longer occurs showing the complete elimination of the groups on the material surface.

Figure 2(a) shows the TG-DTG curves of Ti(OH)$_4$. The TG curve for the Ti(OH)$_4$ showed weight loss in two stages. The first was observed between 25°C and 150°C, and the second was recorded at a temperature range from 250°C to 300°C. The weight loss in the first temperature interval could be attributed to the removal of adsorbed water in the catalyst surface. At temperatures between 250°C and 300°C, there were weight losses of 24%, which could be attributed to the peroxo groups bonded to catalyst particles.

The two weight loss regions observed in the TG are associated with one endothermic DSC peak and a few overlapped exothermic peaks (Figure 2(b)). The first weight loss of 28% was attributed to the loss of physically adsorbed water, and it could be manifested by a clear endothermic DSC peak at

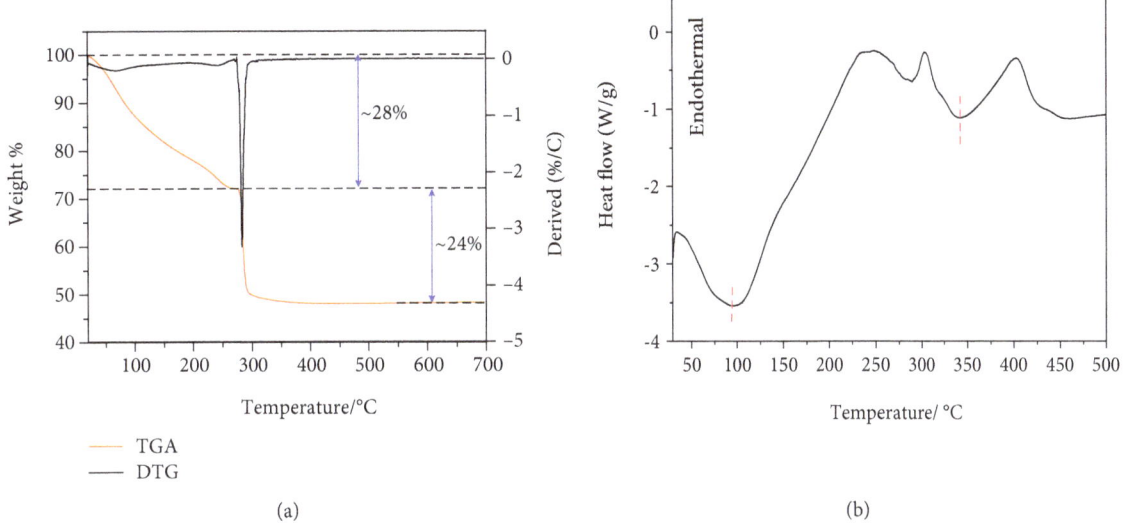

FIGURE 2: (a) TG/DTG and (b) DSC patterns of Ti(OH)$_4$ powder.

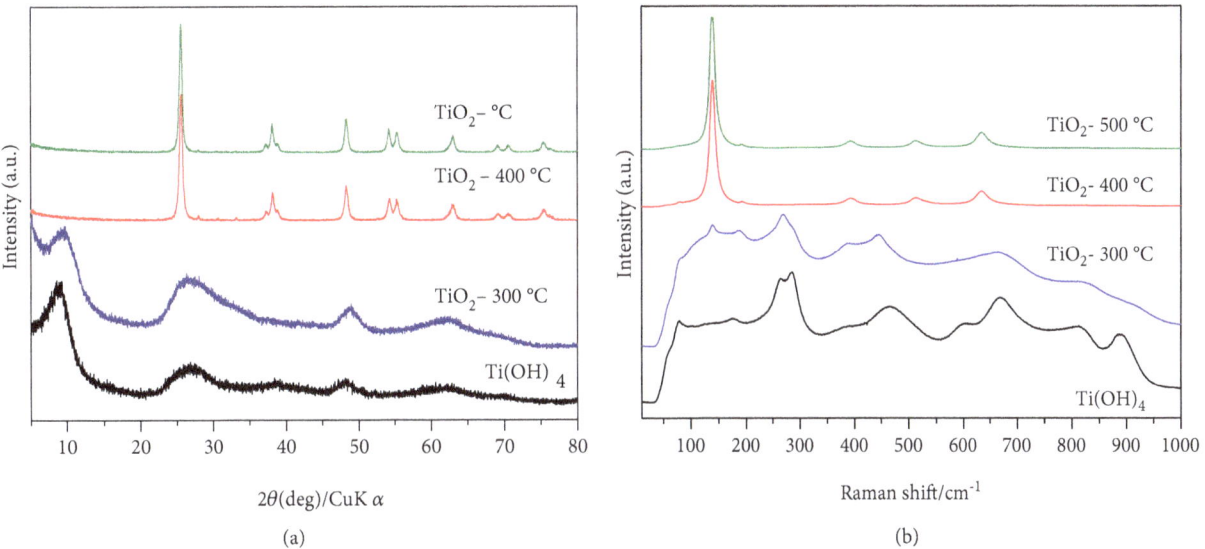

FIGURE 3: XRD patterns (a) and Raman spectroscopy of Ti(OH)$_4$, TiO$_2$-300°C, TiO$_2$-400°C, and TiO$_2$-500°C.

around 100°C. Between 200 and 500°C, a few overlapped exothermic peaks were observed in DSC accompanied by the second weigh loss of 24% in the TG. These exothermic peaks might be originated from the removal of the peroxo group on the Ti(OH)$_4$ surface.

The XRD patterns of the Ti(OH)$_4$, TiO$_2$-300°C, TiO$_2$-400°C, and TiO$_2$-500°C are shown in Figure 3. It can be seen from Figure 3(a) that the Ti(OH)$_4$ and TiO$_2$-300°C patterns show low crystalline with peaks at $2\theta = 26.5°$ related to the rutile phase (JCPDS 21-1276) and $2\theta = 48.7°$ and $62.2°$ related to the anatase phase; however, the materials calcined at 400 and 500°C presented peaks related only to the anatase phase (JCDS 21-1272). The crystallite size of the particles has been estimated from Debye–Scherrer's equation using the XRD line broadening of the (1 0 1) plane diffraction peak. In temperatures higher than 400°C, the TiO$_2$ starts to crystallize resulting in larger crystallites, which can be attributed to

the thermally promoted crystallite growth. The obtained crystallite sizes are shown in Table 1.

The structural properties of the materials were further investigated by Raman spectroscopy (Figure 3(b)). The samples calcined at 400 and 500°C exhibit the characteristic Raman-active modes of the TiO$_2$ anatase phase, in agreement with the phase composition of the materials determined by XRD [22].

In order to provide additional evidence and to confirm the effect of calcination temperatures on the Ti(OH)$_4$, FTIR characterizations were performed. The infrared spectra of Ti(OH)$_4$ and samples calcined at different temperatures are presented in Figure 4. The broad band at 2700–3650 cm^{-1} and the band at 1636 cm^{-1} correspond to the surface-adsorbed water and hydroxyl groups, respectively. The main band at 400–800 cm^{-1} was attributed to Ti-O stretching and Ti-O-Ti bridging stretching modes [23]. Notably, increasing

TABLE 1: Band gap energy, specific surface area, and crystallite size of the materials.

Sample	Band gap (eV)	S_{BET} (m^2 g^{-1})	Crystallite size (nm)
Ti(OH)$_4$	2.15	15	—
TiO$_2$-300°C	2.76	45	—
TiO$_2$-400°C	2.93	23	19
TiO$_2$-500°C	2.88	79	20

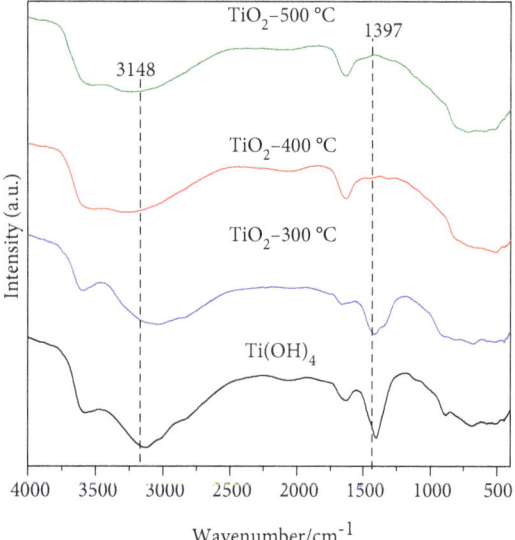

FIGURE 4: FTIR spectra of Ti(OH)$_4$, TiO$_2$-300°C, TiO$_2$-400°C, and TiO$_2$-500°C.

the temperature, the surface-adsorbed water and hydroxyl groups decreased slightly. There is no obvious characteristic band of peroxo groups on the catalyst surface, which may be limited by the detection resolution of IR.

Figure 5 shows the FEGSEM images of Ti(OH)$_4$ and catalysts calcined at 300 to 500°C. It can be seen that the Ti(OH)$_4$ particles are irregular in shape and have a wide particle size distribution. Most are larger than 200 nm, and it is observed that the treatment with temperature did not change the size and morphology.

The optical properties of the Ti(OH)$_4$, commercial TiO$_2$ (TiO$_2$-COM), and catalysts calcined at 300 to 500°C were examined using diffuse reflectance UV-Vis spectroscopy, and the band gap values were calculated by the Tauc method, as shown in Figure 6 [15, 16].

It is observed in the UV-Vis spectra that the increase of the calcination temperature caused a shift of the absorption edge to higher values of energy, as can be observed Figure 6. All the catalysts synthesized presented a lower band gap value than commercial titanium dioxide (Table 1).

This variation of band gap energy with increasing calcination temperature may be related to the different densities of structural and surface defects present in each of the samples. Because defects cannot be controlled in the materials, their presence only provides a change in the amount

and distribution of intermediate energy levels within the band gap region.

Ti(OH)$_4$ presented a band gap energy value of 2.15 eV, with an onset of absorption from 576 nm, indicating that this catalyst is sensitive to visible light, which represents a possible photoactivity under visible radiation.

3.2. Photocatalytic Tests

3.2.1. Photodegradation. Photocatalytic studies were carried out on Ti(OH)$_4$ and materials calcined at 300, 400, and 500°C and were compared with the TiO$_2$-COM nanopowder (Sigma-Aldrich 99.8%) (Figure 7). To test the photocatalytic activity of the catalysts, two organic compounds were used as the pollutant model, the methylene blue dye (maximum absorption at 554 nm), and the amiloride drug (maximum absorption 286 nm) (Figure 8).

The photolysis test of the MB was carried out under UV and visible radiation under the same conditions of the photocatalysis reaction, except without the presence of a catalyst. In this study, we can observe that under UV radiation the dye presents a small degradation by this process, which indicates that the catalyst plays a key role in the degradation of MB. The catalysts were investigated by the adsorption capacities under the same conditions as the photoreduction but without the presence of radiation, in order to distinguish between the adsorption and photocatalytic phenomena (Figure S1). Ti(OH)$_4$ presented the highest adsorption capacity of approximately 19% in relation to other materials; however, it is observed that the photocatalytic effect of Ti(OH)$_4$ is much higher than that of other materials under visible radiation (see Supplementary Materials). It is important to note that the results are due to the combined effect of the adsorption and photocatalysis process.

Figure 7(a) shows that Ti(OH)$_4$ and TiO$_2$-COM were the two catalysts that showed the highest photocatalytic activity, with 100% discoloration of the solution in 60 min of reaction under UV radiation. The degradation rate becomes less pronounced when the calcination temperature used for the production of materials is increased from 300 to 500°C. On the other hand, when the reaction was conducted under visible radiation, the Ti(OH)$_4$ was much more active than the other catalysts, which was expected since it has a higher absorption in the visible region compared to the other catalysts (Figure 7(b)).

One of the problems of testing photocatalytic activity of a catalyst using dye is that the sensitization process can occur [24]. This mechanism occurs when a molecule that has absorption in the visible region, such as dyes, is adsorbed on the surface of a semiconductor and is excited from its ground state (HOMO) to the excited state (LUMO), transferring these electrons to the semiconductor spontaneously, being the first stage of its oxidation. Furthermore, it can be found that amiloride is very stable under visible light radiation (Figure 8(b)), revealing that degradation does not occur without a photocatalyst. It is observed that Ti(OH)$_4$ had the best photocatalytic activity, presenting 90% removal with 20 min of reaction under

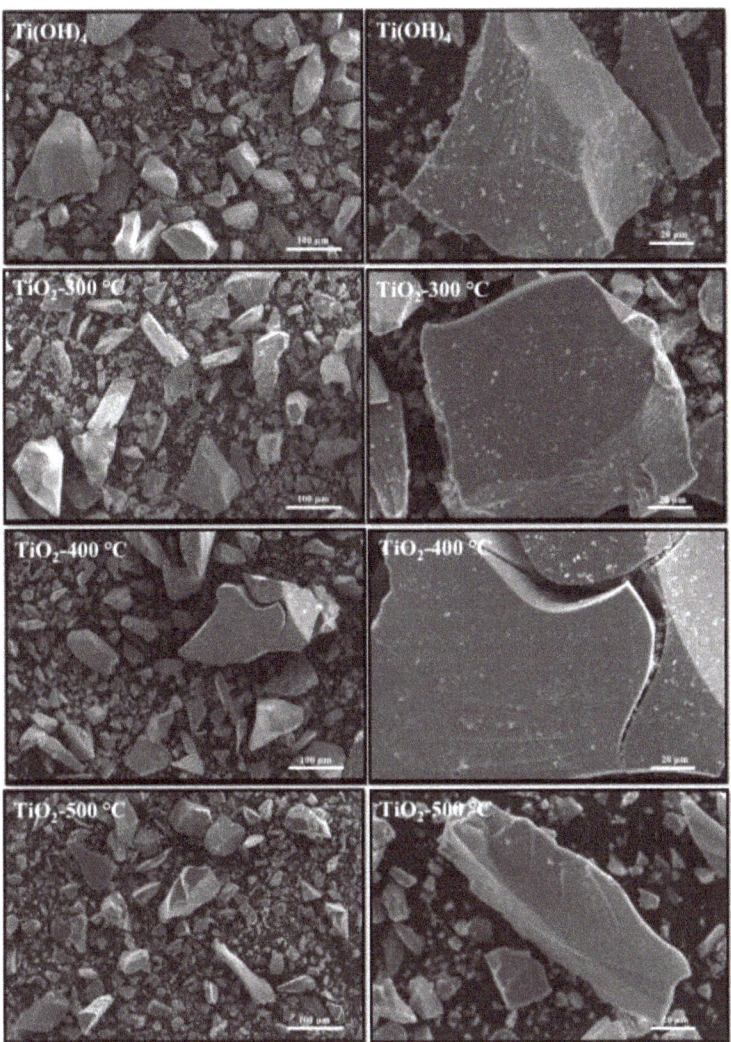

FIGURE 5: FEGSEM image of the Ti(OH)$_4$ powder and the catalyst calcined at different temperatures for 1 h.

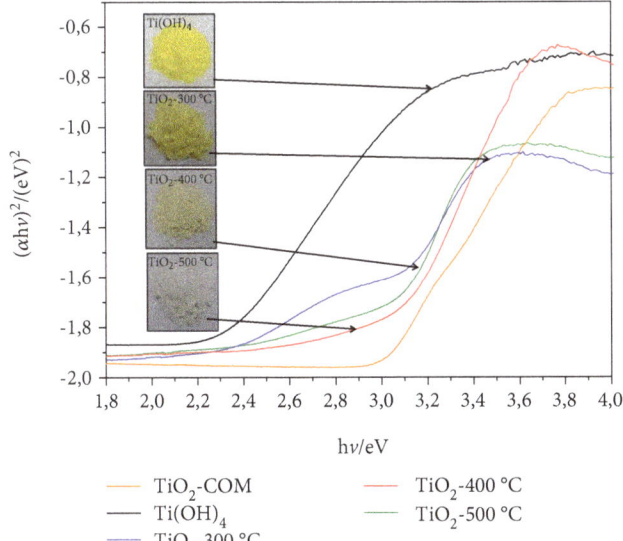

FIGURE 6: UV-Vis diffuse reflectance spectroscopy of catalysts and inset photography of materials.

visible radiation, while TiO$_2$-COM presented only 5% of removal at this same reaction time.

These results show that the absorption edge shift of Ti(OH)$_4$, due to the presence of the peroxo groups on the surface, had a great influence on the catalytic activity of the catalyst under visible radiation, where it was observed that the removal of these peroxo groups in the surface by the calcination process decreased the photocatalytic activity of these materials under visible radiation, probably due to their low absorption capacity in the visible region.

The order of reaction with respect to MB and AML degradation was determined by plotting the reaction time as a function of ln $[C]/[C_0]$ according to the following equation for the materials:

$$\ln \left(\frac{C}{C_0} \right) = -kt, \qquad (2)$$

where $[C_0]$ and $[C]$ represent the concentration of the substrate in solution at time zero and the time of illumination,

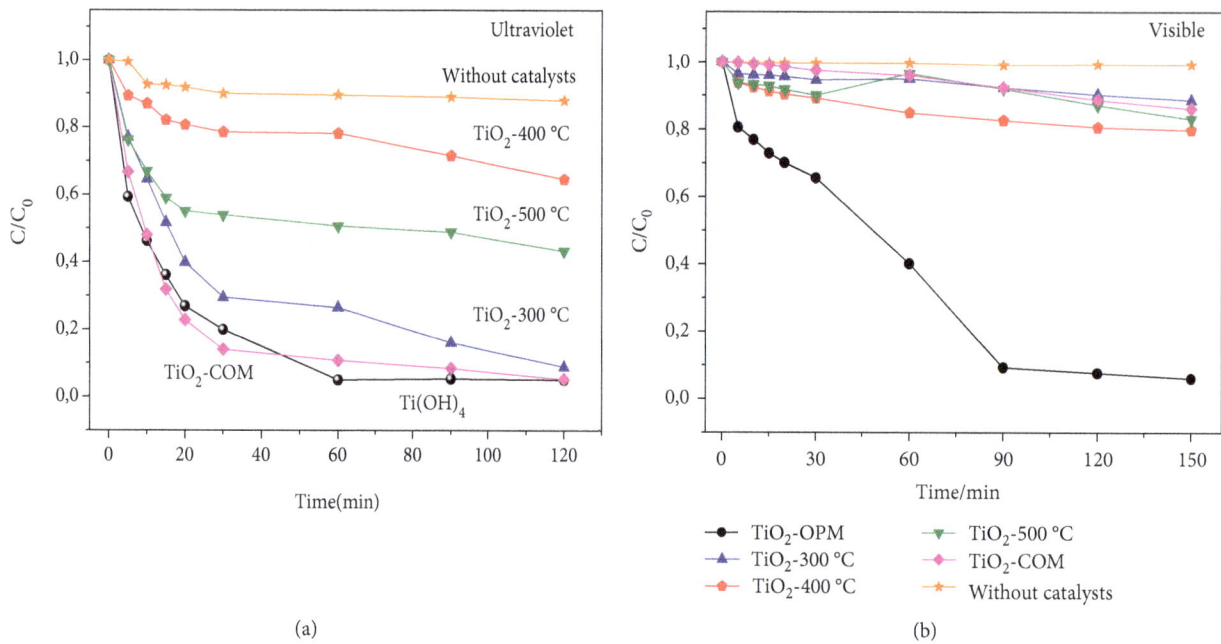

(a)

(b)

FIGURE 7: The kinetic process of MB degradation (a) under UV and (b) visible light radiation.

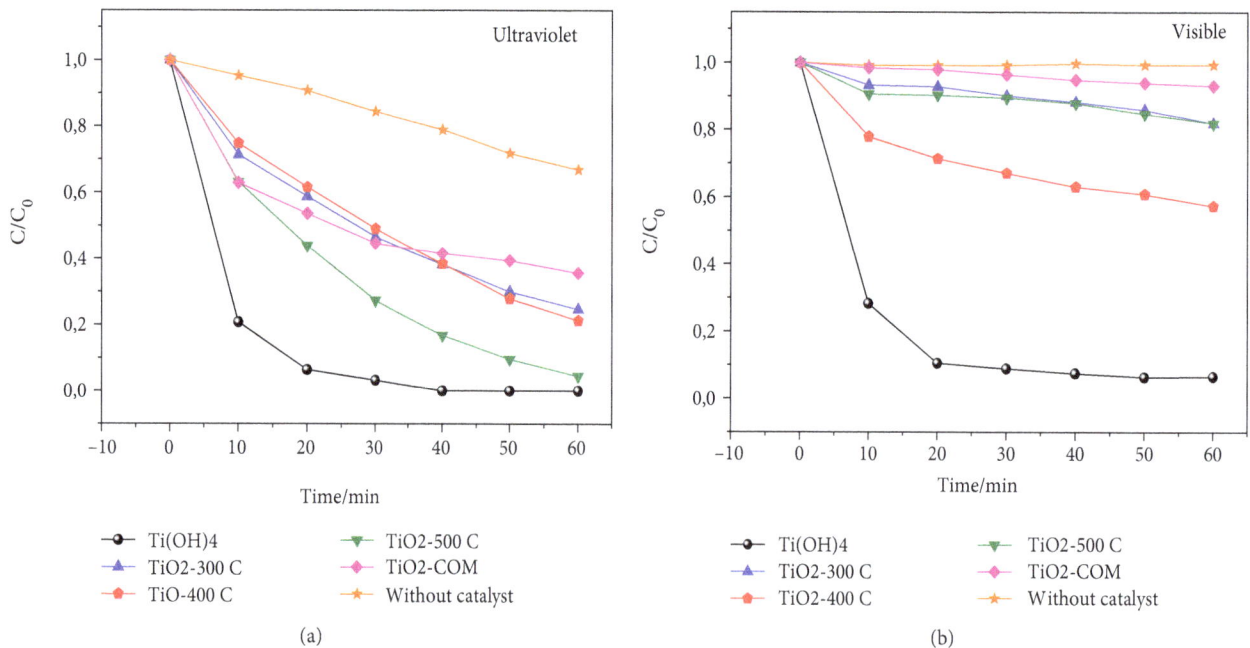

(a)

(b)

FIGURE 8: Degradation kinetics of AML (a) under UV and (b) visible light radiation in the solutions containing $Ti(OH)_4$, TiO_2-COM, TiO_2-300°C, TiO_2-400°C, and TiO_2-500°C.

respectively, and k represents the apparent rate constant (min^{-1}). The apparent rate constants are summarized in Tables 2 and 3. The results show that the reaction followed first-order kinetics.

The stability of the catalyst over a photodegradation is a critical factor for its practical application. Therefore, the photostability of the $Ti(OH)_4$ sample, which showed the best photoactivity, was evaluated by performing recycling

experiments under UV irradiation (Figure S2). After each reaction cycle, the sample was separated from the MB solution by centrifugation and placed immediately in contact with a freshly prepared MB solution. This procedure was repeated four times under the same conditions used in the photocatalytic tests. As shown in Figure S2, a slight decrease (ca. 4%) occurs after the first photocatalysis cycle. In addition, the peroxo groups were analyzed after the

TABLE 2: Photocatalytic reaction constants of the MB degradation.

Material	Ultraviolet		Visible	
	$k \times 10^{-3}$ (min^{-1})	R^2	$k \times 10^{-3}$ (min^{-1})	R^2
$Ti(OH)_4$	44.1	0.9777	8.26	0.9851
TiO_2-COM	63.3	0.9842	0.972	0.9880
TiO_2-300°C	39.5	0.9872	0.752	0.9292
TiO_2-400°C	5.30	0.8844	1.95	0.9588
TiO_2-500°C	20.8	0.9882	1.63	0.9575

TABLE 3: Photocatalytic reaction constants of the AML degradation.

Material	Ultraviolet		Visible	
	$k \times 10^{-3}$ (min^{-1})	R^2	$k \times 10^{-3}$ (min^{-1})	R^2
$Ti(OH)_4$	114	0.9571	82	0.8713
TiO_2-COM	15	0.9937	1	0.9876
TiO_2-300°C	23	0.9949	3	0.9472
TiO_2-400°C	25	0.9856	8	0.8663
TiO_2-500°C	50	0.845	3	0.8490

photocatalysis reaction, in which it was observed that after the reaction the groups were maintained (Figure S3).

The formation of the stable metal-peroxo group, which may be partially in equilibrium with radical metal-$O_2^·$ species, has been recognized to play an important role in the photocatalytic process, as these groups can inhibit the recombination of photogeneration charges and interact with photogenerated holes [25, 26]. Photogenerated valence band holes (h^+) are trapped at the surface oxygen to form the $·OH$ radical. On the other hand, the electrons in the conduction band can react with H_2O to form the $·OOH$ radical (Figure 9) [27].

Zhang and Nosaka studied the formation of the $·OH$ radical from TiO_2 and demonstrated that the mechanism of $·OH$ generation occurs differently depending on the crystalline phase of TiO_2 (Figure 9). They observed that on the surface of the rutile, $·OH$ is generated by the oxidation of the water through the h^+ photogenerated with the peroxo group on the surface; on the other hand, in the anatase phase, the OH is generated by the reaction of the metal-$O_2^·$ radical [28, 29]. This difference in the mechanism of $·OH$ generation occurs due to the packaging of the crystalline structure, in which the rutile phase presents a greater packaging with respect to the anatase phase, better stabilizing the peroxo group in the surface. Since the peroxo group is equivalent to the adsorbed H_2O_2, the presence of the rutile phase in $Ti(OH)_4$ may be favoring an increase in the generation of the $·OH$ radical that consequently increases its photocatalytic activity.

3.2.2. CO_2 Photoreduction. Another aspect of the photocatalysis process is the CO_2 photoreduction also known as artificial photosynthesis. Unlike the photodegradation process, which generates oxidizing radicals ($·OH$ e $·OOH$), the photoreproduction process happens with the electrons

photoexcited to the CB that are captured by the molecules of CO_2 adsorbed on the surface of the catalyst forming carbon dioxide ($CO_2^·$) radicals, which will later lead to the formation of compounds with a higher added value such as CH_4, ethanol, and formic acid [30].

$$CO_2 \xrightarrow{e-} CO_2^- \xrightarrow{H^+ + e-} CH_3 \xrightarrow{H^+ + e-} CH_4 \qquad (3)$$

Thus, to verify the $Ti(OH)_4$ activity in CO_2 photoreduction, tests were performed under UV radiation in the aqueous medium (Figure 10).

A series of control experiments were conducted for quality assurance and to ascertain correct results. The first tests were performed in the dark with and without a catalyst, under the same experimental conditions as the CO_2 photoreduction. No products were detected in the two control experiments, indicating that CO_2 conversion, as described below, is a true photocatalytic reduction process.

Analysis of the gas samples indicated that only CH_4 was formed using $Ti(OH)_4$ and the yield was 6.81 $\mu mol·L^{-1}·g^{-1}$, which was almost 3 times higher than the result obtained using P25 (2.18 $\mu mol·L^{-1}·g^{-1}$), which is the conventional catalyst used in photocatalytic processes. The concentration of CO_2 present in the gaseous medium varied during the reaction (Figure 10(b)); this variation is related to the equilibrium displacement over the photoreduction process. During the photoreduction process, the CO_2 is consumed from the aqueous medium, and in order to maintain the equilibrium, the part of the gas moves to the liquid medium causing that variation in the concentration of CO_2 in the gas phase during the reaction.

It is important to note that the CO_2 concentration variation in the gas phase was approximately 140–130 $mmol·L^{-1}·g^{-1}$ for all materials, indicating that, despite the small variation observed, CO_2 concentration may be considered constant during the reaction time.

4. Conclusions

The OPM route proved to be efficient in obtaining $Ti(OH)_4$ nanoparticles with the surface modified by peroxo groups. These groups had a great influence on the processes of photocatalysis, both in the oxidation process of organic compounds and in the photoreduction of CO_2, not only increasing the absorption of the radiation in the visible region but also providing an effective separation of photogenerated charges. We investigated the effect of the calcination temperature on the photocatalytic activity of $Ti(OH)_4$, and the results showed that the best result was achieved with uncalcined $Ti(OH)_4$, exhibiting 98% degradation of AML and MB under irradiation of visible light, being more than 10 times more active than TiO_2-COM. In addition, $Ti(OH)_4$ with the modified surface proved to be a good candidate for applications in CO_2 photoreduction processes.

It is important to note that the OPM method can be used to obtain other oxides such as niobium oxide, vanadium oxide, and tungsten oxide that are good candidates for applications in photocatalytic processes.

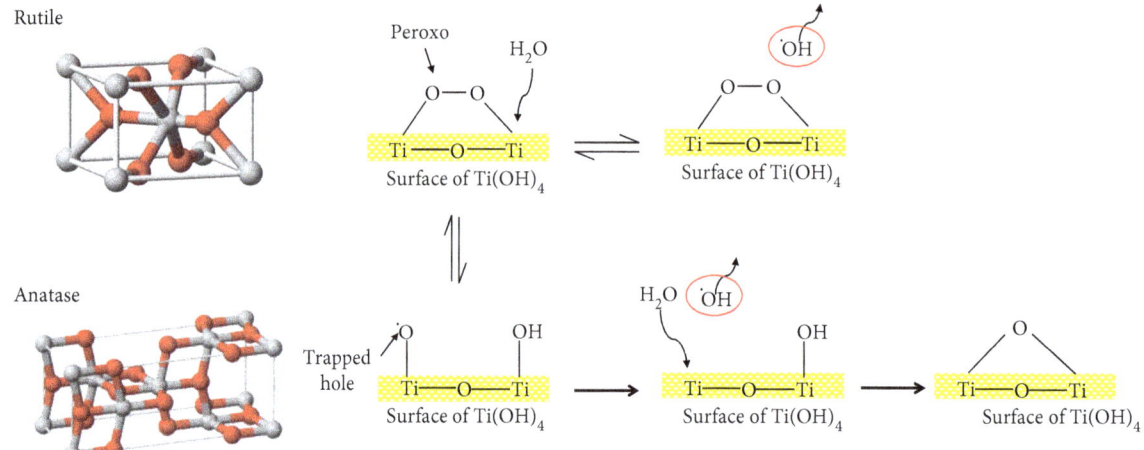

FIGURE 9: Schematic illustration of OH generation at rutile and anatase TiO_2.

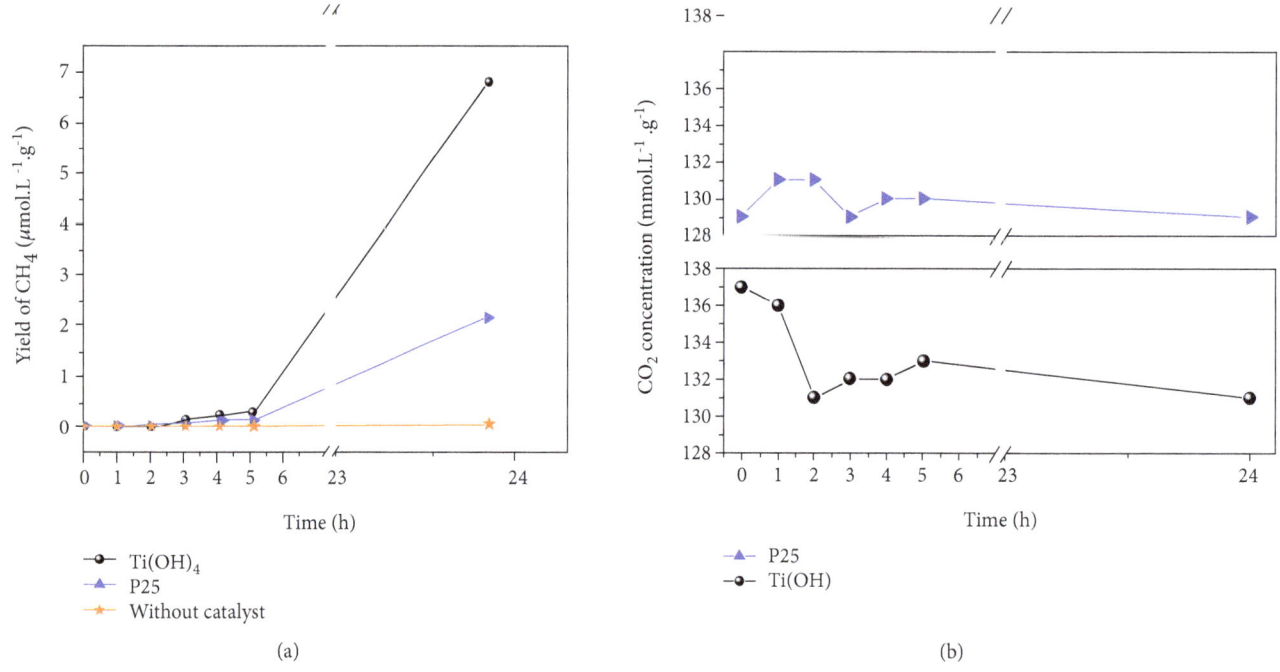

(a)

(b)

FIGURE 10: (a) CO_2 photoreduction kinetics and (b) CO_2 concentration as a function of irradiation time, under UVC-light.

Conflicts of Interest

The authors declare no conflict of interest.

Acknowledgments

The authors would like to thank the financial support given by FAPESP (2014/09014-7), CNPEM/LNNano, SISNANO/ MCTI, and Empresa Brasileira de Pesquisa Agropecuária AgroNano research network. We also acknowledge the CDMF/CEPID.

References

[1] M. Pelaez, N. T. Nolan, S. C. Pillai et al., "A review on the visible light active titanium dioxide photocatalysts for environmental applications," *Applied Catalysis. B, Environmental*, vol. 125, pp. 331–349, 2012.

[2] S. Abu-Bakar and C. Ribeiro, "An insight toward the photocatalytic activity of S doped 1-D TiO_2 nanorods prepared via novel route: as promising platform for environmental leap," *Journal of Molecular Catalysis A:Chemical*, vol. 412, pp. 78–92, 2016.

[3] A. Mills, C. O'Rourke, and K. Moore, "Powder semiconductor photocatalysis in aqueous solution: an overview of kinetics-based reaction mechanisms," *Journal of Photochemistry and Photobiology A: Chemistry*, vol. 310, pp. 66–105, 2015.

[4] S. M. Rodríguez, J. B. Gálvez, M. I. Maldonado Rubio, P. F. Ibáñez, W. Gernjak, and I. O. Alberola, "Treatment of chlorinated solvents by TiO_2 photocatalysis and photo-Fenton: influence of operating conditions in a solar pilot plant," *Chemosphere*, vol. 58, no. 4, pp. 391–398, 2005.

[5] J. Mao, K. Li, and T. Peng, "Recent advances in the photocatalytic CO_2 reduction over semiconductors," *Catalysis Science & Technology*, vol. 3, no. 10, p. 2481, 2013.

[6] A. E. Nogueira, A. R. F. Lima, E. Longo, E. R. Leite, and E. R. Camargo, "Effect of lanthanum and lead doping on the microstructure and visible light photocatalysis of bismuth titanate prepared by the oxidant peroxide method (OPM)," *Journal of Photochemistry and Photobiology A: Chemistry*, vol. 312, pp. 55–63, 2015.

[7] M. C. Neves, J. M. F. Nogueira, T. Trindade, M. H. Mendonça, M. I. Pereira, and O. C. Monteiro, "Photosensitization of TiO_2 by Ag_2S and its catalytic activity on phenol photodegradation," *Journal of Photochemistry and Photobiology A: Chemistry*, vol. 204, no. 2-3, pp. 168–173, 2009.

[8] H. J. Yun, H. Lee, J. B. Joo, N. D. Kim, M. Y. Kang, and J. Yi, "Facile preparation of high performance visible light sensitive photo-catalysts," *Applied Catalysis. B, Environmental*, vol. 94, no. 3-4, pp. 241–247, 2010.

[9] J. Wen, X. Li, W. Liu, Y. Fang, J. Xie, and Y. Xu, "Photocatalysis fundamentals and surface modification of TiO_2 nanomaterials," *Chinese Journal of Catalysis*, vol. 36, no. 12, pp. 2049–2070, 2015.

[10] W. Macyk, K. Szaciłowski, G. Stochel, M. Buchalska, J. Kuncewicz, and P. Łabuz, "Titanium(IV) complexes as direct TiO_2 photosensitizers," *Coordination Chemistry Reviews*, vol. 254, no. 21-22, pp. 2687–2701, 2010.

[11] R. Abe, K. Sayama, and H. Arakawa, "Dye-sensitized photocatalysts for efficient hydrogen production from aqueous I⁻ solution under visible light irradiation," *Journal of Photochemistry and Photobiology A: Chemistry*, vol. 166, no. 1-3, pp. 115–122, 2004.

[12] C. Ribeiro, C. M. Barrado, E. R. de Camargo, E. Longo, and E. . R. Leite, "Phase transformation in titania nanocrystals by the oriented attachment mechanism: the role of the pH value," *Chemistry - A European Journal*, vol. 15, no. 9, pp. 2217–2222, 2009.

[13] E. R. Camargo, J. Frantti, and M. Kakihana, "Low-temperature chemical synthesis of lead zirconate titanate (PZT) powders free from halides and organics," *Journal of Materials Chemistry*, vol. 11, no. 7, pp. 1875–1879, 2001.

[14] E. R. Camargo, M. Popa, J. Frantti, and M. Kakihana, "Wet-chemical route for the preparation of lead zirconate: an amorphous carbon- and halide-free precursor synthesized by the hydrogen peroxide based route," *Chemistry of Materials*, vol. 13, no. 11, pp. 3943–3948, 2001.

[15] D. L. Wood and J. Tauc, "Weak absorption tails in amorphous semiconductors," *Physical Review B*, vol. 5, no. 8, pp. 3144–3151, 1972.

[16] L. Tolvaj, K. Mitsui, and D. Varga, "Validity limits of Kubelka-Munk theory for DRIFT spectra of photodegraded solid wood," *Wood Science and Technology*, vol. 45, no. 1, pp. 135–146, 2011.

[17] E. R. Camargo, M. G. Dancini, and M. Kakihana, "The oxidant peroxo method (OPM) as a new alternative for the synthesis of

lead-based and bismuth-based oxides," *Journal of Materials Research*, vol. 29, no. 01, pp. 131–138, 2014.

[18] F. P. Cardoso, A. E. Nogueira, P. S. O. Patrício, and L. C. A. Oliveira, "Effect of tungsten doping on catalytic properties of niobium oxide," *Journal of the Brazilian Chemical Society*, vol. 23, no. 4, 2012.

[19] D. Bayot and M. Devillers, "Peroxo complexes of niobium(V) and tantalum(V)," *Coordination Chemistry Reviews*, vol. 250, no. 19-20, pp. 2610–2626, 2006.

[20] P. Francatto, F. N. Souza Neto, A. E. Nogueira et al., "Enhanced reactivity of peroxo-modified surface of titanium dioxide nanoparticles used to synthesize ultrafine bismuth titanate powders at lower temperatures," *Ceramics International*, vol. 42, no. 14, pp. 15767–15772, 2016.

[21] J. Zou, J. Gao, and F. Xie, "An amorphous TiO_2 sol sensitized with H_2O_2 with the enhancement of photocatalytic activity," *Journal of Alloys and Compounds*, vol. 497, no. 1-2, pp. 420–427, 2010.

[22] H. C. Choi, Y. M. Jung, and S. B. Kim, "Size effects in the Raman spectra of TiO_2 nanoparticles," *Vibrational Spectroscopy*, vol. 37, no. 1, pp. 33–38, 2005.

[23] A. N. Murashkevich, A. S. Lavitskaya, T. I. Barannikova, and I. M. Zharskii, "Infrared absorption spectra and structure of TiO_2-SiO_2 composites," *Journal of Applied Spectroscopy*, vol. 75, no. 5, pp. 730–734, 2008.

[24] B. Ohtani, "Great challenges in catalysis and photocatalysis," *Frontiers in Chemistry*, vol. 5, 2017.

[25] A. E. Nogueira, T. C. Ramalho, and L. C. A. Oliveira, "Photocatalytic degradation of organic compound in water using synthetic niobia: experimental and theoretical studies," *Topics in Catalysis*, vol. 54, no. 1-4, pp. 270–276, 2011.

[26] T. C. Ramalho, L. C. A. Oliveira, K. T. G. Carvalho, E. F. Souza, E. F. F. da Cunha, and M. Nazzaro, "Catalytic behavior of niobia species on oxidation reactions: insights from experimental and theoretical models," *Journal of Materials Science*, vol. 43, no. 17, pp. 5982–5988, 2008.

[27] A. Ajmal, I. Majeed, R. N. Malik, H. Idriss, and M. A. Nadeem, "Principles and mechanisms of photocatalytic dye degradation on TiO_2 based photocatalysts: a comparative overview," *RSC Advances*, vol. 4, no. 70, pp. 37003–37026, 2014.

[28] J. Zhang and Y. Nosaka, "Mechanism of the OH radical generation in photocatalysis with TiO_2 of different crystalline types," *Journal of Physical Chemistry C*, vol. 118, no. 20, pp. 10824–10832, 2014.

[29] Y. Nosaka, "Surface chemistry of TiO_2 photocatalysis and LIF detection of OH radicals," in *Environmentally Benign Photocatalysts*, Nanostructure Science and Technology, M. Anpo and P. Kamat, Eds., Springer, New York, NY, 2010.

[30] A. Nikokavoura and C. Trapalis, "Alternative photocatalysts to TiO_2 for the photocatalytic reduction of CO_2," *Applied Surface Science*, vol. 391, pp. 149–174, 2017.

6

1D TiO$_2$ Nanostructures Prepared from Seeds Presenting Tailored TiO$_2$ Crystalline Phases and Their Photocatalytic Activity for *Escherichia coli* in Water

Julieta Cabrera,[1] Dwight Acosta,[2] Alcides López,[1] Roberto J. Candal,[3] Claudia Marchi,[4] Pilar García,[1] Dante Ríos,[1] and Juan M. Rodriguez ⓘ[1]

[1]*Universidad Nacional de Ingeniería, Av. TúpacAmaru s/n, Rimac, Lima, Peru*
[2]*Instituto de Física, Universidad Nacional Autónoma de México, 20364 Ciudad de México, Mexico*
[3]*Instituto de Investigación e Ingeniería Ambiental, CONICET, Universidad Nacional de San Martín, Campus Miguelete, 25 de Mayo y Francia, 1650 San Martín, Provincia de Buenos Aires, Argentina*
[4]*Centro de Microscopias Avanzadas, FCEyN, Universidad de Buenos Aires, Ciudad Universitaria, 1428 Buenos Aires, Argentina*

Correspondence should be addressed to Juan M. Rodriguez; jrodriguez@uni.edu.pe

Academic Editor: Joaquim Carneiro

TiO$_2$ nanotubes were synthesized by alkaline hydrothermal treatment of TiO$_2$ nanoparticles with a controlled proportion of anatase and rutile. Tailoring of TiO$_2$ phases was achieved by adjusting the pH and type of acid used in the hydrolysis of titanium isopropoxide (first step in the sol-gel synthesis). The anatase proportion in the precursor nanoparticles was in the 3–100% range. Tube-like nanostructures were obtained with an anatase percentage of 18 or higher while flake-like shapes were obtained when rutile was dominant in the seed. After annealing at 400°C for 2 h, a fraction of nanotubes was conserved in all the samples but, depending on the anatase/rutile ratio in the starting material, spherical and rod-shaped structures were also observed. The photocatalytic activity of 1D nanostructures was evaluated by measuring the deactivation of *E. coli* in stirred water in the dark and under UV-A/B irradiation. Results show that in addition to the bactericidal activity of TiO$_2$ under UV-A illumination, under dark conditions, the decrease in bacteria viability is ascribed to mechanical stress due to stirring.

1. Introduction

TiO$_2$ nanomaterials are well-studied and commonly used photocatalysts for the degradation of organics, water splitting, and solar cells, among others [1–4]. In the last years, several approaches were explored to increase the photoefficiency of TiO$_2$, with the modification of the particle morphology and dimensionality being one of the newest [5]. One-dimensional (1D) nanostructures such as nanotubes, nanorods, nanowires, and nanobelts have attracted great attention because of their unique properties that may be beneficial for photocatalysis: (i) enhanced light absorption due to the high length/diameter ratio, (ii) rapid and long-distance electron transport capability, (iii) large specific surface area, and (iv) ion exchange ability [6]. Hydrothermal treatment of TiO$_2$ particles in alkaline solutions is one of the simplest and cheapest techniques to produce 1D-layered titanate structures. The hydrothermal synthesis of TiO$_2$ nanotubes involves several steps where the structure of the TiO$_2$ precursor changes completely.

Results obtained in our laboratories show that 1D TiO$_2$ nanostructures display photocatalytic activity for dye degradation [7]. Although its performance as a photocatalyst is not as good as other industrially produced TiO$_2$, this form of TiO$_2$ can be easily recuperated from the solution.

Furthermore, since 1985 when Matsunaga et al. [8] published the first report of the photocatalytic biocide effects of TiO$_2$ under metal halide lamp irradiation, there has been increasing interest in photocatalytic disinfection. Use of TiO$_2$ nanoparticles in suspension is an efficient method for decontamination due to the large surface area of catalysts available to perform the reaction. It has, however, some drawbacks before its scaling at the industrial level; for example, the necessity of removing the catalyst from the solution after decontamination using filtration increases the cost and time of the cleaning process [9].

In this sense, considering the advantages that our 1D TiO_2 nanostructures are more easily filterable than nanoparticles and can be easily removed from solutions—in addition to the fact that the efficiency of 1D TiO_2 as bactericide under UV-A irradiation was only briefly explored—in this work, we assess the photocatalytic activity of 1D TiO_2, obtained from nanoparticles with a controlled proportion of anatase and rutile made by the sol-gel method, for *E. coli* ATCC 25922 in water.

2. Materials and Methods

2.1. Materials. Titanium isopropoxide purity 98%, hydrochloric acid fuming 37%, nitric acid 65%, and pure sodium hydroxide pellets were purchased from Merck. All reagents were used as received.

2.2. Synthesis of TiO_2 Nanostructures. TiO_2 nanoparticles (TiO_2 NPs) were synthesized by the sol-gel method (SG). Titanium isopropoxide was added drop by drop to vigorously stirred HNO_3 or HCl solutions at pH 0.5, 0.8, and 1.0. Suspensions were heated at 70°C for 2 h, autoclaved in a stainless-steel chamber at 220°C for 12 h, washed by centrifugation, and dried at 60°C.

1D TiO_2 nanostructures were synthesized by hydrothermal treatment of 1 g TiO_2 NPs obtained by the sol-gel method in 40 mL of 10 M NaOH at 130°C for 24 h. After hydrothermal treatment, the obtained white powder was vacuum filtered, washed with HCl solution for ionic exchange, and then washed with distilled water until a neutral pH was reached. Finally, the samples were annealed at 400°C for 2 h to crystallize the material.

The obtained nanostructures were characterized by X-ray diffraction (XRD) in a Rigaku diffractometer using Cu Kα radiation ($\lambda = 1.54056$ Ãfâ€¦). The morphology was studied by field emission scanning electron microscopy (FE-SEM SUPRA 40, Carl Zeiss) and high-resolution transmission electron microscopy (HRTEM) using a JEOL JEM-2010F transmission electron microscope operating at 200 kV. TEM samples were prepared by dispersing a small amount of the sample in ethanol with the help of an ultrasonic bath. Small droplets of the freshly prepared dispersion were placed onto a copper grid covered with carbon to improve the conduction of the electrons.

2.3. Assessment of Photocatalytic Activity of TiO_2 Nanoparticles and 1D TiO_2 Nanostructures against Escherichia coli in Water. The photocatalytic activity for water disinfection was tested using *E. coli* ATCC 25922. Experiments were performed in a batch reactor, with illumination from above using an Ultra-Vitalux 300 W lamp (30 W/m^2) and, under dark conditions, containing 100 mL aqueous solution with 10^7 CFU/mL bacteria. 1.0 mL aliquots were collected after 0, 20, 40, and 60 min irradiation. Aliquots were diluted 1 : 10 with sterile water to fit in the range 10–500 CFU/mL.

1.0 mL samples of the final dilutions were vacuum filtered through a sterile filter; this results in all bacteria present in the water being retained on the filter. Finally, the filters were placed onto a paper pad soaked in "membrane lauryl sulphate broth" (Oxoid MM0615), which feeds *E. coli* bacteria but inhibits the growth of any other bacteria. The bacterial concentration was determined by counting after 18 h incubation at 37°C.

TiO_2 nanoparticle samples were codified with C or N (for samples made with HCl and HNO_3, resp.) accompanied by 0.5, 0.8, and 1.0, depending on the pH used in the sol-gel synthesis. In a similar way, 1D TiO_2 nanostructures were codified adding 1D to the nanoparticle code (e.g., N0.5 refers to TiO_2 nanoparticles obtained with HNO_3 in pH 0.5, and N0.5-1D refers to a one-dimensional TiO_2 nanostructure obtained for N0.5) resulting in twelve samples.

2.4. Assessment of Stirring in Bacteria Viability. In order to evaluate the mechanical stirring effect in bacteria viability, *E. coli* ATCC 25922 were tested in the dark under stirring (100 rpm) and without stirring at room temperature (20°C). 1.0 mL aliquots were collected after 0, 20, 40, and 60 min stirring.

3. Results and Discussion

3.1. 1D TiO_2 Nanostructures Prepared from Seeds Presenting Tailored TiO_2 Crystalline Phases. Figure 1 shows FE-SEM images of TiO_2 nanoparticles obtained by the sol-gel method (SG-TiO_2 NPs) using HNO_3 and HCl as catalysts in the acid hydrolysis reaction of titanium isopropoxide (pH = 0.5, 0.8, and 1). At pH 1, regardless of the acid, the images show spherical nanoparticles with average diameters of about 15 nm and 13 nm, with HNO_3 and HCl, respectively. Polyhedral structures (60–100 nm) were observed when the pH decreased to 0.8, and octahedral structures with edges of about 140 nm were obtained with HCl at pH 0.5.

XRD patterns (Figures 2(a) and 2(b)) show that the crystalline structures correspond mainly to anatase when acidic solutions with pH 1 were used with both catalysts. Both anatase and rutile were observed with acidic solutions at pH 0.8 and 0.5; a small amount of brookite was detected in most cases. The amount of rutile increased as the pH decreased, and it was the dominant phase when HCl at pH 0.5 was used. The small peak for brookite disappears in this case.

The average crystallite size for anatase and rutile ($D_{anatase}$ and D_{rutile}, resp.), the anatase content, A_p, estimated with the Spurr-Myers equation [10] from the main diffraction peaks, and the pH of the acidic solution (HNO_3 and HCl) are shown in Table 1.

Figure 3 shows the morphology, by FE-SEM, of the 1D nanostructures obtained after alkaline hydrothermal treatment of the sol-gel TiO_2 nanoparticles presented in Figure 1 and Table 1. In the case of SG-TiO_2 synthesized with HNO_3, the particles displayed a tube-like shape, with an average diameter of 11 ± 1 nm in different anatase contents within 18% to 100%, respectively. On the other hand, using as precursor SG-TiO_2 synthesized with HCl acid, flake-like

FIGURE 1: FE-SEM images of TiO$_2$ nanoparticles obtained from the sol-gel method using HNO$_3$ and HCl as catalysts of the titanium isopropoxide hydrolysis reaction (pH = 0.5, 0.8, and 1.0, from left to right, resp.).

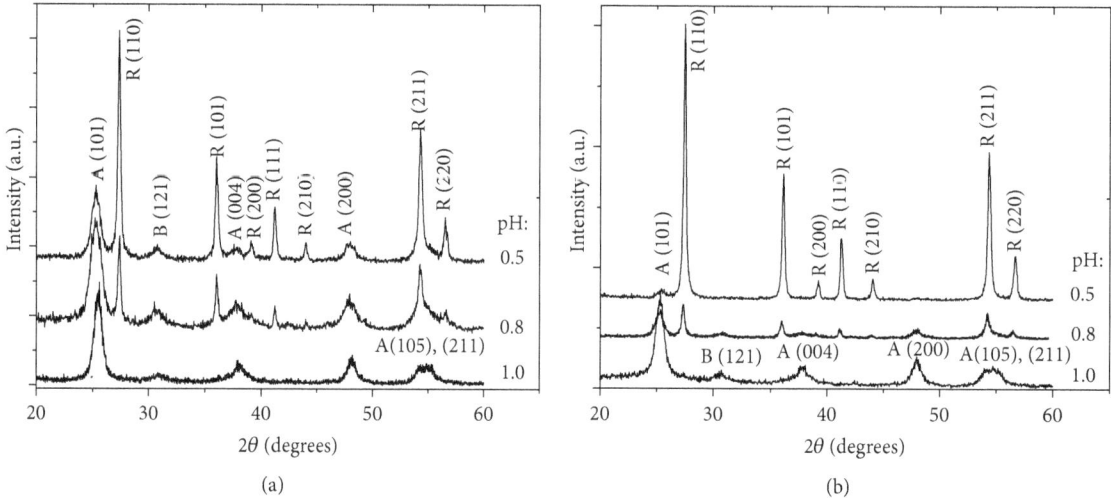

FIGURE 2: XRD patterns from sol-gel TiO$_2$ nanoparticles with a range of crystalline phases obtained with (a) HNO$_3$ and (b) HCl at pH = 0.5, 0.8, and 1.0. (A = anatase; R = rutile; B = brookite).

TABLE 1: Crystallite size, $D_{anatase}$ or D_{rutile}, and anatase proportion, A_p, of the SG-TiO$_2$ powders prepared at the indicated pH using HCl or HNO$_3$ acid solutions.

Catalyst	pH	$D_{anatase}$ (nm)	D_{rutile} (nm)	A_p
HNO$_3$	0.5	9.4 ± 0.2	26.8 ± 0.2	0.18
	0.8	7.4 ± 0.2	29.0 ± 0.2	0.56
	1	9.0 ± 0.2	—	1
HCl	0.5	—	29.7 ± 0.2	~0.03
	0.8	9.5 ± 0.2	25.9 ± 0.2	0.39
	1	9.8 ± 0.2	—	1

particles were identified together with tube-like structures. The proportion of flake-shaped particles increased as the pH decreased.

After the annealing process at 400°C for 2 h, TEM images (Figure 4) show nanotube structures in all the samples; depending on the seed material, some spherical and rod-shaped structures were also present. It can be seen that a sintering-like process took place during the annealing and that, as a consequence, bundles of tube-like structures and cracked structures were produced.

Tube-like structures seemed to be best conserved when obtained from TiO$_2$ nanoparticles with 56% of anatase, synthesized with HNO$_3$. When seed material with lower anatase content (~18%) was employed, large and irregular particles measuring about 80 nm were accompanying the nanotube structures. These might be rutile seed aggregates that could not react in the hydrothermal treatment because of their large particle size. In contrast, needle-like shapes and nanotubes turning to nanorods were observed when anatase-rutile TiO$_2$ nanoparticles synthesized with HCl were used

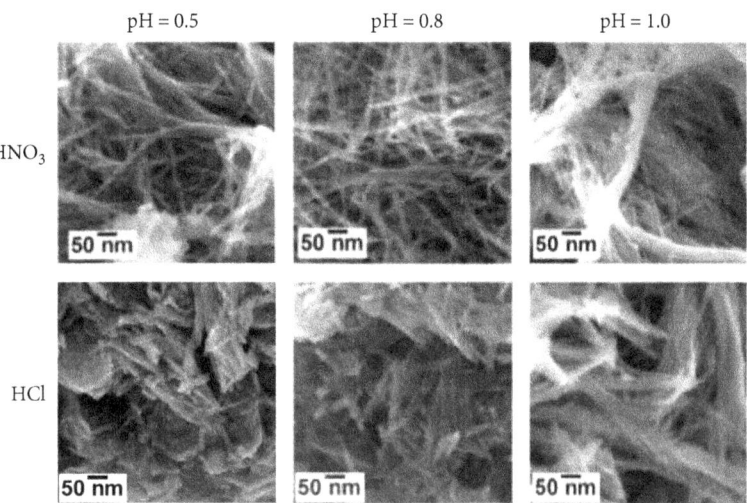

FIGURE 3: FE-SEM images of 1D TiO_2 nanostructures obtained from SG-TiO_2, prepared with HNO_3 or HCl at pH = 0.5, 0.8, and 1, after 24 h of hydrothermal treatment.

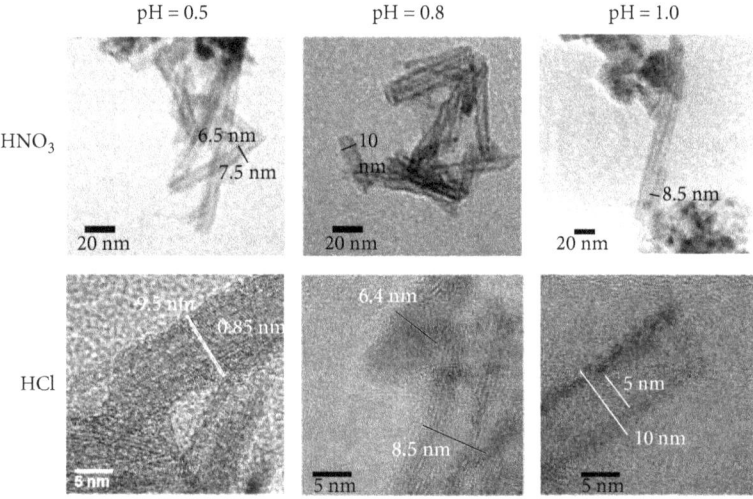

FIGURE 4: TEM images of 1D TiO_2 nanostructures obtained from SG-TiO_2, prepared with HNO_3 or HCl at pH = 0.5, 0.8, and 1.0, after annealing process (400°C/2 h).

as seed materials. It must be mentioned that, because of the lack of homogeneity in the samples, it is difficult to represent the final TiO_2 structure in a single TEM image; the pictures shown represent the most typical structure in each sample.

The XRD analysis of the samples after hydrothermal treatment (Figure 5(a)) shows that the crystalline structure of the seed material changed and displayed peaks around 2θ = 10, 24.5, 28.4, and 48.3°. These peaks represent the diffraction of sodium titanate with the chemical formula $Na_2Ti_nO_{2n+1}$ (n = 3, 6, and 9). This is observed for samples where the anatase content in the seed material was higher than 55%. In other cases, rutile was also present as shown by the reflection peaks around 2θ = 27.5, 36.1, 41.5, 54, and 56°, corresponding to the (110), (101), (111), (211), and (220) planes in agreement with JCPDS No. 21-1276. This confirms that part of the rutile seeds could remain unreacted after the hydrothermal treatment. After the acid

treatment, the features corresponding to titanates almost disappeared, leaving those of the rutile TiO_2 polymorph (not shown).

After the annealing process (Figure 5(b)), a mix of anatase and rutile was observed for samples whose seed had a rutile content larger than 60%. Only peaks corresponding to anatase TiO_2 were observed for samples with anatase higher than 56% in seed. This suggests that when rutile was the dominant phase in the seed material, a portion of it remained unreacted, probably because of the large crystallite size of rutile (~28 nm), compared with the anatase crystallite size (~9 nm). The conditions of the hydrothermal treatment seem insufficient to carry out the dissolution-precipitation process that would be involved in the transformation of TiO_2 to sodium titanate, followed by proton exchange to produce hydrogen titanate and, finally, crystallization to anatase after thermal treatment. On the other hand, it can

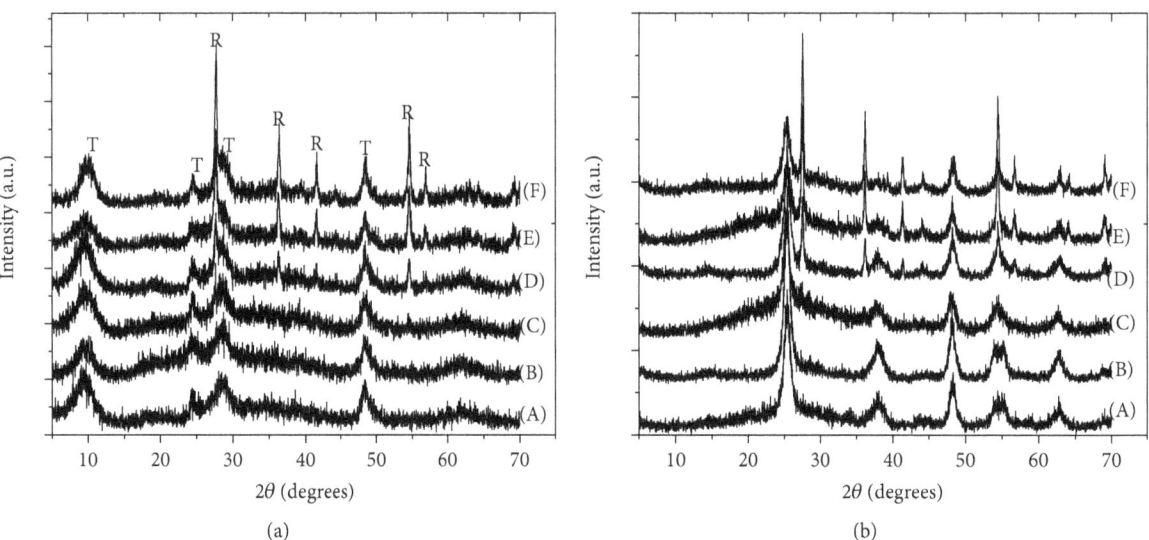

FIGURE 5: XRD patterns of 1D nanostructures obtained after 24 h of hydrothermal treatment of SG-TiO$_2$ NPs (a) and the products obtained after the final annealing process at 400°C for 2 h (b). The anatase contents in seeds were (A) ~100% (HNO$_3$, pH = 1), (B) ~100% (HCl, pH = 1), (C) ~56% (HNO$_3$, pH = 0.8), (D) ~39% (HCl, pH = 0.8), (E) ~18% (HNO$_3$, pH = 0.5), and (F) ~3% (HCl, pH = 0.5) (A = anatase; R = rutile).

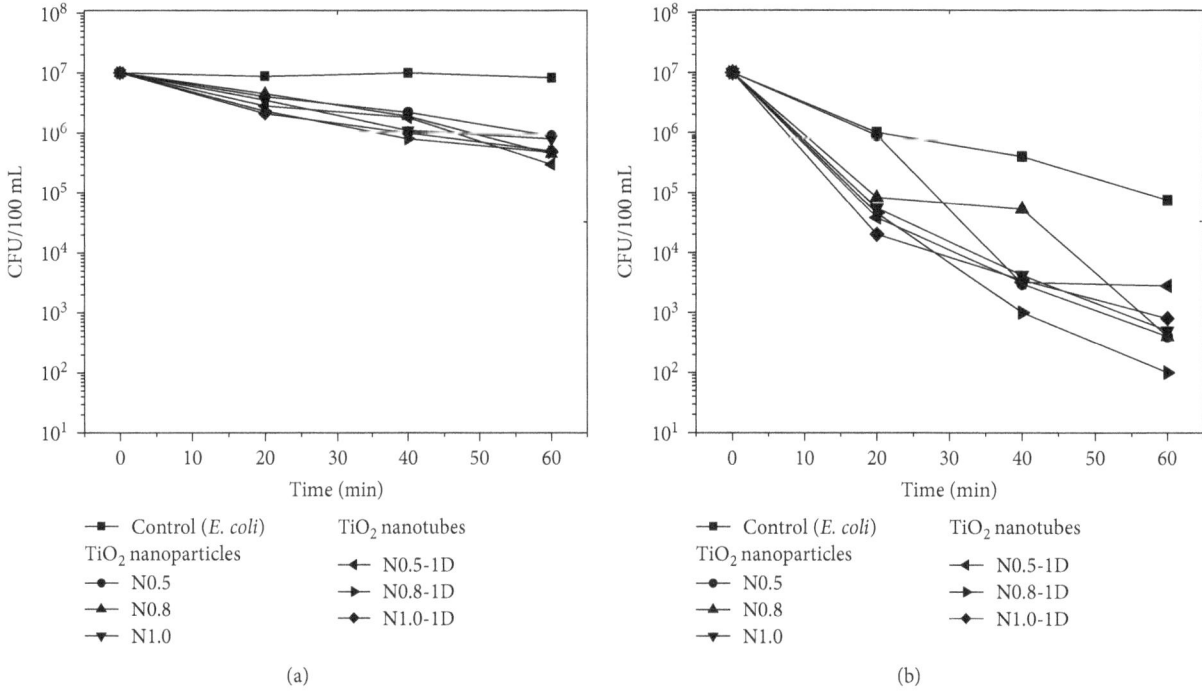

FIGURE 6: *E. coli* bacteria viability under stirring condition in (a) the dark and (b) under UV-A/B irradiation in the presence of TiO$_2$ nanoparticles and nanotubes.

be noted that the seed material obtained with HCl produced samples with the best crystallinity, as the X-ray reflections were well defined, compared to those obtained with HNO$_3$.

3.2. Assessment of the Photocatalytic Activity of TiO$_2$ Nanostructures for Escherichia coli in Water.
Bacteria viability under the stirring process was determined by colony counting after 24 h of incubation. The results showed that the stirring process affects in 1, 3, and 5% (gradually for

20, 40, and 60 minutes, resp.). The assays without stirring were not performed because it was not possible to obtain a homogeneous bacteria distribution.

The effect of stirring in the presence of SG-TiO$_2$ nanoparticles and their corresponding 1D TiO$_2$ nanostructures against *E. coli* was evaluated in the dark and under UV-A/B irradiation. As shown in Figure 6, considering that the initial *E. coli* concentration was 1×10^7 CFU/mL, the presence of TiO$_2$ nanoparticles and nanotubes under stirring conditions in the dark produced a diminution of bacteria viability of around two orders of magnitude (10^5 CFU/mL). It is ascribed to mechanical stress produced by the stirring process.

Also, as is reported in other studies [11], the photolysis is present in our experiments. It contributes to a decrease in bacteria viability at three orders of magnitude. The bactericidal activity of TiO_2 nanostructures is similar to the photolysis in consequence; the catalyst plus irradiation can decrease bacteria viability until five orders of magnitude. However, no major difference was observed for the bactericidal effect of nanoparticles and one-dimensional TiO_2 nanostructures. The latter have an important advantage since 1D TiO_2 nanostructures can be easily removed from solutions.

4. Conclusions

In summary, TiO_2 anatase 1D nanostructures, with different shapes such as tube- and rod-like shapes, were synthesized by hydrothermal treatment of seeds controlling the anatase-rutile proportion. The synthesized 1D TiO_2 nanostructure was effectively used for photocatalytic abatement of *E. coli* in water. Although the 1D TiO_2 nanostructures have a similar photocatalytic activity than the nanoparticles have, the use of one-dimensional TiO_2 nanostructures has an important advantage since the 1D TiO_2 nanostructure can be easily removed from solutions and could be reusable avoiding the necessity of use filtration that increases the cost and time of the cleaning process.

Conflicts of Interest

The authors declare that they have no conflicts of interest.

Acknowledgments

This work was partially supported by the Innovate Peru Project C.133-PNICP-PIAP-2015 and the Concytec Project 223-2015-FONDECYT-DE. The authors are also grateful to the Microscopy Centre of FCEyN, Universidad de Buenos Aires, and Central Microscopy Laboratory of the Institute of Physics (UNAM). Roberto J. Candal is a member of CONICET.

References

[1] A. J. Cowan, J. Tang, W. Leng, J. R. Durrant, and D. R. Klug, "Water splitting by nanocrystalline TiO_2 in a complete photoelectrochemical cell exhibits efficiencies limited by charge recombination," *The Journal of Physical Chemistry C*, vol. 114, no. 9, pp. 4208–4214, 2010.

[2] D. F. Ollis, E. Pelizzetti, and N. Serpone, "Photocatalyzed destruction of water contaminants," *Environmental Science & Technology*, vol. 25, no. 9, pp. 1522–1529, 1991.

[3] Y. Ren, Z. Liu, F. Pourpoint, A. R. Armstrong, C. P. Grey, and P. G. Bruce, "Nanoparticulate TiO_2(B): an anode for lithium-ion batteries," *Angewandte Chemie International Edition*, vol. 51, no. 9, pp. 2164–2167, 2012.

[4] J. Yan and F. Zhou, "TiO_2 nanotubes: structure optimization for solar cells," *Journal of Materials Chemistry*, vol. 21, no. 26, p. 9406, 2011.

[5] C. W. Lai, J. C. Juan, W. B. Ko, and S. B. A. Hamid, "An overview: recent development of titanium oxide nanotubes as photocatalyst for dye degradation," *International Journal of Photoenergy*, vol. 2014, Article ID 524135, 14 pages, 2014.

[6] N. Liu, X. Chen, J. Zhang, and J. W. Schwank, "A review on TiO_2-based nanotubes synthesized via hydrothermal method: formation mechanism, structure modification, and photocatalytic applications," *Catalysis Today*, vol. 225, pp. 34–51, 2014.

[7] J. Cabrera, H. Alarcón, A. López, R. Candal, D. Acosta, and J. Rodriguez, "Synthesis, characterization and photocatalytic activity of 1D TiO_2 nanostructures," *Water Science & Technology*, vol. 70, no. 6, pp. 972–979, 2014.

[8] T. Matsunaga, R. Tomoda, T. Nakajima, and H. Wake, "Photoelectrochemical sterilization of microbial cells by semiconductor powders," *FEMS Microbiology Letters*, vol. 29, no. 1-2, pp. 211–214, 1985.

[9] S. Ponce, E. Carpio, J. Venero et al., "Titanium dioxide onto polyethylene for water decontamination," *Journal of Advanced Oxidation Technologies*, vol. 12, no. 1, pp. 81–86, 2009.

[10] R. A. Spurr and H. Myers, "Quantitative analysis of anatase-rutile mixtures with an X-ray diffractometer," *Analytical Chemistry*, vol. 29, no. 5, pp. 760–762, 1957.

[11] N. Vermeulen, W. J. Keeler, K. Nandakumar, and K. T. Leung, "The bactericidal effect of ultraviolet and visible light on *Escherichia coli*," *Biotechnology and Bioengineering*, vol. 99, no. 3, pp. 550–556, 2008.

7

One-Pot Synthesis of Ru-Doped ZnO Oxides for Photodegradation of 4-Chlorophenol

Maria E. Manríquez,[1] Luis Enrique Noreña (iD),[2] Jin An Wang (iD),[1] Lifang Chen,[1] Jose Salmones,[1] Julio González-García (iD),[1] Carmen Reza,[1] Francisco Tzompantzi (iD),[3] José G. Hernández Cortez,[4] Liqun Ye,[5] and Haiquan Xie[5]

[1]ESIQIE, Instituto Politécnico Nacional, Av. Instituto Politécnico Nacional s/n, Col. Zacatenco, 07738 Mexico City, Mexico
[2]Departamento de Ciencias Básicas, Universidad Autónoma Metropolitana-Azcapotzalco, Av. San Pablo 180, Col. Reynosa-Tamaulipas, 02200 Mexico City, Mexico
[3]Departamento Química, Universidad Autónoma Metropolitana-Iztapalapa, San Rafael Atlixco 186, 09340 Mexico City, Mexico
[4]GDMyPQ, Eje Lázaro Cárdenas 152, Instituto Mexicano del Petróleo, 07730 Mexico City, Mexico
[5]College of Chemistry and Pharmaceutical Engineering, Nanyang Normal University, Nanyang, China

Correspondence should be addressed to Luis Enrique Noreña; lnf@correo.azc.uam.mx and Jin An Wang; jwang@ipn.mx

Academic Editor: Ilya I. Tumkin

The photocatalytic degradation of 4-chlorophenol in water using Ru-doped ZnO mixed oxides (0, 0.5, 1, and 3 wt% RuO_2) synthesized by the one-pot homogeneous coprecipitation method is reported. ZnO with wurtzite structure was present in the mixed oxide as corroborated by Raman spectroscopy and X-ray diffraction analysis. All the samples showed nanorod morphological features. The presence of Ru^{6+}/Ru^{4+} couples on ZnO modified the band gap of the mixed oxides and led to a shift of the band gap energy from 3.20 eV to 3.07 eV. Ru addition increased the surface area and significantly promoted the formation of active surface oxygen species such as hydroradicals evidenced by the fluorescence spectroscopy measurement. In the photodegradation of 4-chlorophenol solution under UV irradiation, a notable increase in photoactivity was obtained as the amount of RuO_2 in the mixed oxides increased to 3 wt%. The charge transfer between Ru^{6+}/Ru^{4+} couples and ZnO nanoparticles together with the formation of free radical oxidant species effectively inhibits electron-hole recombination rate, thus favoring the photodegradation of 4-chlorophenol.

1. Introduction

Although most water effluents are effectively treated by traditional biological and/or physicochemical methods, the presence of chlorinated compounds represents an increasing problem, since these compounds, for example, chlorinated phenol derivatives, are often mutagenic or carcinogenic. 4-Chlorophenol is one of the most dangerous substances as far as toxicity is concerned. Therefore, it is highly necessary to develop more effective processes for the destruction of such contaminants. Actually, the expanding demand for efficient and economical techniques for wastewater treatment is the origin of a wide spectrum of investigations and technological innovation [1–5]. Some new technologies such as the advanced oxidation process (AOP) have been applied for the elimination of organic compounds in water systems. In AOPs, the oxidation of organic compounds by the very reactive free radicals takes place on the surface of heterogeneous catalysts which has been proven to be of interest due to the efficient degradation of recalcitrant organic compounds in water [6–8].

Developed in the 1970s, photocatalytic oxidation has attracted considerable attention, and in recent years, numerous studies have been carried out on the applications of photocatalytic oxidation in the presence of a semiconductor oxide [9–13]. As a good candidate of photocatalyst, ZnO has a band gap energy (E_g) of 3.37 eV, which corresponds to light absorption in the UV region [14]; this E_g value indicates that its photocatalytic properties are almost comparable with TiO_2. Li et al. reported that the total photodegradation of chlorophenol and dichlorobenzene can be realized by using ZnO nanoparticles [15]. On the other hand, RuO_2 has been studied as a photocatalyst showing strong oxidation properties [16]. The catalytic role in the N_2O decomposition over the Ru-doped FAU zeolite was also investigated [17]. Zhuiykova et al. reported that the electrocatalytic activity of ZnO-RuO_2 catalysts appears to be governed by the redox behavior of the active surface sites [18].

In the present work, we report the one-pot homogeneous coprecipitation method for the synthesis of a set of Ru-doped ZnO oxides with different Ru contents (0.5, 1, and 3 wt% of RuO_2). These catalysts were characterized by X-ray diffraction (XRD), scanning electron microscopy (SEM), Raman spectroscopy, X-ray photoelectron, and UV-Vis spectroscopic techniques. The textural properties including the surface area and pore volume were determined by the N_2 adsorption-desorption isotherms method. The crystalline structure of the samples was refined by the Rietveld refinement method, and the surface chemical composition was analyzed by the electron dispersive spectroscopic (EDS) technique. The formation of the surface $^\bullet$OH radicals was monitored by using fluorescence emission spectroscopy. The photocatalytic performance of the RuO_x-ZnO catalysts with various RuO_2 contents for the degradation of 4-chlorophenol was investigated, and the surface photoreaction mechanism was discussed.

2. Experimental

2.1. Sample Preparation. The RuO_x-ZnO semiconductors were synthesized by the one-pot homogeneous coprecipitation method as follows: appropriate amounts of $Zn(CH_3COO)_2$ (Baker, 99.99%) and $RuCl_3 \cdot 3H_2O$ (Aldrich, 99.9%) solutions were mixed at a pH of ~8.5 adjusted with an ammonia solution. The amount of $RuCl_3 \cdot 3H_2O$ was calculated in order to obtain final solids with 0, 0.5, 1.0, and 3.0 wt% of RuO_2. Afterwards, the mixture containing both metallic precursors was heated at 80°C under constant stirring for 2 h. The resultant precipitate was collected by filtration and washed with deionized water until a neutral pH in the effluent was reached, and finally, the wet solid was dried at 100°C in air for 6 h and calcined at 400°C in a muffle furnace for 12 h. These solids will be hereinafter referred to as $x$wt% RuO_2-ZnO, where $x = 0$, 0.5, 1, and 3.

2.2. Characterization. X-ray diffraction (XRD) patterns of the RuO_2-ZnO solids were obtained by an Empyrean multipurpose research X-ray diffractometer (PANalytical) with a Cu-Kα radiation (1.5418 Å) source. Diffraction intensity was measured between 4° and 70° with a 2θ step of 0.02°. The crystalline structures of the catalysts were refined using the Rietveld method. The JAVA-based software named Materials Analysis Using Diffraction (MAUD) was employed for refining each XRD pattern [19]. The atomic fraction coordinates for ZnO are reported in Table 1.

TABLE 1: Atomic fractional coordinates in the crystalline structure of ZnO. Space group: P63mc; SG number: 186; crystal system: hexagonal.

Atom	Site	x	y	z
Zn	2b	0.3333	0.6667	0
O	2b	0.3333	0.6667	0.3825

The textural properties of the samples were obtained by nitrogen adsorption-desorption isotherms at −196°C in a Quantachrome Autosorb-1C instrument. The surface area was calculated by the Brunauer–Emmett–Teller (BET) equation. All samples were degassed at 300°C for 6 h before N_2 adsorption.

Raman spectra were recorded at room temperature on previously calcined samples using an ISA Labram micro-Raman apparatus. The excitation line of a He-Ne laser was 632 nm. To avoid thermal effects, the laser power on the samples was kept at ~1 nW.

X-ray photoelectron spectroscopy (XPS) was used to analyze the surface metal oxidation state of the samples. The XPS spectra were acquired with a THERMO Scientific K-Alpha spectrometer, equipped with a hemispherical electron analyzer and Al Kα X-ray source (1486.6 eV).

Scanning electron microscopy analysis (SEM) was carried out on a Quanta 3D FEG microscope. The microscope has detectors for surface chemical analysis by X-ray energy dispersive spectroscopy (EDS).

The band gap energy (E_g) of the catalysts was measured using a Cary 100 spectrophotometer with an integration sphere, and MgO was used as 100% reflectance patron. The band gap energy values were determined using the Kubelka-Munk function by the method.

Fluorescence emission spectra of 2-hydroxyterephthalic acid were measured on a SCINCO fluorescence spectrometer FS-2. The $^\bullet$OH radicals generated by the semiconductor material in the absence of 4-chlorophenol were measured with the following procedure: terephthalic acid (TA) ($5 \times 10{-}4$ M) was dissolved in a water/NaOH solution (2×10^{-3} M). Then 200 mg of the photocatalysts was added and the suspension was stirred for 60 min under dark conditions. Afterwards, it was irradiated by a UV light Pen-Ray lamp with a wavelength of 254 nm ($I0 = 4.4$ mW cm^{-2}), during 60 min, and aliquots were taken each 5 min. The fluorescence emission spectra of the irradiated solution were analyzed by PL (excited at 320 nm).

2.3. Photocatalytic Activity. The photocatalytic degradation of 4-chlorophenol in water was performed at room temperature (~25°C) using a solution of 4-chlorophenol in water (80 ppm) employing a Pyrex reactor (500 mL). 200 mg of the catalyst with particle size 60–80 meshes was used for the photodegradation reaction. The mixture was under stirring and bubbling with air (1 mL/s) generated by a magnetic stirring and electric pump at room temperature (approximately 30°C). The mixture prepared was exposed to UV radiation for 6 h. The irradiation source was a UV Pen-Ray lamp ($\lambda = 254$ nm and intensity $= 4400\,\mu W/cm^2$), protected with a quartz tube whose outer and inner diameters were 0.5 and 0.35 in nm, respectively. The 4-chlorophenol photooxidation reaction procedure was monitored via following the variation of the characteristic absorption band at 221 nm using the UV-Vis spectrophotometer (Varian Cary 100 UV-Vis). The 4-chlorophenol concentration was monitored by taking aliquots of 3 mL every 15 min of reaction. The aliquots were filtered with a 0.45 μm nylon filter to eliminate any suspended solids.

3. Results and Discussion

3.1. Crystalline Structure and Rietveld Refinements. The XRD patterns of the samples calcined at 400°C are shown in Figure 1. All the diffraction peaks can be indexed to the hexagonal Wurtzite structure of ZnO (JCPDS card 36-1451) [20, 21]. XRD peaks corresponding to RuO_2 and changes in the crystalline structure of ZnO are not observed; thus, it can be assumed that the ruthenium oxide like RuO_2 is not detectable since its quantity is very small; perhaps, it is dispersed uniformly on the surface of ZnO with a crystallite size smaller than 4 nm.

The crystalline structures of the solids were refined with the Rietveld refinement technique. A plot of the Rietveld refinement for the 3 wt% RuO_2-ZnO catalyst is shown in Figure 2. The crystallite size, lattice cell parameters, and R_{wp} are reported in Table 2. The crystallite sizes determined by Scherrer's equation and obtained by the Rietveld refinement were 60.6, 75.9, 69.5, and 61.8 nm for the $0RuO_2$-ZnO, $0.5RuO_2$-ZnO, $1RuO_2$-ZnO, and $3RuO_2$-ZnO samples, respectively. It seems that Ru addition led to a slightly greater ZnO crystallite size, even if this effect becomes weak for the higher Ru content sample. Ru doping probably promotes the formation of surface hydroxyls which may gain linkage degree with the hydroxide units during the synthesis, leading to a slight enlargement of the crystallite size. The lattice cell parameters of ZnO with a hexagonal wurtzite structure varied: $a = b$ from 3.2903 to 3.4098 Å and c from 5.2219 to 5.4039 Å, depending on the Ru content.

3.2. Raman Spectroscopy. The Raman spectra of the samples were obtained in the wavelength interval 300–550 cm^{-1}, and they are shown in Figure 3(a). The Raman bands at 330, 380, and 437 cm^{-1} are assigned to ZnO with a wurtzite (hexagonal) structure belonging to the P63mc space group. The dominant and sharp peak at 332 cm^{-1} is due to the second-order Raman spectrum, and the peak labeled as E_2 at 437 cm^{-1} is known as the Raman active phonon mode, which is a characteristic of the Wurtzite hexagonal phase of ZnO.

In Figure 3(b), the Raman modes of E_g, A_{1g}, and B_{2g} at 528, 646, and 716 cm^{-1}, respectively, are shown, corresponding to RuO_2 in $1RuO_2$-ZnO and $3RuO_2$-ZnO samples. The intensity of Raman peaks increases with the increasing of the RuO_2 content [22, 23]. Although Ru oxides were not detectable by XRD analysis, Ru-O bonds are clearly present in the surface of the catalysts. Therefore, the Raman spectroscopy is more sensitive for identifying the surface Ru-O bonds.

3.3. XPS Characterization. The chemical oxidation states of Ru and Zn in the Ru-doped ZnO samples were analyzed by the XPS technique. The Ru 3d core levels in the XPS spectra of all samples are presented in Figure 4. By deconvoluting with software (XPS PEAK 4.1), two doublets in XPS spectra were determined: the peaks at 280.9 and 285.0 eV are assigned to Ru 3d 5/2 and Ru 3d 3/2, respectively, indicating that Ru is in the oxidation state Ru^{4+} (RuO_2) [24]. The other two small peaks at 282.5 and 286.3 eV correspond to Ru^{6+} (ReO_3). The intensity of these peaks increases with increasing content of RuO_2.

The XPS spectra of Zn 2p core levels are shown in Figure 5, and they exhibited two symmetric peaks centered at 1022.0 and 1044.7 eV, which are attributed to Zn 2p 3/2 and Zn 2p 1/2, respectively, indicating the Zn^{2+} oxidation state [25].

Figure 6 shows the asymmetric O1s spectra on the surface of $0RuO_2$-ZnO, $0.5RuO_2$-ZnO, $1RuO_2$-ZnO, and $3RuO_2$-ZnO samples. For the pure ZnO sample, the O1s signal was deconvoluted into two peaks at 530.6 eV and 532.6 eV, corresponding to the lattice O^{2-} in the Zn-O bond and surface oxygen in hydroxyls (OH) [26]. For the other three Ru-doped samples, four peaks at approximately 529.0, 530.3, 530.6, and 532.6 eV were observed. These are assigned to lattice oxygen in RuO_2 (529.0 eV, Ru^{4+}-O) and RuO_3 (530.3 eV, Ru^{6+}-O), lattice oxygen in ZnO (530.6 eV, Zn-O), and surface oxygen in the hydroxyl group (532.6 eV, O-H) in the RuO_2 and ZnO solids, respectively [27]. The XPS analyses confirm that different kinds of oxygen species exist on the surface of the RuO_2-ZnO solids and Ru ions have two oxidation states: Ru^{4+} (major) and Ru^{6+} (minor).

The surface atomic concentrations were obtained from the XPS quantitative analysis and are reported in Table 3. The surface Ru/Zn atomic ratio increases from 0.081 to 0.093 and 0.207 as the RuO_2 nominal content increases from 0.5 wt% to 1 and 3 wt%. The O/(Zn + Ru) atomic ratio varies from 1.12 for 0.5RuZnO to 1.40 for 1RuZnO and 2.84 for 3RuZnO. Therefore, the surface oxygen concentration clearly gains as the Ru content increases. Ru doping led to oxygen species rich in the surface of the RuO_2-ZnO solids.

3.4. SEM Observation and EDS Analysis. Figure 7 shows the SEM images and the selective EDS spectra for ZnO and RuO_2-ZnO solids. For the pure ZnO, nanorods with uniform size were obtained (Figure 7(a)). These nanorods did not present a preferential alignment but had a random distribution. The length of the rods varied from 300 to 500 nm. In the sample with 0.5 wt% of RuO_2 (Figure 7(c)), the morphology is slightly modified as the nanorods become thinner and shorter. When the RuO_2 content increases to 1 wt% (Figure 7(e)), the length of nanorods ranges between 100 and 500 nm in length. Finally, with a greater amount of RuO_2 (Figure 7(g)), the morphology was more homogeneous and the nanorods had a length of approximately 100 nm. So doping with a proper amount of Ru may lead to the formation of short and homogeneous ZnO nanorods.

The surface elemental composition analysis of each sample was performed by the EDS technique. Signals belonging

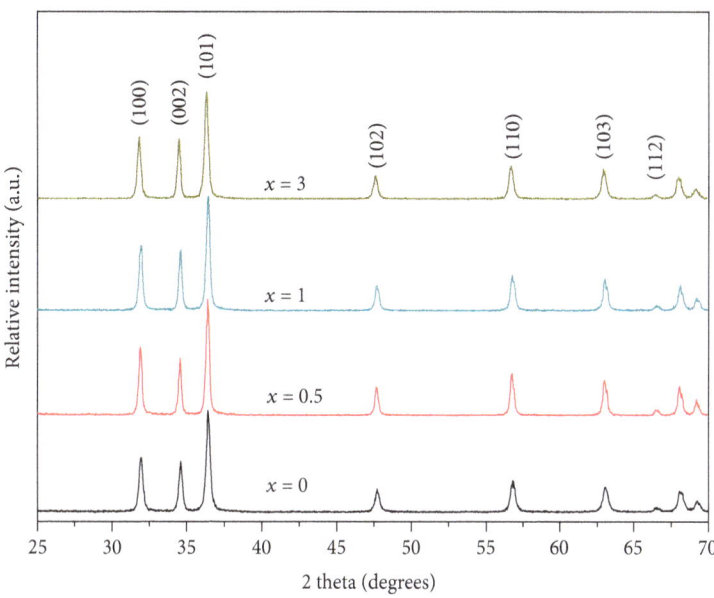

FIGURE 1: XRD patterns of the $x$wt% RuO_2-ZnO samples. $x = 0$, $x = 0.5$, $x = 1$, and $x = 3$.

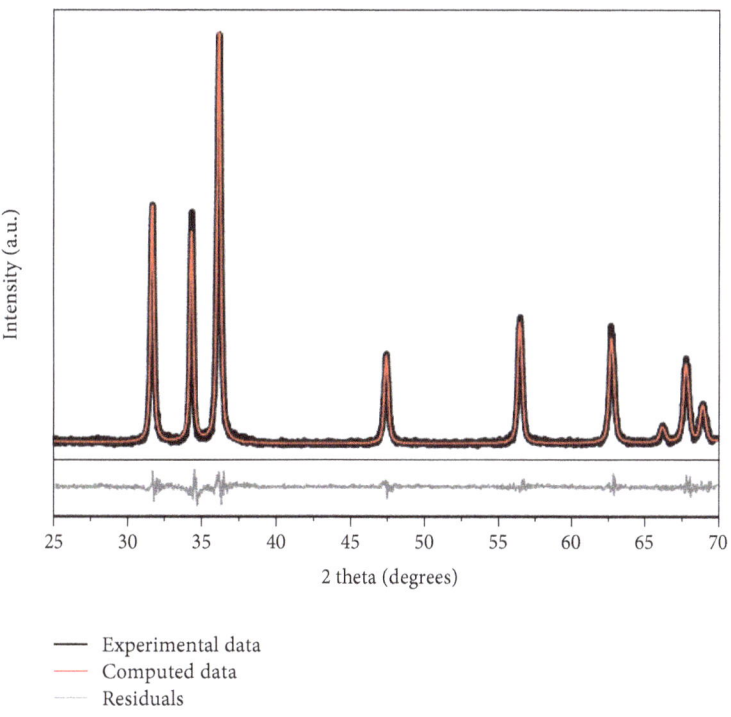

— Experimental data
— Computed data
······ Residuals

FIGURE 2: A plot of Rietveld refinement of 3 wt% RuO_2-ZnO. The black line corresponds to the experimental data, the red line corresponds to the calculated data, and the grey line indicates residuals between experimental and calculation data.

to Zn, Ru, C, and O elements are observed (see Figures 7(b), 7(d), 7(f), and 7(h)). Figure 7(b) only shows the signals of Zn and O with carbon traces. Carbon was probably formed from the acetate precursor used in the ZnO synthesis. Figures 7(d), 7(f), and 7(h) show the chemical composition of the xRuO$_2$-ZnO. A small peak for Ru appears which increases in intensity as the content of RuO_2 increases. It is also noted that as the RuO_2 content increases, the peak corresponding

to oxygen increases in agreement with the results of XPS analysis. The increment of the surface oxygen content is important as the oxygen species usually take part in the oxidation reactions.

3.5. Textural Properties. The textural properties of the RuO_2-ZnO samples are reported in Table 4. The surface area of the samples with different RuO_2 loadings varies in the range of

TABLE 2: Crystalline size, lattice cell parameters, and R_{wp} data obtained from Rietveld refinements.

Sample	Crystallite size (nm)	Lattice cell parameters			Rwp (%)
		a (Å)	b (Å)	c (Å)	
3 wt% RuO_2-ZnO	61.81	3.2603	3.2603	5.2219	0.43
1 wt% RuO_2-ZnO	69.46	3.4588	3.4588	5.5412	0.63
0.5 wt% RuO_2-ZnO	75.93	3.4298	3.4298	5.4939	1.15
0 wt% RuO_2-ZnO	60.58	3.4258	3.4258	5.4873	2.17

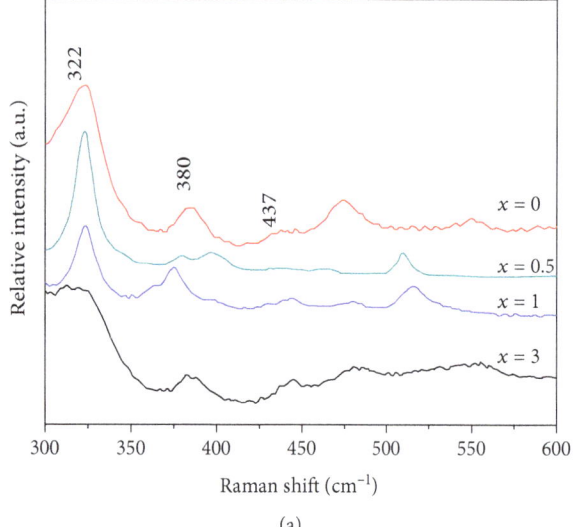

(a)

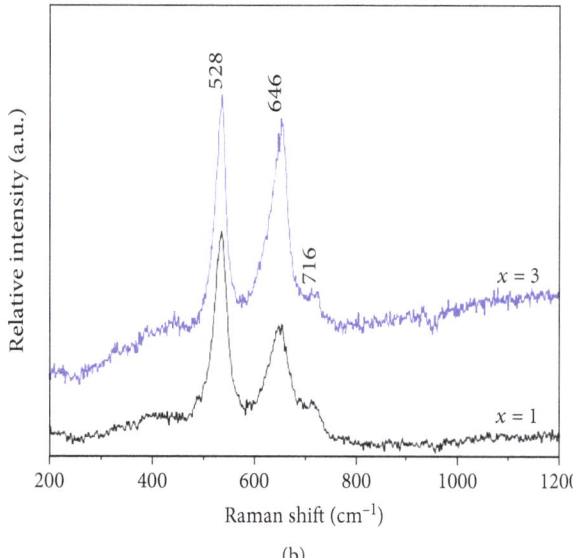

(b)

FIGURE 3: (a) Raman spectra of the $x$wt% RuO_2-ZnO samples in the region between 200 and 600 cm^{-1}. $x = 0$, $x = 0.5$, $x = 1$, and $x = 3$. (b) Raman spectra of $x$wt% RuO_2-ZnO samples in the region between 200 and 1200 cm^{-1}. $x = 1$ and $x = 3$.

11~38 m^2/g. When the RuO_2 content is greater, the specific surface area is bigger. The 3 wt% RuO_2-ZnO sample shows the biggest specific surface area, three times greater than that obtained for pure ZnO. According to the above results, we can assume that the variations on the BET surface area are due to the changes of the structure resulted from the modification of RuO_2 doping.

3.6. Band-Gap Energy Measurement. Figure 8 shows the plots of the Kubelka-Munk function ($[F(R_1)E]^{1/2}$) against the absorbed light energy (E). The optical band gap energies (E_g) of each sample can be estimated by extrapolating the slope to cut the abscissa axis [28–30]. The presence of RuO_2 (0.5, 1, and 3 wt %) produces to a red-shift of the absorption band towards the visible region. The band gap energy for pure ZnO was approximately 3.20 eV, which is close to report in the literature [31, 32]. Several previous investigations suggest that the visible emission is due to the surface defects and oxygen vacancies in the crystalline structure of the ZnO nanomaterials [33, 34]. These RuO_2-ZnO catalysts show a continuous absorption in the visible range and the absorption is stronger when the RuO_2 content is greater. The estimated band gap energies decrease from 3.20 eV for pure ZnO to 3.07 eV for 3 wt% RuO_2-ZnO. The E_g data obtained from Figure 8 are reported in Table 4.

3.7. Photocatalytic Activity Test. To investigate the photocatalytic activity, the RuO_2-ZnO samples were evaluated in the photocatalytic degradation of 4-chlorophenol in aqueous solution. Figure 9 illustrates a set of UV-Vis absorption spectra of 4-chlorophenol molecule at different intervals of reaction time for the 3 wt% RuO_2-ZnO catalyst. The maximum absorption bands of 4-chlorophenol appear at 221 and 277 nm under UV irradiation. These absorption bands gradually decrease in intensity within 150 min, indicating that 4-chlorophenol was photodegradated.

The influence of the content of RuO_2 on the photodegradation efficiency (noted as C/C_0) of 4-chlorophenol is presented in Figure 10. The Ru-doped ZnO catalysts exhibit an increased photocatalytic activity compared to pure ZnO. After 225 min of reaction time, the photodegradation efficiency of 4-chlorophenol pollutant is about 46%, 61%, 70%, and 92% for 0RuO_2-ZnO, 0.5RuO_2-ZnO, 1RuO_2-ZnO, and 3RuO_2-ZnO, respectively.

3.8. Determination of •OH Radical Generation and Reaction Mechanism. In order to understand the mechanism of the photodegradation of 4-chlorophenol, it is important to obtain information about the formation of free radicals, especially the formation of •OH radicals, which have a high oxidative power and great capacity for degradation of organic contaminants in an aqueous medium. The •OH radicals can be monitored by the fluorescence spectrum resulting from the formation of 2-hydroxy terephthalic acid by the •OH radicals reacting with terephthalic acid when the solid was irradiated with ultraviolet light [35, 36]. The most active catalyst 3 wt%RuO_2 was measured by comparison with TiO_2-P25 in the absence of 4-chlorophenol. The

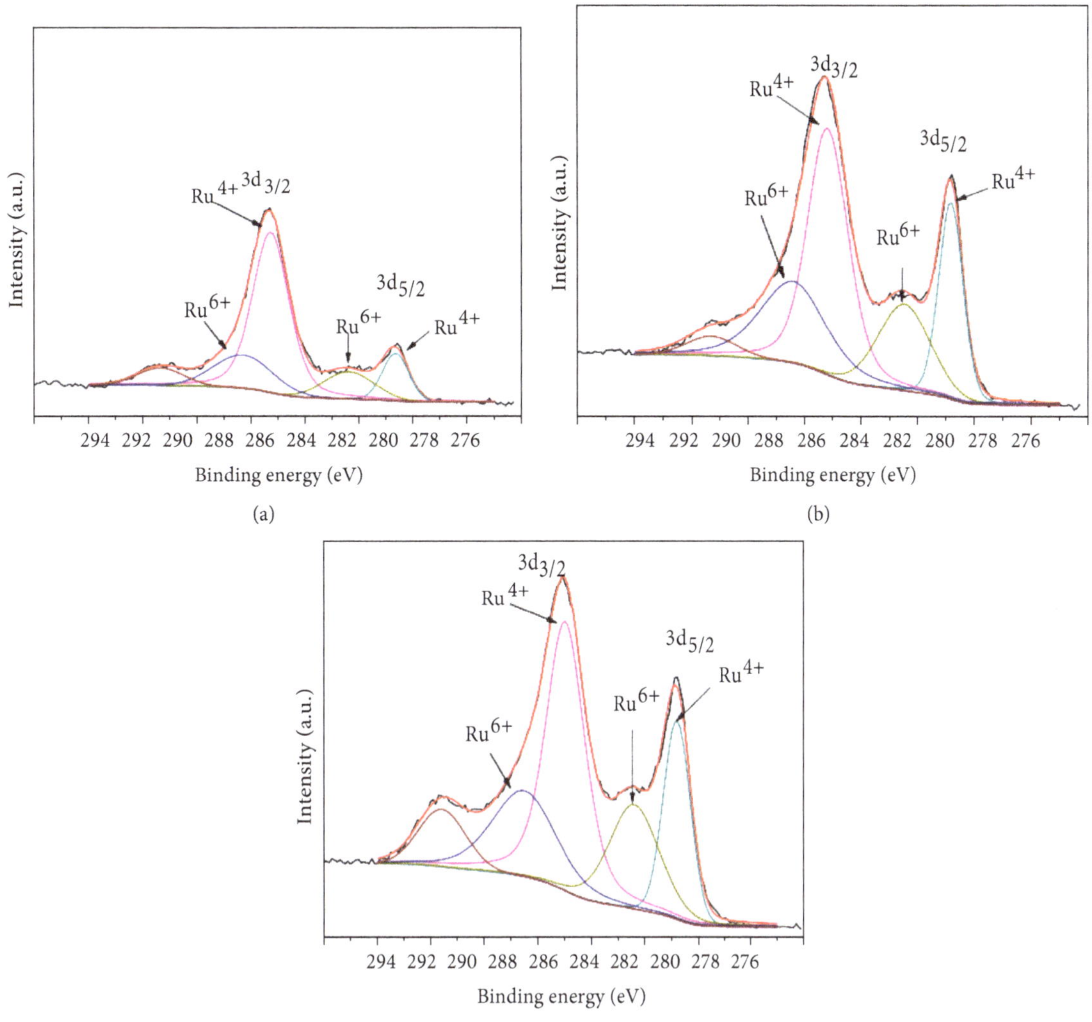

FIGURE 4: Ru 3d core levels of the XPS spectra for the RuO_2-ZnO catalysts. (a) 0.5 wt% RuO_2-ZnO, (b) 1 wt% RuO_2-ZnO, and (c) 3 wt% RuO_2-ZnO.

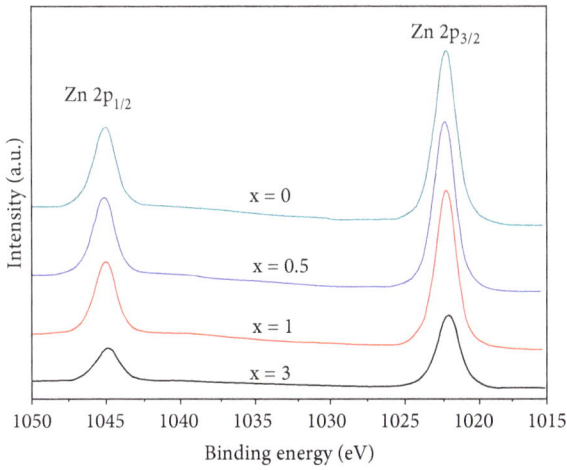

FIGURE 5: Zn 2p core levels of the XPS spectra for the $x$wt% RuO_2-ZnO catalysts, where x indicates the RuO_2 concentration.

photolysis performance was also obtained in the absence of any catalyst.

Figure 11(a) shows the fluorescence spectra obtained with the 3RuO_2-ZnO and TiO_2 P25 catalysts. The peak at 454 nm indicated the amount of produced 2-hydroxy terephthalic acid or the amount of •OH radicals produced. The peak intensity clearly increases as a function of irradiation time. Both 3RuO_2-ZnO and TiO_2 P25 show almost the linear relationship between the band intensity and the irradiation time, but 3RuO_2-ZnO exhibits a greater concentration of •OH radicals on the surface in comparison with TiO_2 P25 during the irradiation procedure. Without the catalyst, the photolysis produces a rather low concentration of hydroxyl terephthalic acid, indicating the low concentration of •OH radicals.

The reaction mechanism considered for heterogeneous photocatalysis usually consists of two steps: the fast adsorption of the reactants on the photocatalyst surface, followed by the slow surface reaction of the adsorbed reactants with photo-generated radicals (e.g., O_2^-, •OH) to generate the

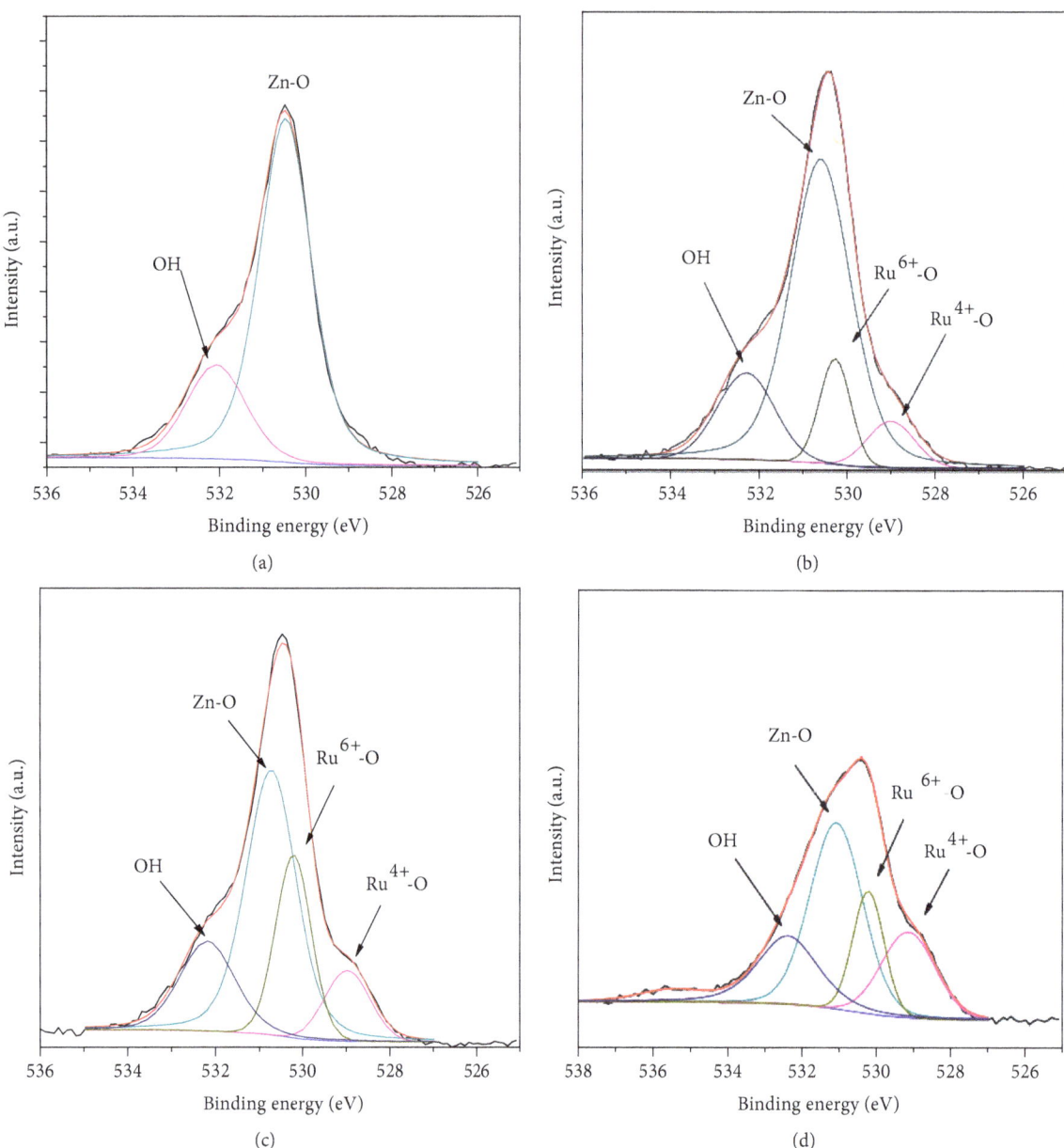

FIGURE 6: O1s core level of the XPS spectra for the $x$wt% RuO_2-ZnO samples. (a) $x = 0$, (b) $x = 0.5$, (c) $x = 1$, and (d) $x = 3$.

TABLE 3: Surface chemical composition of the Ru-doped samples obtained from XPS analysis.

Catalysts	Zn : O	O/(Zn + Ru)	Ru/Zn	Ru : Zn : O
0.5RuO$_2$-ZnO	1 : 1.27	1.18	0.081	1 : 12.32 : 15.65
1RuO$_2$-ZnO	1 : 1.29	1.40	0. 093	1 : 10.00.13.89
3RuO$_2$-ZnO	1 : 2.72	2.85	0.207	1 : 4.83 : 13.13

photooxidized active species. In the presence of dissolved oxygen in aqueous solution and the surface hydroxyl radicals in the solid catalysts, oxygen species may exist in several forms such as O_2, $^{\bullet}OH$, O_2^{2-}, and O_2^{-}, depending on the adsorption capability of the structure and temperature. The formation of surface oxygen species in the RuO_2-ZnO

catalysts can be explained by the following reactions on the surface of the catalysts:

$$RuO_2 - ZnO + h\nu \rightarrow e^{-}(ZnO) + h^{+}(RuO_2)$$

$$O_2 + e^{-} \rightarrow O_2^{-} \;\frac{and}{or}\; O_2 + 2e^{-} \rightarrow O_2^{2-} \qquad (1)$$

$$O_2^{-} + H_2O \rightarrow HO_2^{-} + {}^{\bullet}OH$$

The surface of RuO_2-ZnO solids presents a high degree of surface hydroxylation. For instance, the catalyst with 3 wt%RuO_2 presents a high concentration of surface hydroxyl species as evidenced by the XPS spectra in Figure 6 and by the fluorescence spectroscopy in Figure 11, and therefore, it shows the highest concentration of surface active oxygen

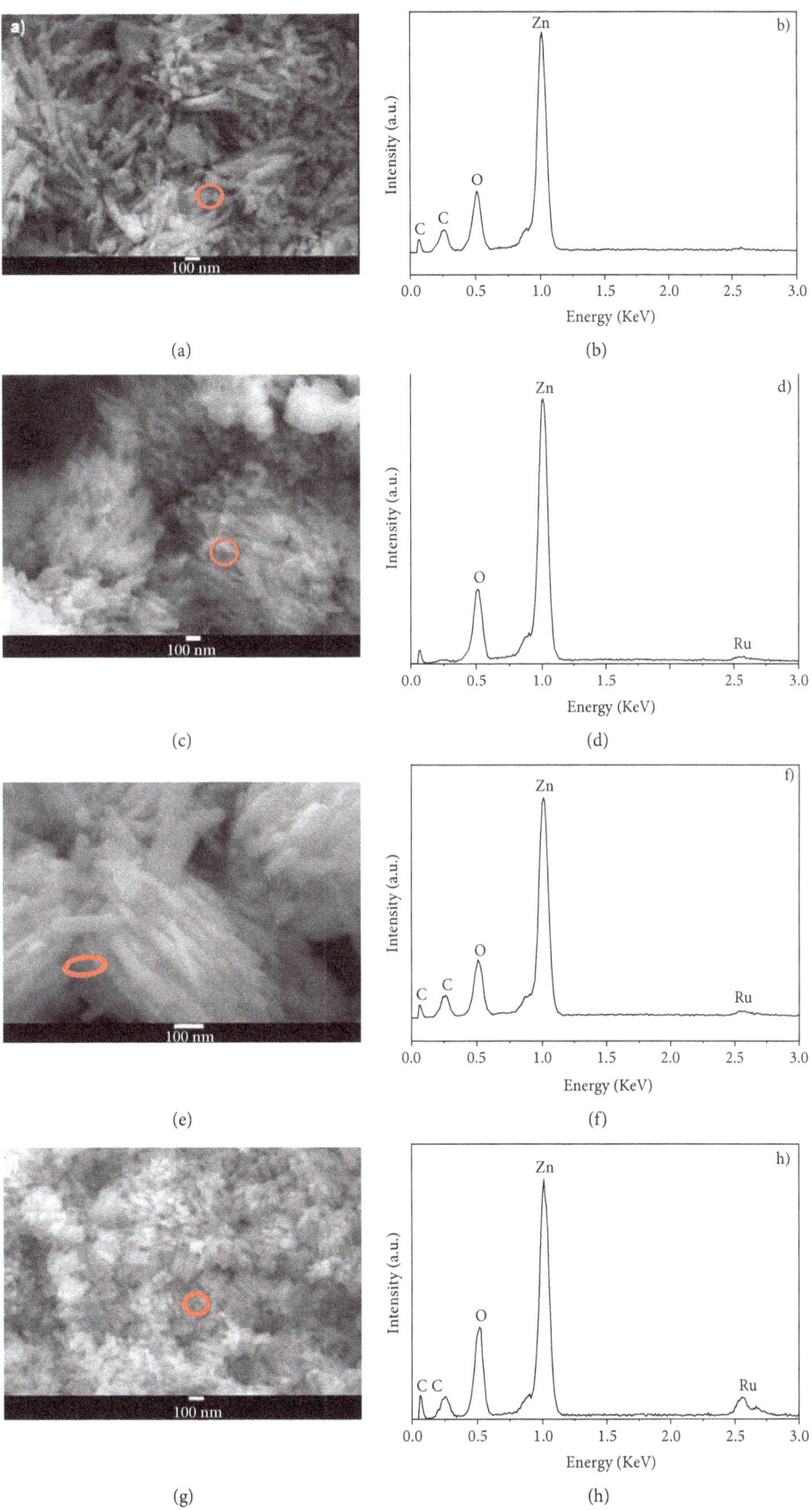

FIGURE 7: SEM images and EDS spectra of the $x$wt% RuO$_2$-ZnO samples. (a, b) $x = 0$; (c, d) $x = 0.5$; (e, f) $x = 1$; and (g, h) $x = 3$.

TABLE 4: Textural properties and band gap of the RuO_2-ZnO solids.

Sample	S_{BET} (m^2/g)	Pore volume (cm^3/g)	Eg (eV)
$0RuO_2$-ZnO	11	0.38	3.20
$0.5RuO_2$-ZnO	19	0.16	3.12
$1RuO_2$-ZnO	32	0.15	3.11
$3RuO_2$-ZnO	38	0.22	3.07

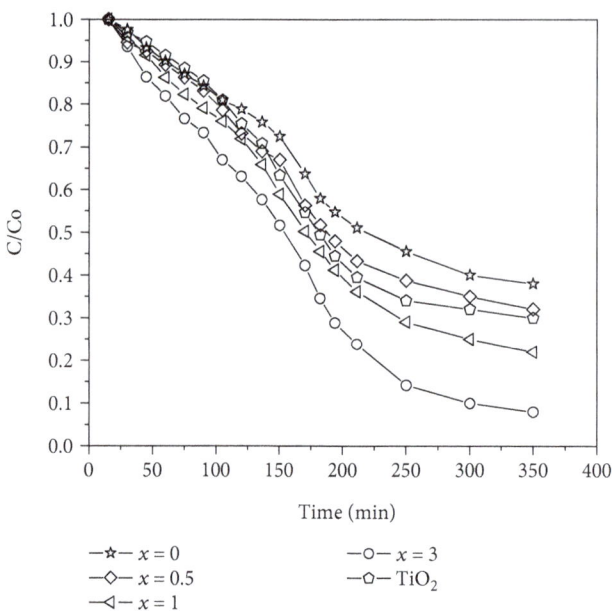

FIGURE 10: Photodegradation of 4-chlorophenol with $x$wt% RuO_2-ZnO and in 350 min of reaction. The activity of TiO_2 P25 was evaluated as reference.

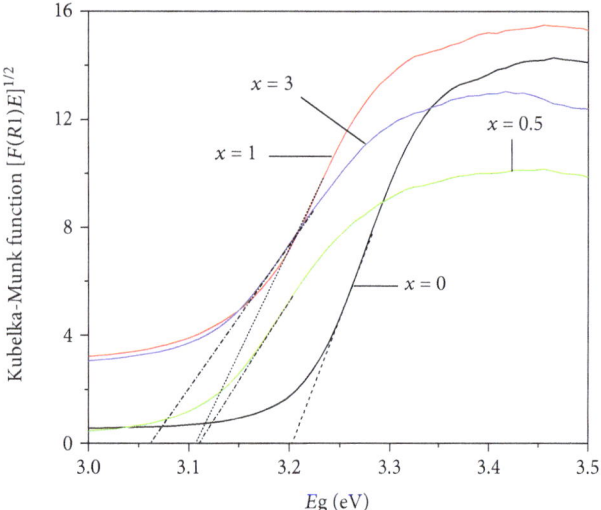

FIGURE 8: A plot of the Kubelka-Munk function versus energy (hν) for the $x$wt% RuO_2-ZnO samples. $x = 0$, $x = 0.5$, $x = 1$, and $x = 3$.

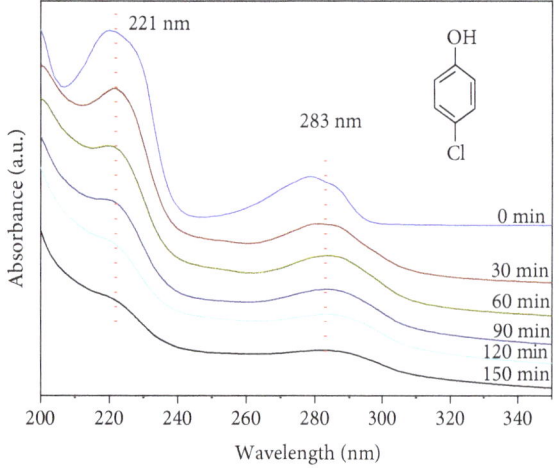

FIGURE 9: A set of UV-Vis spectra of 4-chlorophenol solution recorded in the presence of the 3 wt% RuO_2-ZnO sample at different irradiation times.

species and thus the best photoactivity. All catalysts with RuO_2 exhibit higher reaction rates than the pure ZnO. Perhaps, there is a synergistic interaction between RuO_2 and ZnO that favors the improvement of photocatalytic activity. It is assumed that a heterojunction structure was created in the interface of the ZnO and RuO_2 nanoparticles. Ru doping

strongly affects the surface properties of semiconductors by generating a "barrier" in the ZnO which acts as an electron trap of the Ru metal in contact with the ZnO semiconductor surface. The electrons in the Ru conduction band (CB) transferred to the ZnO CB under the force of the electronic field of the heterojunction structure [37]. Thus, the surface of the ZnO was more negative and favors the formation of more hydroradicals. Therefore, Ru doping promotes the active surface oxygen formation via electron transfer. All these may inhibit the recombination rate of the electron-holes and benefit the photodegradation of 4-chlorophenol. A proposed reaction mechanism is illustrated in Scheme 1.

4. Conclusions

This work showed that Ru doping for ZnO reduced the band gap energy of ZnO solid, generated more surface active oxygen species such as hydroxyls, O_2^{2-}, and O_2^-, and increased the surface area by modification of the crystalline structure and morphological features. In the photodegradation of 4-chlorophenol solution, the RuO_x-ZnO catalysts could effectively photodegrade the 4-chlorophenol contaminant with a much faster reaction rate than the pure ZnO. This can be explained by rich active oxygen species and high concentration of free radicals on the surface of the Ru-doped catalyst that may inhibit the electron-hole recombination rate and promotes the Ru^{6+}/Ru^{4+} ox-red cycles due to the electron transfer from the RuO_x (RuO_3 and RuO_2) nanoparticles to the Zn-O bond. These results show that the one-pot homogeneous coprecipitation synthesis reported herein is a simple but very effective method for obtaining RuO_2-ZnO nanocatalysts with high photocatalyic activity in the photodegradation of 4-chlorophenol.

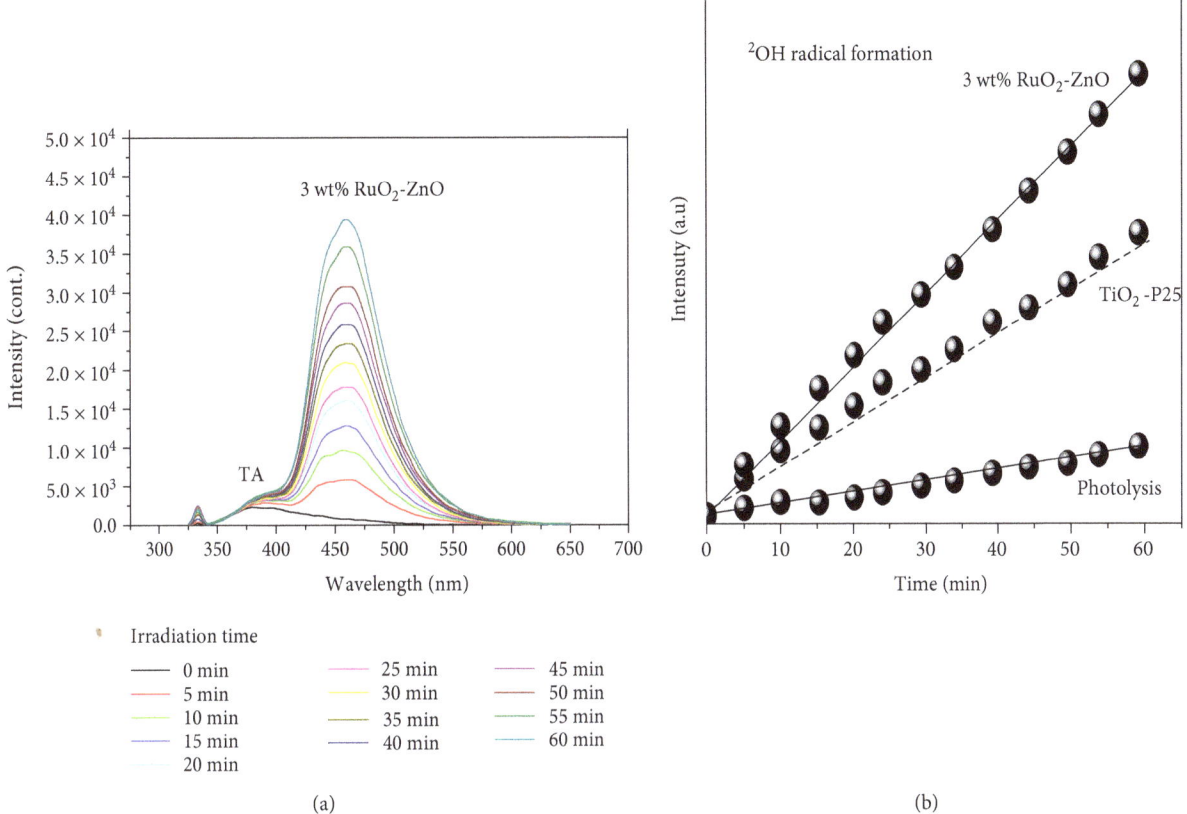

FIGURE 11: (a) Fluorescence spectrum of hydroxyl terephthalic acid with 3 wt% Ru-ZnO catalyst at various irradiation times. (b) Profiles of the •OH radical formation on the 3 wt%Ru-ZnO catalyst and TiO$_2$-P25 in the absence 4 chlorophenol. The •OH radical formation was also measured in the absence of catalyst (photolysis).

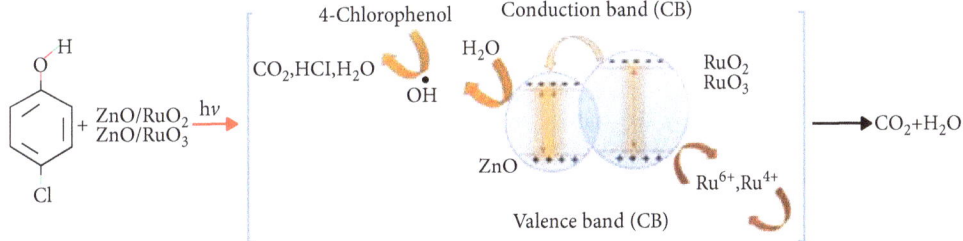

SCHEME 1: A proposed photodegradation mechanism of 4-chlorophenol on the surface of RuO$_2$-ZnO catalysts.

Nanotecnologías del Instituto Politécnico Nacioonal. Jin An Wang thanks the Universidad Autónoma Metropolitana-Azcapotzalco for supporting his sabbatical year.

Conflicts of Interest

The authors declare herein that this article currently has no conflict of interests.

Acknowledgments

The authors are thankful to the financial support from projects SIP-20180199 and SIP-20181280 and the technical support from Centro de Nanociencias y Micro y

References

[1] Y. Li, J. Niu, and L. Yin, "Photocatalytic degradation kinetics and mechanism of pentachlorophenol based on superoxide radicals," *Journal of Environmental Sciences*, vol. 23, no. 11, pp. 1911–1918, 2011.

[2] J. Saien and S. Khezrianjoo, "Degradation of the fungicide carbendazim in aqueous solutions with UV/TiO$_2$ process: optimization, kinetics and toxicity studies," *Journal of Hazardous Materials*, vol. 157, no. 2-3, pp. 269–276, 2008.

[3] B. Krishnakumar and M. Swaminathan, "Influence of operational parameters on photocatalytic degradation of a genotoxic azo dye Acid Violet 7 in aqueous ZnO suspensions," *Spectrochimica Acta Part A: Molecular and Biomolecular Spectroscopy*, vol. 81, no. 1, pp. 739–744, 2011.

[4] K. Vasanth Kumar, K. Porkodi, and A. Selvaganapathi, "Constrain in solving Langmuir–Hinshelwood kinetic expression for the photocatalytic degradation of Auramine O aqueous solutions by ZnO catalyst," *Dyes and Pigments*, vol. 75, no. 1, pp. 246–249, 2007.

[5] B. Ohtani, "Preparing articles on photocatalysis—beyond the illusions misconceptions and speculation," *Chemistry Letters*, vol. 37, no. 3, pp. 216–229, 2008.

[6] N. Daneshvar, A. Aleboyeh, and A. R. Khataee, "The evaluation of electrical energy per order (EEo) for photooxidative decolorization of four textile dye solutions by the kinetic model," *Chemosphere*, vol. 59, no. 6, pp. 761–767, 2005.

[7] M. Sleiman, D. Vildozo, C. Ferronato, and R. F. P. M. Moreira, "Photocatalytic degradation of azo dye Metanil Yellow: optimization and kinetic modeling using a chemometric approach," *Applied Catalysis B: Environmental*, vol. 77, no. 1-2, pp. 1–11, 2007.

[8] Y. Dongju, Z. Lizhong, Z. Xiufeng, C. Han, and Z. Qian, "Iron-glutamate-silicotungstate ternary complex as highly active heterogeneous Fenton-like catalyst for 4-chlorophenol degradation," *Chinese Journal of Catalysis*, vol. 36, no. 12, pp. 2203–2210, 2015.

[9] S. Valizadeh, M. H. Rasoulifard, and M. S. S. Dorraji, "Modified Fe$_3$O$_4$- hydroxyapatite nanocomposites as heterogeneous catalysts in three UV, Vis and Fenton like degradation systems," *Applied Surface Science*, vol. 319, pp. 358–366, 2014.

[10] Z. J. Huang, P. X. Wu, B. N. Gong, Y. H. Lu, N. W. Zhu, and Z. X. Hu, "Preservation of Fe complexes into layered double hydroxides improves the efficiency and the chemical stability of Fe complexes used as heterogeneous photo-Fenton catalysts," *Applied Surface Science*, vol. 286, pp. 371–378, 2013.

[11] Y. Huang, D. K. Sarkar, and X. Chen, "Superhydrophobic nanostructured ZnO thin films on aluminum alloy substrates by electrophoretic deposition process," *Applied Surface Science*, vol. 327, pp. 327–334, 2015.

[12] S. Parra, J. Olivero, and C. Pulgarin, "Relationships between physicochemical properties and photoreactivity of four biorecalcitrant phenylurea herbicides in aqueous TiO$_2$ suspension," *Applied Catalysis B: Environmental*, vol. 36, no. 1, pp. 75–85, 2002.

[13] T. Sauer, G. Cesconeto Neto, H. J. José, and R. F. P. M. Moreira, "Kinetics of photocatalytic degradation of reactive dyes in a TiO$_2$ slurry reactor," *Journal of Photochemistry and Photobiology A: Chemistry*, vol. 149, no. 1-3, pp. 147–154, 2002.

[14] H. D. Mansilla, J. Villaseñor, G. Maturana, J. Baeza, J. Freer, and N. Durán, "Homogeneous and heterogeneous advanced oxidation of a bleaching effluent from the pulp and paper industry," *Water Science and Technology*, vol. 35, no. 4, pp. 273–278, 1997.

[15] X. Li, J. W. Cubbage, and W. S. Jenks, "Photocatalytic degradation of 4-chlorophenol. 2. The 4-chlorocatechol pathway," *The Journal of Organic Chemistry*, vol. 64, no. 23, pp. 8525–8536, 1999.

[16] L. Mančić, S. G. Sipka, V. M. Djinović, Z. Marinković, T. Sabo, and O. Milosević, "Fine nanophased ZnO:Ru and ZnO:Pt powder synthesis through aerosols," *Materials Science Forum*, vol. 494, pp. 149–154, 2005.

[17] D. Chen and A. K. Ray, "Photocatalytic kinetics of phenol and its derivatives over UV irradiated TiO$_2$," *Applied Catalysis B: Environmental*, vol. 23, no. 2-3, pp. 143–157, 1999.

[18] S. Zhuiykova, E. Katsa, V. Plashnitsab, and N. Miurac, "Toward selective electrochemical "E-tongue": potentiometric DO sensor based on sub-micron ZnO–RuO$_2$ sensing electrode," *Electrochimica Acta*, vol. 56, no. 15, pp. 5435–5442, 2011.

[19] L. Lutterotti, "MAUD version 2.55," May 2015, http://www.maud.radiographema.com.

[20] L. Yuan-Chang, M. Y. Tsaib, H. Chiem-Lum, H. Chia-Yen, and C. S. Hwang, "Structural and optical properties of electrodeposited ZnO thin films on conductive RuO$_2$ oxides," *Journal of Alloys and Compounds*, vol. 509, no. 8, pp. 3559–3565, 2011.

[21] S. Zhuiykov and E. Kats, "Electrochemical DO sensor based on sub-micron ZnO-doped RuO$_2$ sensing electrode: Influence of sintering temperature on sensing performance," *Sensors and Actuators B: Chemical*, vol. 187, pp. 12–19, 2013.

[22] T. C. Damen, S. P. Porto, and B. Tell, "Raman effect in zinc oxide," *Physical Review*, vol. 142, no. 2, pp. 570–574, 1966.

[23] C. K. Xu, G. D. Xu, Y. K. Liu, and G. H. Wang, "A simple and novel route for the preparation of ZnO nanorods," *Solid State Communications*, vol. 122, no. 3-4, pp. 175–179, 2002.

[24] M. T. Uddin, Y. Nicolas, C. Olivier et al., "Preparation of RuO$_2$/TiO$_2$ mesoporous heterostructures and rationalization of their enhanced photocatalytic properties by band alignment investigations," *The Journal of Physical Chemistry C*, vol. 117, no. 42, pp. 22098–22110, 2013.

[25] R. Al-Gaashania, S. Radimana, A. R. Dauda, N. Tabetc, and Y. Al-Dourid, "XPS and optical studies of different morphologies of ZnO nanostructures prepared by microwave methods," *Ceramics International*, vol. 39, no. 3, pp. 2283–2292, 2013.

[26] K. Kotsis and S. Volker, "Ab initio calculations of the O1s XPS spectra of ZnO and Zn oxo compounds," *Physical Chemistry Chemical Physics*, vol. 8, no. 13, pp. 1490–1498, 2006.

[27] K. S. Kim and N. Winograd, "X-ray photoelectron spectroscopic studies of ruthenium-oxygen surfaces," *Journal of Catalysis*, vol. 35, no. 1, pp. 66–72, 1974.

[28] K. Vanheusden, C. H. Seager, W. L. Warren, D. R. Tallant, and J. A. Voigt, "Correlation between photoluminescence and oxygen vacancies in ZnO phosphors," *Applied Physics Letters*, vol. 68, no. 3, pp. 403–405, 1996.

[29] H. J. Egelhaaf and D. Oelkrug, "Luminescence and nonradiative deactivation of excited states involving oxygen defect centers in polycrystalline ZnO," *Journal of Crystal Growth*, vol. 161, no. 1-4, pp. 190–194, 1996.

[30] S. A. Studenikin and M. Cocivera, "Time-resolved luminescence and photoconductivity of polycrystalline ZnO films," *Journal of Applied Physics*, vol. 91, no. 8, pp. 5060–5065, 2002.

[31] S. B. Zhang, S. H. Wei, and A. Zunger, "Intrinsic *n*-type versus *p*-type doping asymmetry and the defect physics of ZnO," *Physical Review B*, vol. 63, no. 7, article 075205, 2001.

[32] Z. Wang, S. C. Su, M. Younas, F. C. C. Ling, W. Anwand, and A. Wagner, "The Zn-vacancy related green luminescence and donor–acceptor pair emission in ZnO grown by pulsed laser deposition," *RSC Advances*, vol. 5, no. 17, pp. 12530–12535, 2015.

[33] F. D. Auret, S. A. Goodman, M. J. Legodi, W. E. Meyer, and D. C. Look, "Electrical characterization of vapor-phase-grown single-crystal ZnO," *Applied Physics Letters*, vol. 80, no. 8, pp. 1340–1342, 2002.

[34] K. Ozoemena, N. Kuznetsova, and T. Nyokong, "Comparative photosensitised transformation of polychlorophenols with different sulphonated metallophthalocyanine complexes in aqueous medium," *Journal of Molecular Catalysis A: Chemical*, vol. 176, no. 1-2, pp. 29–40, 2001.

[35] G. Mendoza-Damián, F. Tzompantzi, A. Mantilla, R. Pérez-Hernández, and A. Hernández-Gordillo, "Improved photocatalytic activity of SnO_2–ZnAl LDH prepared by one step Sn^{4+} incorporation," *Applied Clay Science*, vol. 121-122, pp. 127–136, 2016.

[36] N. Milan-Segovia, Y. Wang, F. S. Cannon, R. C. Voigt, C. James, and J. C. Furness, "Comparison of hydroxyl radical generation for various advanced oxidation combinations as applied to foundries," *Ozone-Science & Engineering*, vol. 29, no. 6, pp. 461–471, 2007.

[37] M. Tamez Uddin, Y. Nicolas, C. Olivier et al., "Improved photocatalytic activity in RuO_2–ZnO nanoparticulate heterostructures due to inhomogeneous space charge effects," *Physical Chemistry Chemical Physics*, vol. 17, no. 7, pp. 5090–5102, 2015.

8

Photocatalytic Activity of Ag-TiO$_2$ Composites Deposited by Photoreduction under UV Irradiation

Carlos Díaz-Uribe ⓘ,[1] Jose Viloria,[1] Lorraine Cervantes,[1] William Vallejo ⓘ,[1]
Karen Navarro,[1,2] Eduard Romero,[3] and Cesar Quiñones[4]

[1]Grupo de Fotoquímica y Fotobiología, Universidad del Atlántico, Barranquilla, Colombia
[2]Universidad Nacional de Córdoba, Cordoba, Argentina
[3]Departamento de Química, Universidad Nacional de Colombia, Bogotá, Colombia
[4]Institución Universitaria Politécnico Gran Colombiano, Bogotá, Colombia

Correspondence should be addressed to Carlos Díaz-Uribe; carlosdiaz@mail.uniatlantico.edu.co

Academic Editor: Mohammad Muneer

In this work, we synthesized Ag nanoparticles on TiO$_2$ thin films deposited on soda lime glass substrates. Ag nanoparticles were synthesized by photoreduction under UV irradiation silver nitrate solution. X-ray diffraction, Raman spectroscopy, scanning electron microscopy (SEM), and X-ray photoelectron spectroscopy (XPS) measurements were used for physicochemical characterization. The structural study showed that all samples were polycrystalline, main phases were anatase and rutile, and no additional signals were detected after surface modification. Raman spectroscopy suggested that silver aggregates deposited on the TiO$_2$ films could exhibit the surface plasmon resonance (SPR) phenomenon; XPS and SEM analysis confirmed TiO$_2$ film morphological modification after photoreduction process. Photocatalytic degradation of methylene blue (MB) was studied under UV irradiation in aqueous solution, and, besides, pseudo-first-order model was used to obtain kinetic information about photocatalytic degradation. Results indicated that Ag-TiO$_2$ showed an important increase in photocatalytic activity under UV (from 20% to 35%); finally, Ag-TiO$_2$ thin films had k_{app} value $2.4 \times 10^{-3} \pm 0.003$ min^{-1} of 1.8 times greater than the k_{app} value $1.3 \times 10^{-4} \pm 0.0004$ min^{-1} of TiO$_2$ thin films.

1. Introduction

Nowadays, titanium dioxide (TiO$_2$) is one of the most important semiconductors in photocatalytic applications because of its prominent properties (e.g., it is harmful, stable, resistant to photocorrosion, inexpensive, and has high photocatalytic properties in the degradation of different pollutants) [1, 2]. However, despite all their characteristics, TiO$_2$ has three main drawbacks: (a) fast recombination rate of photogenerated electron-hole pair, (b) low quantum yield in the photocatalytic reactions in aqueous solutions, and (c) a high bandgap value (3.2 eV, photocatalytic active only under UV irradiation) [3–5]. Due to energy transfer that must be efficient during electron-hole pair separation, fast recombination of the photogenerated charge pair (e$^-$/h$^+$) after activation of semiconductor is an important drawback for using

titanium dioxide (TiO$_2$) as a photocatalyst, because it inhibits redox processes and reduces strongly photocatalytic efficiency [6–8]. Two strategies have been explored to solve this drawback: (a) TiO$_2$ synthesis methods can be modified and optimized and (b) the TiO$_2$ surface can be modified. TiO$_2$ surface modification by transition metal deposition (e.g., silver, platinum, ruthenium, and palladium) has been reported to decrease the recombination process by forming heterostructures and new interfaces that improve semiconductor photocatalytic properties [9–11]. Noble metal cocatalysts can improve photocatalytic properties of semiconductor because they act as electron traps near to conduction bands and/or they improve surface electron excitation by surface plasmonic resonance [12–14]. Nalbandian et al. fabricated Ag-TiO$_2$ nanofibers by electrospinning, and they showed that the presence of the Ag cocatalyst enhanced reactivity

[15]. Yilmaz et al. deposited photochemically Pd nanoparticles on TiO_2 nanorods, and they corroborated that electron transfer between Pd and TiO_2 was improved after modification process [16]. Currently, different reports showed advances for the fabrication of Ag/TiO_2 nanoheterostructure catalysts (e.g., Ag-modified TiO_2 nanoparticles, nanowires, and nanorods) [17, 18]; however, several practical problems arise from the use of powder during the photocatalytic process: (a) separation of insoluble catalyst from suspension is harder, (b) particles in suspensions create aggregates especially at high concentrations, and (c) suspensions are difficult to apply on continuous flow systems. The TiO_2 thin film is an efficient alternative to solve these drawbacks.

In this contribution, we evaluate the photocatalytic activity of Ag-TiO_2 composite thin films in the degradation of methylene blue as a potential alternative in photocatalysis.

2. Experimental

2.1. Photocatalyst Synthesis. TiO_2 films were deposited on soda lime glass (SLG) by doctor blade method. TiO2-Degussa powder (P25) was used as reagent, and after this process, $1.5\,\mu m$ TiO_2 thin film thickness was obtained [19]. Silver particle deposition on TiO_2 films was carried out by chemical photoreduction: first, TiO_2 films were immersed in an aqueous silver nitrate solution (8.0×10^{-2} M) for 30 and 60 minutes; after the photoreduction process, TiO_2 film colors changed from white to grayish brown.

2.2. Photocatalyst Characterization. We study thin film physicochemical properties through measurements of Raman spectroscopy, X-ray diffraction (XRD), and scanning electronic microscopy (SEM); furthermore, the thickness of the thin films was measured through a Veeco Dektak 150 profilometer. Raman spectra were recorded on DRX Raman microscope with laser at 680 nm; SEM images were carried out using a Cambridge S360 microscope operating at acceleration voltage of 20 kV with a scanned area $11\,\mu m \times 10\,\mu m$, and XRD X-ray diffraction patterns of the samples were recorded in Shimadzu 6000 diffractometer with a source of Cu$K\alpha$ radiation ($\lambda = 0.15418$ nm) in a range diffraction angle 2θ between 20° and 70°. Finally, XPS measurements were performed on X-ray photoelectronic spectrometer (NAP-XPS; brand Specs) with a PHOIBOS 150 1D-DLD analyzer, using a monochromatic source of Al-K$_\alpha$ (1486.7 eV, 13 kV, 100 W) with energy from the 90 eV for the general spectra and 20 eV for the high-resolution spectra. The step was 1 eV for the general spectra and 0.1 eV for the high spectra. We performed 20 measurement cycles for the high-resolution spectra and 3 for the general spectra.

2.3. Photocatalytic Activity. Thin films were immersed in blue methylene solution (10 ppm was used as target solution), and prior to irradiation, the system was magnetically stirred in the dark by 1 hour to ensure the equilibrium of dye adsorption-desorption on thin film surface. The system was irradiated by UV lamp with an emission maximum of 260 nm and power 30 W during 180 minutes. All the tests

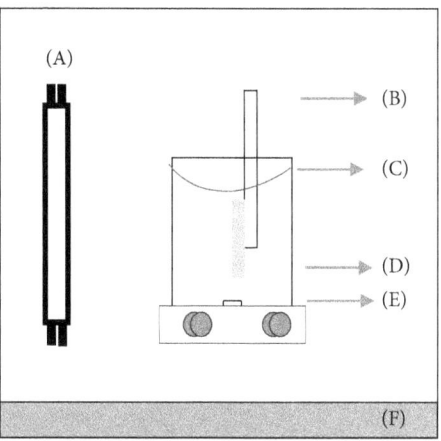

FIGURE 1: General scheme of photoreactor: (A) UV light source, (B) holder, (C) the beaker container and methylene blue solution, (D) photocatalyst thin film, (E) magnetic stirring, and (F) INOX container.

of catalytic reactivity were carried out in aerobic conditions (magnetically stirred: 150 rpm); the irradiated surface area was $3.0\,cm^2$, and the volume of the suspension was 0.015 l. The incident photon flow per unit volume ($I_o = 1.3 \times 10^{-6}$ Einstein*L^{-1}*s^{-1}; Figure 1) shows a general scheme of photoreactor. We chose methylene blue as the model compound because of, since 2010, International Organization for Standardization (ISO) published standard: 10678:2010, namely, the "Determination of photocatalytic activity of surfaces in an aqueous medium by degradation of methylene blue" [20]. Finally, the concentration of dye was determined by spectrophotometry at $\lambda = 665$ nm for using calibration curve (correlation coefficient $R = 0.998$).

3. Results

3.1. SEM Analysis. SEM and EDS were used for morphological compound characterization. Figure 2 shows SEM images for both TiO_2 and Ag-TiO_2 thin films. Figures 2(a) and 2(b) show that the microaggregates compose TiO_2 films: microaggregates have a size range around 50 nm; these results are according to the typical morphology of TiO_2-Degussa P25 [21]. Figures 2(c) and 2(d) show the morphological results to Ag-TiO_2 thin films; the SEM image allows to differentiate the deposited silver particles. Figure 2 shows the photoreduced particles agglomerated reaching sizes of the order of 200 nm. The EDS analysis showed that the agglomerates observed in the surface of TiO_2 corresponded to silver particles (Figures 2(e)).

3.2. Raman Assay. Figure 3 shows the Raman spectrum for TiO_2 films and the Ag-TiO_2 composite thin films deposited after 30 minutes of UV light irradiation. TiO_2 films showed characteristic vibration Raman mode signals to anatase phase: signals located at $144\,cm^{-1}$, $398\,cm^{-1}$, and $520\,cm^{-1}$ are assigned to E_{1g} vibration mode, for the signal located at $639\,cm^{-1}$, it is attributed to $398\,cm^{-1}$ E_{3g} vibration mode, and results are according to other reports [22]. Raman

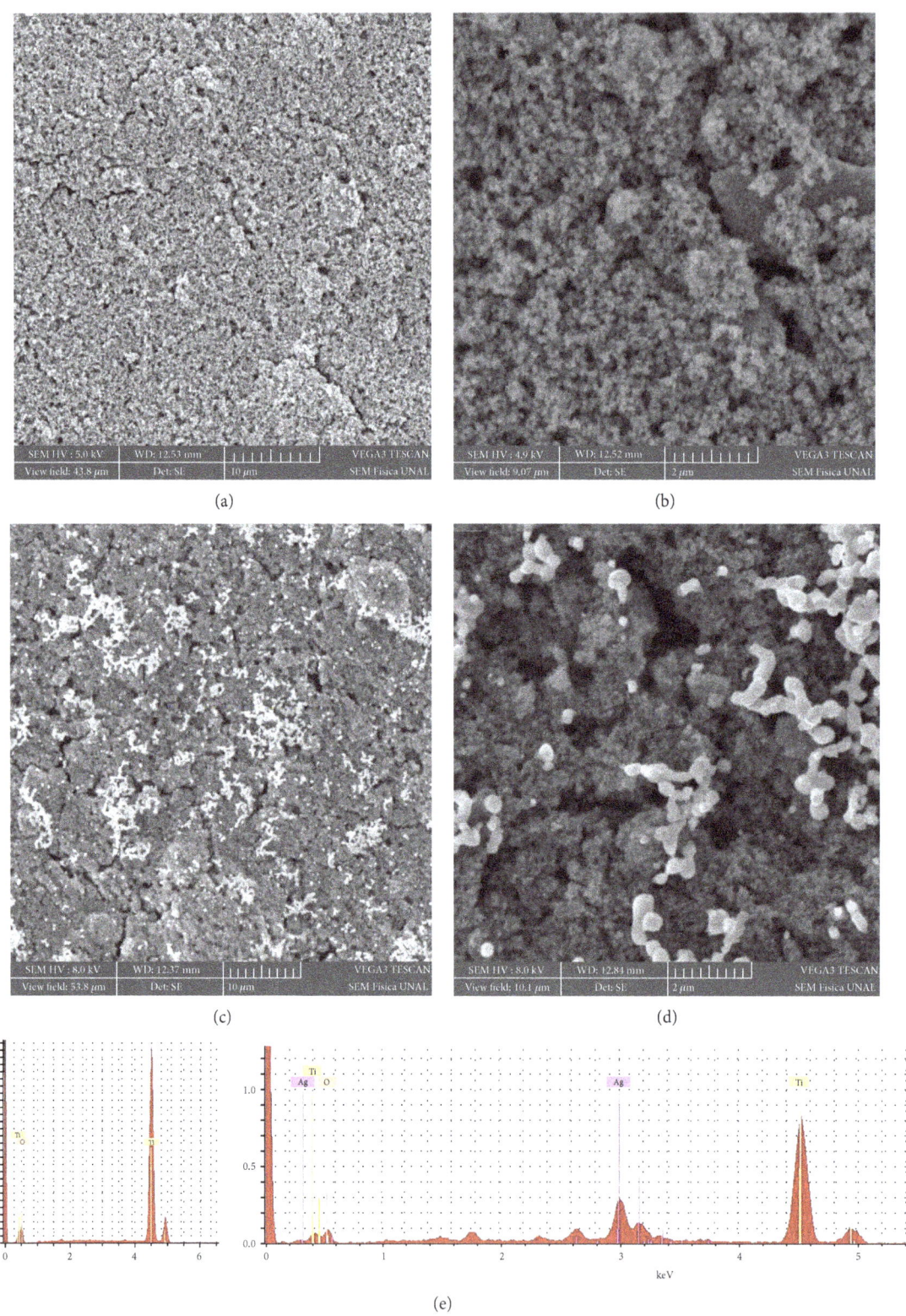

FIGURE 2: SEM images: (a) TiO_2 thin films (10 μm), (b) TiO_2 thin films (2 μm), (c) Ag-TiO_2 (10 μm), (d) Ag-TiO_2 (2 μm), and (e) TiO_2 and Ag-TiO_2Ag thin film EDS assay.

spectrum of the modified Ag-TiO_2 thin films shows same signals than TiO_2 thin films; however, the intensity of all signals is increasing; this behavior is associated to phenomenon known as surface plasmon resonance (SPR). This process is

characterized by the increase in several orders of magnitude of the intensity of the signals in the Raman spectrum due to the effect of metal nanometric scale species adsorbed on semiconductor surface [23–25]. Different reports indicated

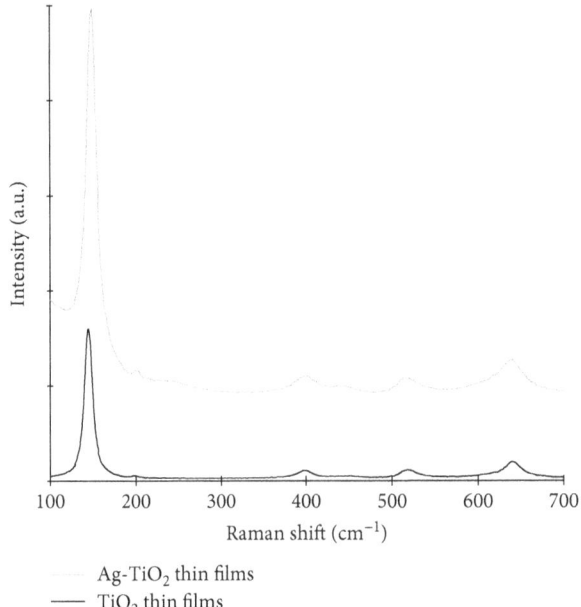

FIGURE 3: Raman spectrum to TiO$_2$ thin films and Ag-TiO$_2$ composite thin films.

--- Ag-TiO$_2$ thin films
— TiO$_2$ thin films

that some noble metal nanoparticles (gold, silver, palladium, and platinum) can generate an intense surface plasmon resonance when adsorbed onto semiconductors; finally, SPR process could affect optical, electrical, and photocatalytic properties of TiO$_2$ [26, 27].

3.3. XRD Assay. Figure 4 shows the experimental XRD pattern for both TiO$_2$ and Ag-TiO$_2$ thin films. XRD pattern showed peaks at $2\theta = 25.4°$, $2\theta = 38.2°$, $2\theta = 48.2°$, and $2\theta = 62.5°$; they correspond to planes (101), (112), (200), and (204). These signals are typical of the TiO$_2$ anatase phase (JCPDS #071-1166); diffraction pattern also shows additional diffraction signals at $2\theta = 27.8°$ and $2\theta = 55.0°$; these reflections can be associated to the planes (110) and (220) of the rutile crystalline phase (JCPDS #021-1276). The presence of the rutile is due to the composition of the material used as TiO$_2$ source (Degussa P25). Furthermore, after Ag photoreduction on TiO$_2$ surface, the intensity of main signal (located at $2\theta = 25.4°$) decreases indicating that the anchoring process did not change the characteristic crystal surface structure of TiO$_2$; these results are according to SEM assay [28].

3.4. XPS Assay. XPS measurements were conducted to verify the surface components and valence states of Ag-TiO$_2$ thin film. Figure 5(a) shows profile XPS spectrum to Ag-TiO$_2$ thin film. The profile of the sample shows the typical signals of binding energy corresponding to Ag, Ti, O, and C elements.

The signals at 574.0 eV, 374.0 eV, and 368.0 eV correspond to the electronic transitions of Ag 3p$_{3/2}$, Ag 3d$_{3/2}$, and Ag 3d$_{3/2}$, respectively; Figure 5(c) plots the HRXPS spectrum of Ag 3d$_{5/2}$ and Ag 3d$_{3/2}$ double peaks, which are centered at 367.8 and 373.8 eV, respectively; and the splitting

of the 3d doublet was 6.0 eV. This binding energy indicated that silver was of metallic nature. These results are according to other reports [29–31]. Figure 5(b) shows the peaks at 458.6 and 464.3 eV; these signals correspond to Ti 2p$_{3/2}$ and Ti 2p$_{1/2}$ indicating the formation of Ti^{4+} in TiO$_2$ [32–34]. Figure 5(a) also shows an important signal at 530.2 eV, and this corresponds to the phototransition O1s; finally, the signal located at 285 eV corresponds to electronic transition of C 1s. This signal is typical of atmospheric CO$_2$ absorbed on the surface of the simple.

We performed quantitative assay of Ag-TiO$_2$ thin films for using XPS survey spectra shown in Figure 5; it mainly consists in determination of the relative concentration of respective surface atoms like Ti, O, C, and Ag as follows:

$$n_i = \frac{(I_i/\text{ASF}_i)}{\sum_i^n (I_i/\text{ASF}_i)}, \tag{1}$$

where I_i corresponds to relative intensity height of the O$_{1s}$, Ti$_{2p}$, Ag$_{3d}$, and C$_{1s}$ core-level line peaks and ASF$_i$ is the atomic sensitivity factors related to the height of specific peaks [35–37]. Table 1 summarizes the corresponding partial concentration of specific elements in the Ag-TiO$_2$ thin films.

3.5. Photocatalytic Assay. Figure 6 shows the C_t/C_o (methylene blue) as a function of time under UV irradiation, in the presence of TiO$_2$ and Ag-TiO$_2$ thin films for 180 min. Figure 6 shows similar profile to behavior observed for the photocatalytic degradation of different dyes; previous studies showed that photocatalytic degradation rate of textile dyes in heterogeneous photocatalytic oxidation systems under UV light illumination followed Langmuir–Hinshelwood (L–H); the Langmuir–Hinshelwood expression that explains the kinetics of heterogeneous catalytic systems is given by [38–44]

$$r = -\frac{dC}{dt} = \frac{k_t K[C]}{1 + K[C]}, \tag{2}$$

where r is the rate of dye mineralization, k_t is the rate constant, $[C]$ is the methylene blue concentration, and K is the adsorption coefficient. Equation (1) can be solved explicitly for (t) by using discrete changes in [MB] from the initial concentration to a zero reference point; however, when the concentration of substrate is in the scale of μmoles, an apparent first-order model can be assumed; in our case,

$$r = -\frac{dC}{dt} = k_t K[C]. \tag{3}$$

Integration of (2):

$$[C] = [C]_o e^{-k_{app}*t},$$
$$\ln\left(\frac{[C]_t}{[C]_o}\right) = -k_{app}*t, \tag{4}$$

where (t) represents time in minutes, k_{app} $(k_t K)$ is the apparent reaction rate constant (min^{-1}), and $[C]_t$ is the

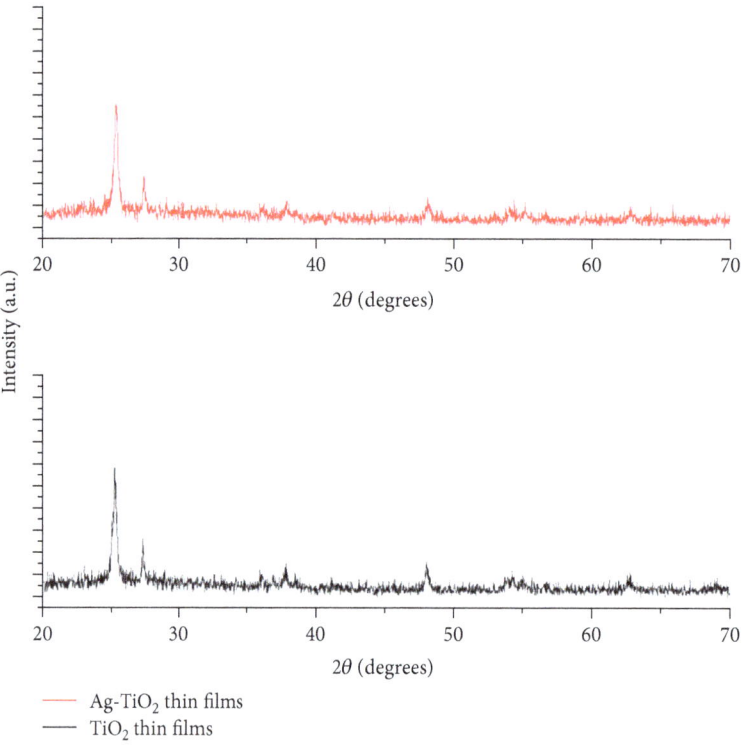

FIGURE 4: XRD pattern for both TiO_2 and Ag-TiO_2 thin films.

concentration of MB at each time. (L–H) kinetics is commonly used to explain the kinetics of the heterogeneous catalytic processes; the assumption of a pseudo-first-order kinetic model was used in several studies to characterize the effect of different experimental conditions [45–47]. Figure 6 shows that when the MB solution was exposed to UV radiation on TiO_2 thin films, a small reduction in the concentration of MB was observed (near 20%). These results are according to expected because TiO_2 thin films are photocatalytic active under this kind of radiation. Visible is a typical behavior for TiO_2 due to their well-known wide bandgap value [48].

Furthermore, Figure 6 shows that Ag-TiO_2 thin films showed an important increase in the photocatalytic activity under UV (near 35%); these results could be due to the silver aggregates deposited on the TiO_2 surface that can accumulate charge and accept electrons inside the semiconductor. The accumulation of electrons can reduce the carrier's recombination increasing generation of reactive oxygen species, and besides, Ag particle on TiO_2 can change and a collective oscillation can be generated, which is known as localized surface plasmon resonance (SPR) [49]; according to other reports, silver particles exhibit collective oscillation of their electrons in the conduction band, and it is suggested that the association between the semiconductor and the metal can increase photocatalytic activity, which coincides with the results obtained in this work [50]. Pseudo-first-order model was applied to kinetic data of Figure 6. The k_{app} values for each test were determined from the slope of the linear fitting of $\ln (C_t/C_o)$ vs time (Figure 7).

Ag-TiO_2 thin films had a k_{app} value $2.4 \times 10^{-3} \, min^{-1}$; this value was 1.8 times greater than the k_{app} value $1.3 \times 10^{-4} \, min^{-1}$ of TiO_2 thin films. k_{app} values confirmed that the photoreduction process was effective and demonstrated that this method could be used as an alternative to increasing the photocatalytic activity of TiO_2 under UV light irradiation. We reported in a previous work that Ag-TiO_2 thin films had antimicrobial activity against *Staphylococcus aureus*; same material demonstrated both photocatalytic and antimicrobial activities [51]. Reactive oxygen species generated in advance oxidation process are endogenous, highly reactive, oxygen molecules, and these species can originate oxidation reactions of many organic and biological compounds (e.g., singlet oxygen and oxygen free radicals such as hydroxyl radical, superoxide anion, hydroperoxyl (HOO˙), or other similar radicals); our results corroborate high reactivity of these chemical species [52].

4. Conclusions

In this work, we deposited Ag microaggregates on TiO_2 thin films. SEM characterization demonstrated microaggregate formation on TiO_2 surface, and besides, Raman spectroscopy indicated that surface modification generated SPR process; furthermore, XRD assay indicated that all thin films were crystalline and XPS assay indicates that silver was of metallic nature after photoreduction on TiO_2 thin films. Photocatalytic activity of Ag-TiO_2 thin films was higher than that of TiO_2 thin films; kinetic results showed that the k_{app} for

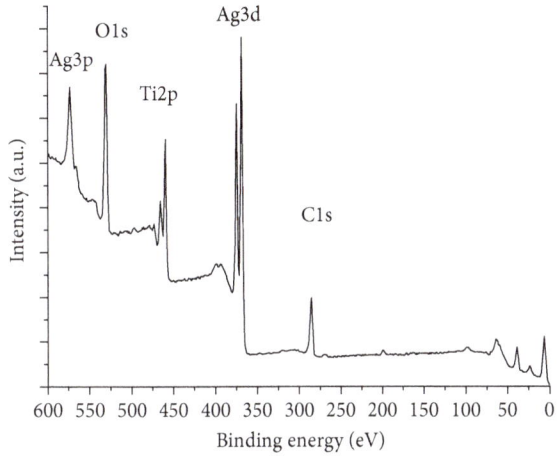

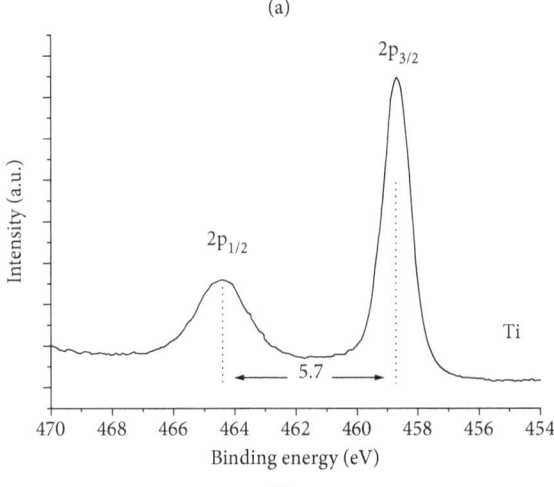

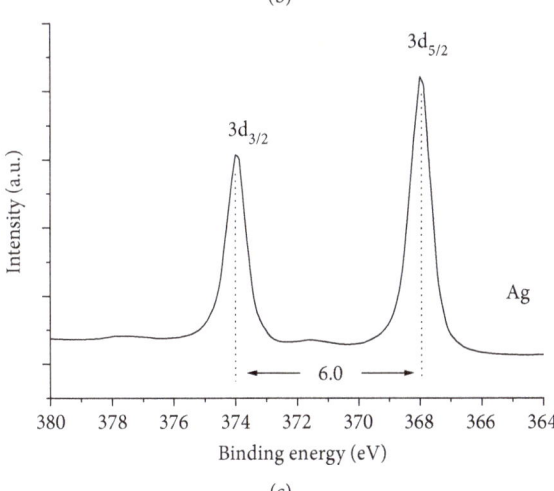

(a)

(b)

(c)

FIGURE 5: (a) XPS spectrum to Ag-TiO$_2$ thin films, (b) HRXPS Ti 2p spectrum to Ag-TiO$_2$ thin films, and (c) HRXPS Ag 3d spectrum to Ag-TiO$_2$ thin films. All binding energies of the XPS spectra are calibrated with reference to the C1s peak at 285 eV.

TABLE 1: Partial concentration for elements of Ag-TiO$_2$ thin films.

Element	Oxygen	Titanium	Silver	Carbon
%	39.1	15.9	8.5	36.6

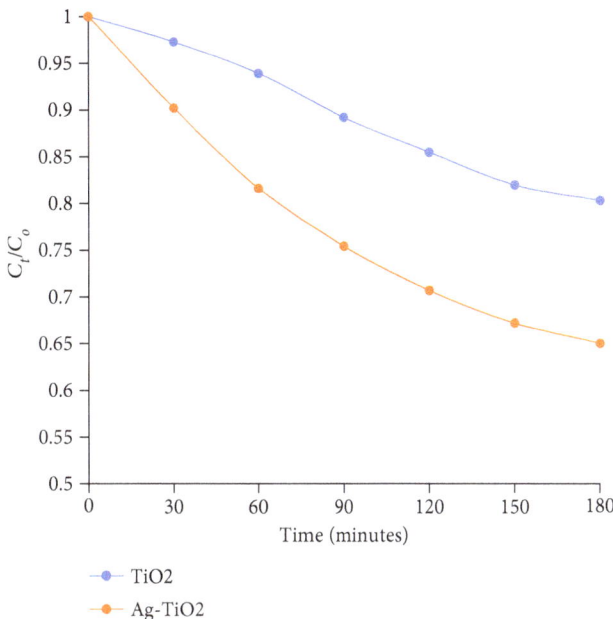

FIGURE 6: C_t/C_o vs time under UV light irradiation, where C_t refers to the concentration at time t of MB and C_o refers to initial concentration.

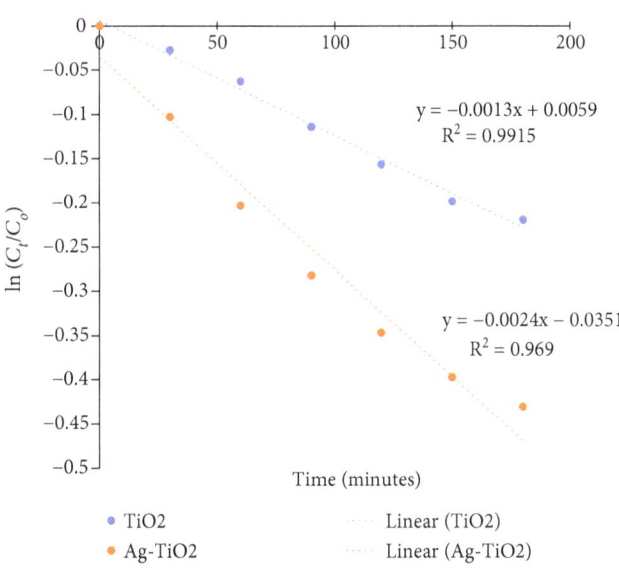

FIGURE 7: Fitting $\ln(C_t/C_o)$ vs (t) for tests, where C_t refers to the concentration of MB at time (t) and C_o refers to the initial concentration of MB; inside figure is the R value and equation of linear fitting.

Ag-TiO$_2$ thin films was greater $(2.4 \times 10^{-3}\,\text{min}^{-1})$ than the k_{app} for TiO$_2$ thin films $(1.3 \times 10^{-3}\,\text{min}^{-1})$.

Conflicts of Interest

The authors declare that they have no conflicts of interest.

Acknowledgments

This research was supported by Universidad del Atlántico. Furthermore, the authors thank Universidad de Antioquia for supporting the XPS measurements.

References

[1] J. Schneider, M. Matsuoka, M. Takeuchi et al., "Understanding TiO$_2$ photocatalysis: mechanisms and materials," *Chemical Reviews*, vol. 114, no. 19, pp. 9919–9986, 2014.

[2] M. J. G azquez, J. P. Bolívar, R. Garcia-Tenorio, and F. Vaca, "A review of the production cycle of titanium dioxide pigment," *Materials Sciences and Applications*, vol. 5, no. 7, pp. 441–458, 2014.

[3] Y. Du and J. Rabani, "The measure of TiO$_2$ photocatalytic efficiency and the comparison of different photocatalytic titania," *The Journal of Physical Chemistry. B*, vol. 107, no. 43, pp. 11970–11978, 2003.

[4] C. Ming, Y. Ku, F. Tien, and Y. L. Kuo, "Effect of platinum on the decomposition efficiency and apparent quantum yield of photocatalysis of isopropanol in an annular photoreactor," *Journal of Advanced Oxidation Technologies*, vol. 10, no. 2, pp. 36–368, 2007.

[5] B. Kılıç, N. Gedik, S. Pıravadıllı, A. Serhan, and E. Gürd, "Band gap engineering and modifying surface of TiO$_2$ nanostructures by Fe$_2$O$_3$ for enhanced-performance of dye sensitized solar cell," *Materials Science in Semiconductor Processing*, vol. 31, pp. 363–371, 2015.

[6] P. Zawadzki, A. B. Laursen, K. W. Jacobsen, S. Dahl, and J. Rossmeisl, "Oxidative trends of TiO$_2$ – hole trapping at anatase and rutile surfaces," *Energy & Environmental Science*, vol. 5, no. 12, pp. 9866–9869, 2012.

[7] L. Cavigli, F. Bogani, A. Vinattieri et al., "Carrier recombination dynamics in anatase TiO$_2$ nanoparticles," *Solid State Sciences*, vol. 12, no. 11, pp. 1877–1880, 2010.

[8] D. Punnoose, C. H. S. S. Pavan, H. Woong et al., "Reduced recombination with an optimized barrier layer on TiO$_2$ in PbS/CdS core shell quantum dot sensitized solar cells," *New Journal of Chemistry*, vol. 40, no. 4, pp. 3423–3431, 2016.

[9] N. Rahimi, R. A. Pax, and E. M. Gray, *Progress in Solid State Chemistry*, vol. 44, no. 3, pp. 86–105, 2016.

[10] A. Bumajdad and M. Madkour, "Understanding the superior photocatalytic activity of noble metals modified titania under UV and visible light irradiation," *Physical Chemistry Chemical Physics*, vol. 16, no. 16, pp. 7146–7158, 2014.

[11] J. Choi, H. Park, and M. R. Hoffmann, "Effects of single metal-ion doping on the visible-light photoreactivity of TiO$_2$," *The Journal of Physical Chemistry C*, vol. 114, no. 2, pp. 783–792, 2009.

[12] M. Kotesh, K. Bhavani, G. Naresh, B. Srinivas, and A. Venugopala, "Plasmonic resonance nature of Ag-Cu/TiO$_2$ photocatalyst under solar and artificial light: synthesis, characterization and evaluation of H$_2$O splitting activity," *Applied Catalysis B: Environmental*, vol. 199, pp. 282–291, 2016.

[13] S. Shuang, R. Lv, Z. Xie, and Z. Zhang, "Surface plasmon enhanced photocatalysis of Au/Pt-decorated TiO$_2$ nanopillar arrays," *Scientific Reports*, vol. 6, pp. 1–8, 2016.

[14] S. Kochuveedu, D. Kim, and D. Ha Kim, "Surface-plasmon-induced visible light photocatalytic activity of TiO$_2$ nanospheres decorated by Au nanoparticles with controlled configuration," *The Journal of Physical Chemistry C*, vol. 116, no. 3, pp. 2500–2506, 2012.

[15] M. J. Nalbandian, M. Zhang, J. Sanchez et al., "Synthesis and optimization of Ag–TiO$_2$ composite nanofibers for photocatalytic treatment of impaired water sources," *Journal of Hazardous Materials*, vol. 299, pp. 141–148, 2015.

[16] P. Yilmaz, A. Lacerda, I. Larrosa, and S. Dunn, "Photoelectrocatalysis of rhodamine B and solar hydrogen production by TiO$_2$ and Pd/TiO$_2$ catalyst systems," *Electrochimica Acta*, vol. 231, pp. 641–649, 2017.

[17] C. Su, L. Liu, M. Zhang, Y. Zhanga, and C. Shao, "Fabrication of Ag/TiO$_2$ nanoheterostructures with visible light photocatalytic function via a solvothermal approach," *CrystEngComm*, vol. 14, no. 11, pp. 3989–3999, 2012.

[18] K. Kusdianto, D. Jiang, M. Kubo, and M. Shimada, "Fabrication of TiO$_2$–Ag nanocomposite thin films via one-step gas-phase deposition," *Ceramics International*, vol. 43, no. 6, pp. 5351–5355, 2017.

[19] C. Quiñones, Y. Ayala, and W. Vallejo, "Methylene blue photoelectrodegradation under UV irradiation on Au/Pd-modified TiO$_2$ films," *Applied Surface Science*, vol. 257, no. 2, pp. 367–371, 2010.

[20] A. Mills, "An overview of the methylene blue ISO test for assessing the activities of photocatalytic films," *Applied Catalysis B: Environmental*, vol. 128, pp. 144–149, 2012.

[21] I. Ahmed, Z. Khan, and M. Ali, "Photocatalytic degradation of nitro and chlorophenols using doped and undoped titanium dioxide nanoparticles," *Journal of Nanomaterials*, vol. 2011, Article ID 589185, 8 pages, 2011.

[22] N. Mahdjoub, N. Allen, P. Kelly, and V. Vishnyakov, "SEM and Raman study of thermally treated TiO$_2$ anatase nanopowders: influence of calcination on photocatalytic activity," *Journal of Photochemistry and Photobiology A: Chemistry*, vol. 211, no. 1, pp. 59–64, 2010.

[23] M. Torrell, R. Kabir, L. Cunha et al., "Tuning of the surface plasmon resonance in TiO$_2$/Au thin films grown by magnetron sputtering: the effect of thermal annealing," *Journal of Applied Physics*, vol. 109, no. 7, article 074310, 2011.

[24] A. Tanaka, S. Sakaguchi, K. Hashimoto, and H. Kominami, "Preparation of Au/TiO$_2$ with metal cocatalysts exhibiting strong surface plasmon resonance effective for photoinduced hydrogen formation under irradiation of visible light," *ACS Catalysis*, vol. 3, no. 1, pp. 79–85, 2013.

[25] A. Tanaka, K. Teramura, S. Hosokawa, H. Kominamic, and T. Tanaka, "Visible light-induced water splitting in an aqueous suspension of a plasmonic Au/TiO$_2$ photocatalyst with metal co-catalysts," *Chemical Science*, vol. 8, no. 4, pp. 2574–2580, 2017.

[26] K. H. Leong, H. Y. Chu, S. Ibrahim, and P. Saravanan, "Palladium nanoparticles anchored to anatase TiO$_2$ for enhanced surface plasmon resonance-stimulated, visible-light-driven photocatalytic activity," *Beilstein Journal of Nanotechnology*, vol. 6, pp. 428–437, 2015.

[27] M. Zhou, J. Zhang, B. Cheng, and H. Yu, "Enhancement of visible-light photocatalytic activity of mesoporous Au-TiO$_2$

nanocomposites by surface plasmon resonance," vol. 2012, Article ID 532843, 10 pages, 2012.

[28] Z. Huang, B. Zheng, S. Zhu et al., "Photocatalytic activity of phthalocyanine-sensitized TiO_2–SiO_2 microparticles irradiated by visible light," *Materials Science in Semiconductor Processing*, vol. 25, pp. 148–152, 2014.

[29] J. Yu, J. Xiong, B. Cheng, and S. Liu, "Fabrication and characterization of Ag–TiO_2 multiphase nanocomposite thin films with enhanced photocatalytic activity," *Applied Catalysis B: Environmental*, vol. 60, no. 3-4, pp. 211–221, 2005.

[30] J. F. Moulder, W. F. Stickle, P. E. Sobol, and K. D. Bomben, *Handbook of X-Ray Photoelectron Spectroscopy*, Perkin-Elmer Corporation Physical Electronics Division, 1992.

[31] E. Stathatos and P. Lianos, "Photocatalytically deposited silver nanoparticles on mesoporous TiO_2 films," *Langmuir*, vol. 16, no. 5, pp. 2398–2400, 2000.

[32] L. Ma, Y. Huang, M. Hou, Z. Xie, and Z. Zhang, "Ag nanorods coated with ultrathin TiO_2 shells as stable and recyclable SERS substrates," *Scientific Reports*, vol. 5, no. 1, article 15442, 2015.

[33] J. Chen, H. Su, X. You, J. Gao, W. M. Lau, and D. Zhang, "3D TiO_2 submicrostructures decorated by silver nanoparticles as SERS substrate for organic pollutants detection and degradation," *Materials Research Bulletin*, vol. 49, pp. 560–565, 2014.

[34] B. Erdem, R. A. Hunsicker, G. W. Simmons, E. D. Sudol, V. L. Dimonie, and M. S. El-Aasser, "XPS and FTIR surface characterization of TiO_2 particles used in polymer encapsulation," *Langmuir*, vol. 17, no. 9, pp. 2664–2669, 2001.

[35] M. Kwoka, V. Galstyan, E. Comini, and J. Szube, "Pure and highly Nb-doped titanium dioxide nanotubular arrays: characterization of local surface properties," *Nanomaterials*, vol. 7, no. 12, p. 456, 2017.

[36] J. F. Watts and J. Wolstenholme, *An introduction to surface analysis by XPS and AES*, John Wiley & Sons, Ltd, 2003.

[37] J. C. Vickerman and I. S. Gilmore, "Surface analysis—the principal techniques," John Wiley & Sons, Chichester, UK, 2nd edition, 2009.

[38] S. Ghasemi, S. Rahimnejad, S. Rahman Setayesh, M. Hosseini, and M. R. Gholami, "Kinetics investigation of the photocatalytic degradation of acid blue 92 in aqueous solution using nanocrystalline TiO_2 prepared in an ionic liquid," *Progress in Reaction Kinetics and Mechanism*, vol. 34, no. 1, pp. 55–76, 2009.

[39] I. Konstantinou and T. Albanis, "TiO_2-assisted photocatalytic degradation of azo dyes in aqueous solution: kinetic and mechanistic investigations: a review," *Applied Catalysis B: Environmental*, vol. 49, no. 1, pp. 1–14, 2004.

[40] K. V. Kumar, K. Porkodi, and F. Rocha, "Langmuir–Hinshelwood kinetics – a theoretical study," *Catalysis Communications*, vol. 9, no. 1, pp. 82–84, 2008.

[41] A. R. Khataee, M. Fathinia, and S. Aber, "Kinetic modeling of liquid phase photocatalysis on supported TiO_2 nanoparticles in a rectangular flat-plate photoreactor," *Industrial & Engineering Chemistry Research*, vol. 49, no. 24, pp. 12358–12364, 2010.

[42] V. Djokic, J. Vujovic, A. Marinkovic et al., "A study of the photocatalytic degradation of the textile dye CI basic yellow 28 in water using a P160 TiO_2-based catalyst," *Journal of the Serbian Chemical Society*, vol. 77, no. 12, pp. 1747–1757, 2012.

[43] L. Elsellami, F. Vocanson, and F. Dappozzeetal, "Kinetic of adsorption and of photocatalytic degradation of phenylalanine

effect of pH and light intensity," *Applied Catalysis A: General*, vol. 380, no. 1-2, pp. 142–148, 2010.

[44] N. P. Xekoukoulotakis, C. Drosou, C. Brebou et al., "Kinetics of UV-A/TiO_2 photocatalytic degradation and mineralization of the antibiotic sulfamethoxazole in aqueous matrices," *Catalysis Today*, vol. 161, no. 1, pp. 163–168, 2011.

[45] C. S. Turchi and D. Ollis, "Photocatalytic degradation of organic water contaminants: mechanisms involving hydroxyl radical attack," *Journal of Catalysis*, vol. 122, no. 1, pp. 178–192, 1990.

[46] T. Zhang, T. Oyama, A. Aoshima, H. Hidaka, J. Zhao, and N. Serpone, "Photooxidative N-demethylation of methylene blue in aqueous TiO_2 dispersions under UV irradiation," *Journal of Photochemistry and Photobiology A: Chemistry*, vol. 140, no. 2, pp. 163–172, 2001.

[47] J. W. Rodriguez-Acosta, M. Á. Mueses, and F. Machuca-Martínez, "Mixing rules formulation for a kinetic model of the Langmuir-Hinshelwood semipredictive type applied to the heterogeneous photocatalytic degradation of multicomponent mixtures," *International Journal of Photoenergy*, vol. 2014, Article ID 817538, 9 pages, 2014.

[48] C. Xu, G. P. Rangaiah, and X. S. Zhao, "Photocatalytic degradation of methylene blue by titanium dioxide: experimental and modeling study," *Industrial and Engineering Chemistry Research*, vol. 53, no. 38, pp. 14641–14649, 2014.

[49] P. Liu, P. Zhang, D. Liang, and W. Xue, "Preparation, characterization and photodegradation of methylene blue based on TiO_2 microparticles modified with thiophene substituents," *Chinese Science Bulletin*, vol. 57, no. 33, pp. 4381–4386, 2012.

[50] R. S. André, C. A. Zamperini, E. G. Mima et al., "Antimicrobial activity of TiO_2: Ag nanocrystalline heterostructures: experimental and theoretical insights," *Chemical Physics*, vol. 459, pp. 87–95, 2015.

[51] W. Vallejo, C. Díaz-Uribe, K. Navarro, R. Valle, J. W. Arboleda, and E. Romero, "Estudio de la actividad antimicrobiana de películas delgadas de dióxido de titanio modificado con plata," *Revista de la Academia Colombiana de Ciencias Exactas, Físicas y Naturales*, vol. 40, no. 154, pp. 69–74, 2016.

[52] K. Krumova and G. Cosa, "Chapter 1: overview of reactive oxygen species," in *Singlet Oxygen: Applications in Biosciences and Nanosciences, Volume 1*, pp. 1-2, 2016.

9

Photoinduced Synthesis of Hierarchical Flower-Like Ag/Bi$_2$WO$_6$ Microspheres as an Efficient Visible Light Photocatalyst

Shuisheng Wu ⓘ, Nianyuan Tan, Donghui Lan, and Bing Yi ⓘ

College of Chemistry and Chemical Engineering, Hunan Institute of Engineering, Xiangtan 411104, China

Correspondence should be addressed to Shuisheng Wu; wuss2005@126.com and Bing Yi; 70356@hnie.edu.cn

Academic Editor: Andrey S. Mereshchenko

A series of three-dimensional microflower-like Ag/Bi$_2$WO$_6$ composites were synthesized through a simple and practical photoreduction process with different photoreduction times. The UV-visible diffuse reflectance spectra indicate that the spectrum of Ag/Bi$_2$WO$_6$ is significantly red-shifted compared to pure Bi$_2$WO$_6$ microspheres in the visible light region. The photocatalytic activities of the as-prepared samples were evaluated by the decolorization of rhodamine B under visible light irradiation. The photocatalytic reaction rate constants of the Ag/Bi$_2$WO$_6$ with a photoreduction time of 20 min was 3.60 times bigger than those of pure Bi$_2$WO$_6$. The enhanced photocatalytic activity could be attributed to the synergistic effect of increased light absorption range and the effective separation of photogenerated carriers caused by Ag nanoparticles.

1. Introduction

In the past few decades, photocatalytic degradation of environmental pollutants and photocatalytic hydrogen production have been the concern [1]. Titanium dioxide (TiO$_2$) has been extensively studied by scholars all over the world for its advantages of low cost, nontoxicity, and high chemical stability. However, TiO$_2$ has the disadvantage of lower quantum efficiency and inability to absorb visible light; TiO$_2$-based photocatalysis has not been used for industrial wastewater treatment [2, 3]. It has become a challenge for researchers to develop a visible light-responsive photocatalyst that has both a wide range of light absorption and a low recombination rate of photogenerated charges [4, 5]. Recently, it has been reported that the preparation of a semiconductor heterojunction photocatalyst by coupling another narrow-bandgap semiconductor material can effectively reduce the recombination rate of photogenerated electrons and holes and has been widely used to increase the activity of the photocatalyst [6–8].

With a bandgap width of 2.75 eV, bismuth tungstate (Bi$_2$WO$_6$) has received more and more attention due to its excellent photocatalytic properties [9]. However, pure Bi$_2$WO$_6$ has disadvantages of narrower response range of visible light (less than 450 nm) and rapid recombination of photogenerated electron and hole pairs. Therefore, to broaden the response range of visible light and to enhance the photogenerated carriers, separation efficiency of Bi$_2$WO$_6$ is very important to improve their photocatalytic properties. In recent years, there have been many attempts to improve the photocatalytic activity of Bi$_2$WO$_6$, such as the preparation of a large surface area of the 3D nanostructure Bi$_2$WO$_6$ [10] and element doping [11–13]. Due to its high Schottky barrier at the metal-semiconductor interface, the noble metal on the semiconductor surface has been used as an electron acceptor to separate photogenerated electron-hole pairs and to promote interfacial charge transfer processes [14]. Most of the noble metal-semiconductor composite photocatalysts are prepared by in situ photoreduction. The researchers mainly discuss the influence of the precious metal loading, but the influence of photoreduction time is seldom reported. In this dissertation, a series of Ag/Bi$_2$WO$_6$ composites with different photoreduction times were synthesized by a simple visible light-driven photoreduction method. The synthesized samples were characterized by X-ray powder diffraction (XRD), scanning electron microscopy (SEM) and transmission electron microscopy (TEM), UV-Vis DRS, and photoluminescence (PL) spectroscopic analysis. The photocatalytic activity and degradation mechanism of the Ag/Bi_2WO_6 composite were studied in detail under visible light irradiation.

2. Materials and Methods

2.1. Preparation of Bi_2WO_6 Microspheres. 2 mmol $Bi(NO_3)_3 \cdot 5H_2O$ and 1 mmol $Na_2WO_4 \cdot 2H_2O$ were dissolved in 35 mL deionized water, respectively. After the magnetic stirring for 30 min, the Na_2WO_4 solution was slowly added to the solution of $Bi(NO_3)_3$ dropwise. Magnetic stirring was continued for 30 min, and the solution was transferred to a 100 ml polytetrafluoroethylene hydrothermal vessel. The hydrothermal reaction was carried out at 180°C for 12 h and formed a pale yellow precipitate and was filtered, washed (deionized water and ethanol each three times), and dried in an oven at 60°C.

2.2. Preparation of Ag/Bi_2WO_6 Composites. The Ag/Bi_2WO_6 material with a silver content of 1% by mass was photoreduction synthesized as follows: the weighed amount of $AgNO_3$ was added to 50 mL of deionized water and stirred in the dark until completely dissolved. Then, 1.0 mmol Bi_2WO_6 was added to the $AgNO_3$ solution, stirred at room temperature, and irradiated with a 500 W Xenon lamp (irradiance intensity 100 mW/cm^2; Institute of Electric Light Source, Beijing) for a certain period of time, and silver ions were photoreduced in situ to form silver nanoparticles on the surface of the Bi_2WO_6 by light irradiation. The reaction was filtered, washed, and dried at 80°C for 12 h to obtain the Ag/Bi_2WO_6 photocatalyst. In order to study the effect of photoreduction time on the photocatalytic activity of Ag/Bi_2WO_6 composites, Ag/Bi_2WO_6 samples with different photoreduction times of 10, 20, 40, and 60 min were labeled as AB-10, AB-20, AB-40, and AB-60, respectively.

2.3. Characterization. Phase analysis of the Ag/Bi_2WO_6 photocatalyst was carried out using an X-ray powder diffractometer (Bruker D8, Germany). The morphology and particle size were analyzed by scanning electron microscopy (Hitachi S-4800, Japan) and transmission electron microscopy (JEOL 2010F, Japan). UV-visible diffuse reflectance spectra were measured on a UV-2550 (Shimadzu, Japan). Photoluminescence fluorescence spectra were measured using a fluorescence spectrometer (Hitachi F-4500, Japan).

2.4. Photocatalytic Activity Test. The photocatalytic activities of the as-synthesized Ag/Bi_2WO_6 were carried out in terms of the degradation of RhB solution. A 500 W Xenon lamp (irradiance intensity 100 mW/cm^2; Institute of Electric Light Source, Beijing) with a cutoff filter ($\lambda > 420$ nm) was used as the visible light source. 50 mg of catalyst was dispersed in a beaker of 50 ml RhB solution (1.5×10^{-5} mol/L) or parachlorophenol (4-CP) (1 mg·L^{-1}) and continuously stirred for 30 min in the dark to reach the equilibrium of adsorption-desorption before visible light irradiation. During the irradiation, 3 mL of solution was sampled at certain intervals and centrifuged for analysis. The RhB concentrations in

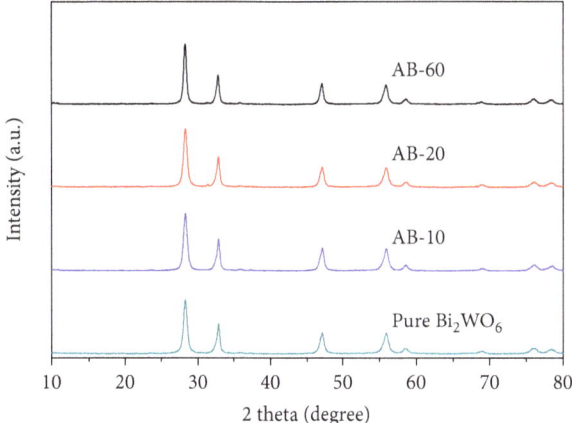

FIGURE 1: XRD patterns of the pure Bi_2WO_6 and Ag/Bi_2WO_6 composites with different photoreduction times.

the solution were measured by a UV-visible spectrophotometer at the maximum absorption wavelength of RhB of 553 nm (Shanghai Meipuda Instrument Co. Ltd. UV-1600). The 4-CP concentrations in the solution were analyzed by high-performance liquid chromatography (HPLC) using an Agilent 1100 series (Santa Clara, CA, USA) equipped with a UV detector at 280 nm. The mobile phase was composed of 80% methanol and 20% water at a flow rate of 0.5 mL·min^{-1}.

3. Results and Discussion

To characterize the crystalline structure of the samples, the XRD patterns of pure Bi_2WO_6 and Ag/Bi_2WO_6 composites with different photoreduction times were obtained (Figure 1). All of the diffraction peaks observed in the XRD patterns belong to the orthorhombic Bi_2WO_6 (JCPDS number 39-0256) with the unit cell parameters which are $a = 5.456$ Å, $b = 16.435$ Å, and $c = 5.438$ Å [15]. The diffraction angles for pure Bi_2WO_6 and Ag/Bi_2WO_6 composites at 2θ that are equal to 28.3°, 32.8°, 47.0°, 55.8°, and 58.5° can be assigned to 113, 200, 113, 220, 331, and 226 reflections, respectively, of the orthorhombic Bi_2WO_6. The diffraction peaks are strong and sharp, and no other impurity peaks appear, indicating that Ag doping does not change the phase structure of Bi_2WO_6. The absence of Ag diffraction peaks can be attributed to too little Ag nanoparticle content below the detection limit of the instrument.

Figures 2(a) and 2(b) show the SEM images of pure Bi_2WO_6 and Ag/Bi_2WO_6 (20 min). It can be clearly seen that the prepared samples have a three-dimensional flower-like microsphere structure with good dispersion and the diameter of flower-like microspheres is about 2.5–3.5 μm. It is indicated that the nano-Ag particles introduced by photoreduction do not affect the macroscopic morphology of Bi_2WO_6 microspheres. Silver nanoparticles were not observed for the Ag/Bi_2WO_6 microspheres (Figure 2(b)), probably due to the light-induced preparation of silver nanoparticles too small in size. In order to further confirm the presence of silver nanoparticles on the surface of flower-like Bi_2WO_6

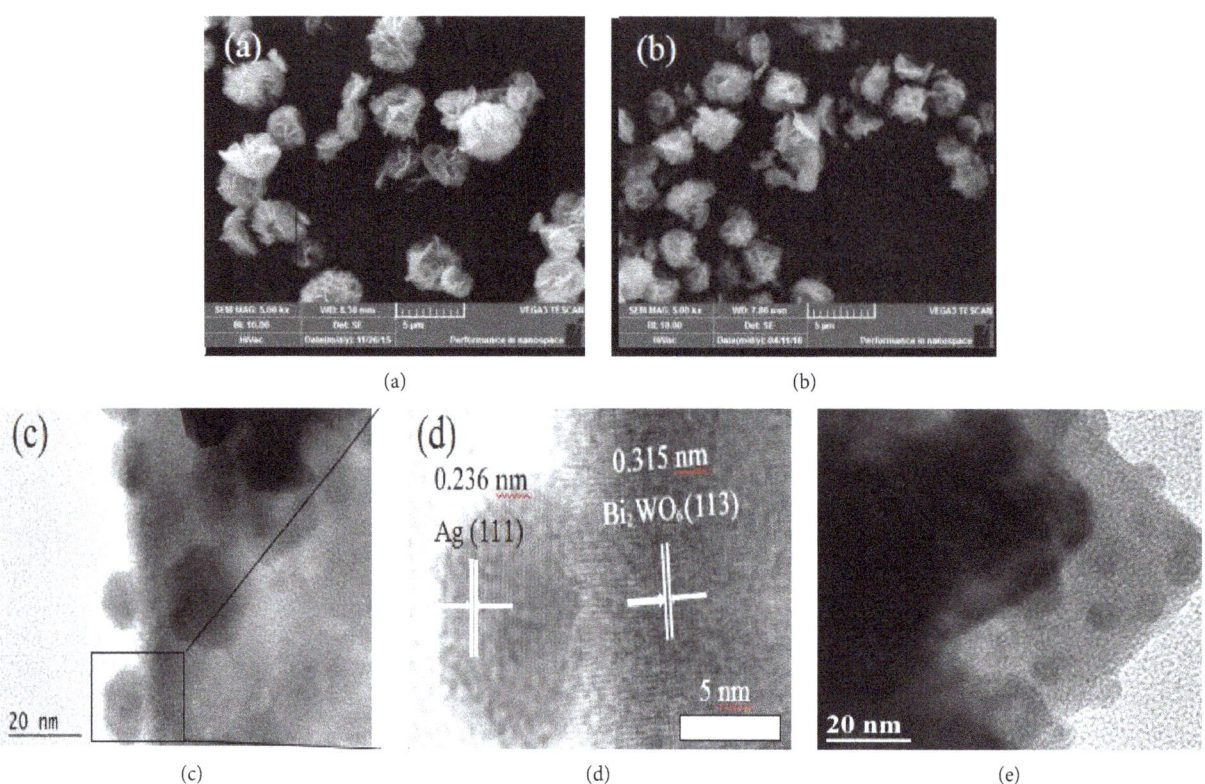

FIGURE 2: SEM images of synthetic samples: (a) Bi_2WO_6; (b) Ag/Bi_2WO_6; (c and e) TEM images of AB-20 and AB-40; (d) HRTEM images of AB-20.

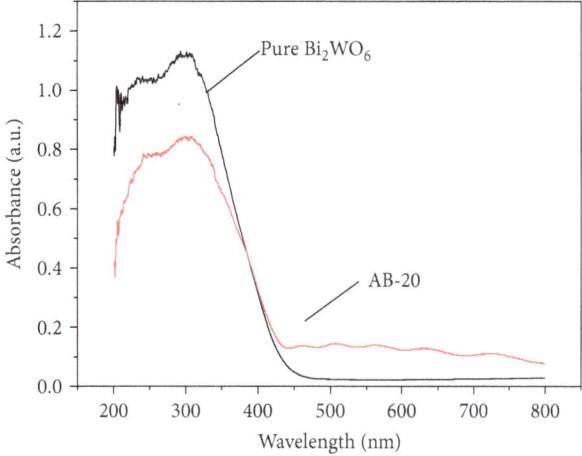

FIGURE 3: UV-visible diffuse reflectance spectroscopy of the synthetic sample.

microspheres, AB-20 was analyzed using the TEM and HRTEM analysis, which shows that Ag nanoparticles of about 10 nm were distributed on Bi_2WO_6 nanosheets. Figure 2(d) shows that the lattice fringes of Bi_2WO_6 and Ag are clearly visible and correspond to the (113) plane of Bi_2WO_6 and the (111) plane of Ag, respectively. Figure 2(e) reveals that prolonged photoreduction results in excessive silver nanoparticle deposition, which may be detrimental to the rapid separation of photogenerated carriers.

UV-visible diffuse reflectance spectroscopy of pure Bi_2WO_6 and AB-20 was measured using a UV-Vis spectrophotometer with an integrating sphere (Shimadzu, Japan), as shown in Figure 3. All samples have a strong absorption band in the UV-visible region, and absorption is steep up to the edge; this is attributed to the transition of the bandgap level of Bi_2WO_6 rather than to the transition of impurity levels [16]. The band gap energy (E_g) was calculated based on the absorption spectra by the formula [17] $(\alpha h\nu)^{1/n} = A(h\nu - E_g)$, where h is Planck's constant, ν is the frequency of vibration, α is the absorption coefficient, E_g is the band gap, and A is the proportionality constant. The value of the exponent n denotes the nature of the sample transition, and $n = 0.5$ when assuming a direct allowed transition. According to the results of this analysis, the band gap values of bare Bi_2WO_6 and AB-20 samples were 2.76 and 2.74 eV, respectively. The Ag/Bi_2WO_6 nanocomposite exhibited similar E_g to pristine Bi_2WO_6, which demonstrated that the introduction of Ag had almost no influence on the band gap energy of Bi_2WO_6. However, the silver-loaded Bi_2WO_6 composite exhibits enhanced light absorption in the visible light region. The enhanced light absorption is due to the surface plasmon resonance (SPR) effect of silver [18]. The plasma peak of silver nanoparticles at about 500 nm is not significantly attributed to the low Ag loading [19]. The silver-supported Bi_2WO_6 composites are more conducive to the visible light absorption compared with the pure Bi_2WO_6, which maybe contributes to the improvement of the photocatalytic activity.

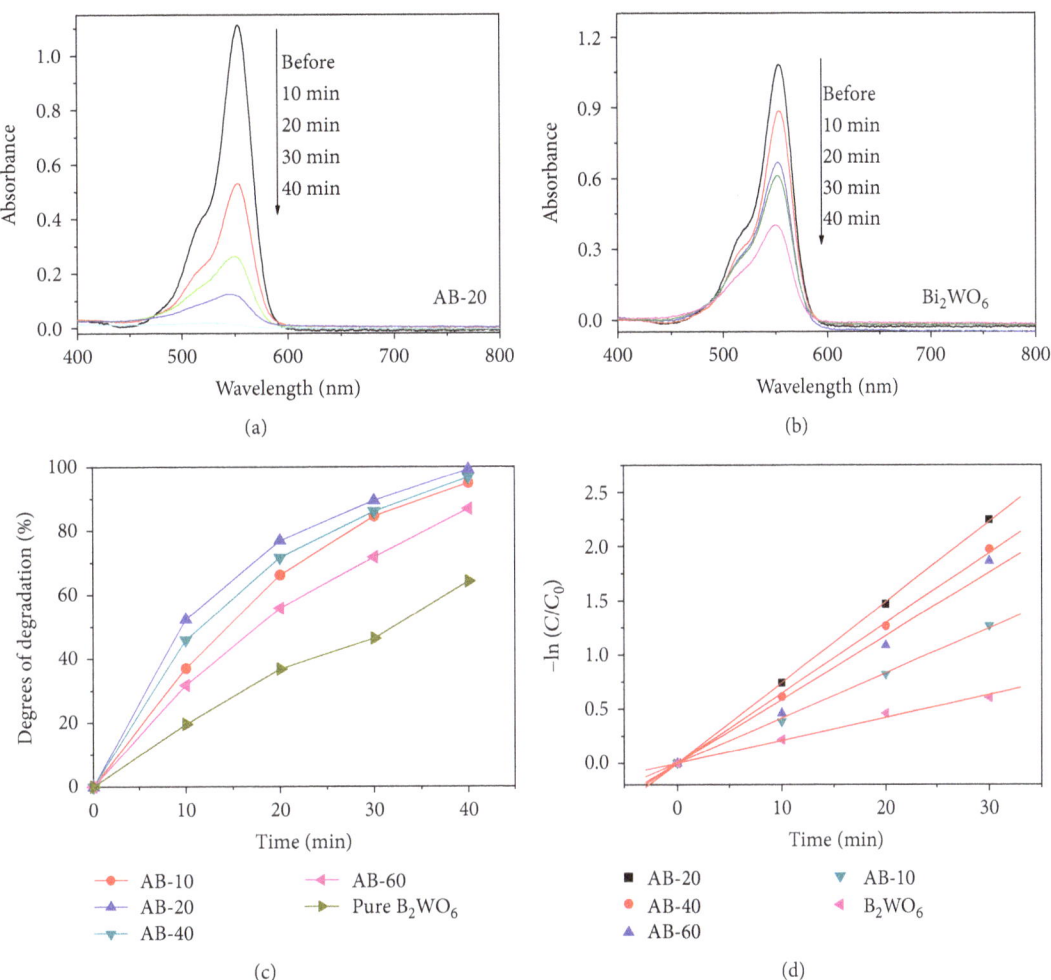

FIGURE 4: (a, b) Changes in the visible absorption spectra of RhB of AB-20 and pure Bi_2WO_6. (c) Photocatalysis degradation profiles of RhB under visible light irradiation. (d) Kinetic plot of the photocatalysts.

The photocatalytic activities of the pure Bi_2WO_6 and Ag/Bi_2WO_6 were investigated by the photocatalytic oxidation of the cationic dye rhodamine B (RhB). Figures 4(a) and 4(b) show that the photodegradation spectra under visible light were acquired for AB-20 and the pure Bi_2WO_6 after different reaction times. The direct degradation of the conjugated system of the dye can be observed through the decrease in the RhB absorbance at 553 nm. Figure 4(c) shows the RhB photocatalytic degradation curve of the different synthetic samples. It can be seen that the photocatalytic activity of Ag/Bi_2WO_6 prepared at different photoreduction times is superior to that of pure Bi_2WO_6. The photocatalytic activity of Ag/Bi_2WO_6 was also enhanced gradually when the photoreduction time was extended from 10 min to 20 min. This is because the Ag nanoparticle promoted the photoelectron-hole pair separation and interfacial electron transfer. In addition, the surface plasmon resonance of Ag nanoparticles under visible light irradiation also contributes to the absorption of light to enhance the photocatalytic activity [17, 18]. The photocatalytic activity of AB-20 was the best when the photoreduction time was 20 min and the degradation rate of RhB reached 99.2% after 40 min of visible light irradiation. The photocatalytic activity of

Ag/Bi_2WO_6 decreases gradually with the photoreduction time more than 20 min. This may be attributed to the fact that excessive silver nanoparticle deposition promotes the recombination of photogenerated carriers [20] and the excess silver may also cover the active sites of Bi_2WO_6, leading to a decrease in photocatalytic activity.

Because the initial concentration of the RhB dye is very small, the photocatalytic degradation of RhB by Ag/Bi_2WO_6 is in accordance with the description of quasi-first-order kinetic equations [21], and the formula is $\ln (C_t/C_0) = -kt$ (C_t and C_0 are the time t RhB solution concentration and initial RhB solution concentration, respectively, and k is the quasi-primary reaction kinetic rate constant). The plot of $-\ln (C_t/C_0)$ versus t is fitted to a straight line, and the fitted slope is the quasi-first-order reaction rate constant k. Figure 4(d) shows the pseudo-first-order linear fit of the photocatalytic degradation reaction of pure Bi_2WO_6 and Ag/Bi_2WO_6 composites synthesized at different photoreduction times. The reaction rate constants of Ag/Bi_2WO_6 composites synthesized at photoreduction times of 10, 20, 40, and 60 min and pure Bi_2WO_6 were 0.0669, 0.08476, 0.07587, 0.04357, and 0.02353 min^{-1}, respectively. The reaction rate constant of Ag/Bi_2WO_6 was the highest when the

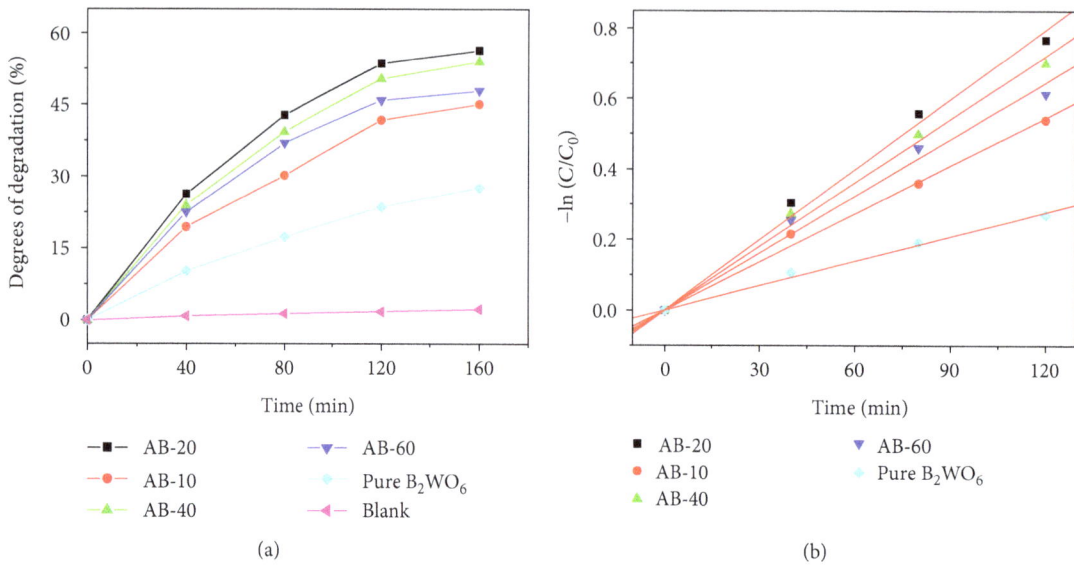

FIGURE 5: (a) Photocatalysis degradation profiles of 4-CP under visible light irradiation. (b) Kinetic plot of the photocatalysts.

photoreduction synthesis time was 20 min, which was 3.60 times higher than that of pure Bi_2WO_6. The photocatalytic results were compared with the already-present related literature; the reaction rate constant of Ag/Bi_2WO_6 synthesized by sonochemical methods [22] or ultrasonic vibration processes [23] was 0.0332 or 0.0279 min^{-1}, respectively. The reaction rate constant of Pt/Bi_2WO_6 [24] using the photoreduction process similar to our synthesis method was 0.035 min^{-1}. Obviously, AB-20 was much more active than Ag/Bi_2WO_6 synthesized by other methods or other metals such as Pt deposited on Bi_2WO_6 using the same method, demonstrating that AB-20 displayed excellent photocatalytic activity.

To further confirm that the photocatalytic activity of Ag/Bi_2WO_6 composites originates from the excitation of the catalysts rather than from the dye sensitization mechanism, the photocatalytic degradation of colorless neutral parachlorophenol (4-CP) over Ag/Bi_2WO_6 composites was also performed (Figure 5). The degradation of 4-CP is extremely slow without photocatalysts after 160 min of reaction, while the degradation efficiency approaches 44.96%, 56.29%, 53.86%, and 47.79% by using AB-10, AB-20, AB-30, or AB-40 as the photocatalyst, respectively (Figure 5(a)). Among these composites, AB-20 exhibits the highest photodegradation efficiency (78.3%), which is much higher than that from Bi_2WO_6 (27.6%). The photodegradation rate constant using AB-20 (0.00664 min^{-1}) is about 2.87 times higher than that using Bi_2WO_6 (0.00231 min^{-1}) (Figure 5(b)).

Photoluminescence (PL) spectra are used to determine the photogenerated carrier recombination, migration, and separation efficiency, as photoluminescence is mainly caused by the recombination of photogenerated electron-hole pairs [25]. The lower the photoluminescence intensity under visible light irradiation, the lower the recombination rate of charge carriers [26]. Figure 6 shows the photoluminescence spectra of pure Bi_2WO_6 and AB-20 samples with an excitation wavelength of 300 nm. Pure Bi_2WO_6 has a broad

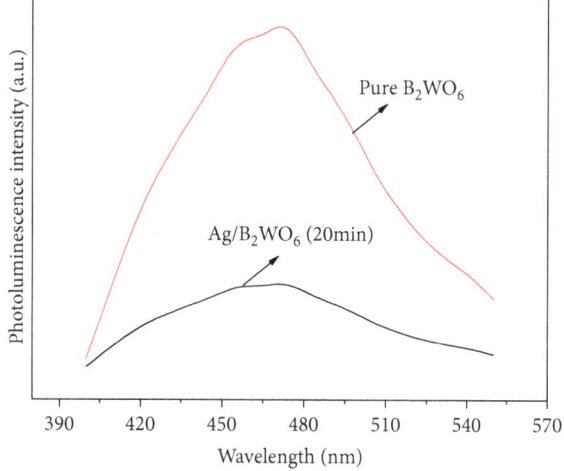

FIGURE 6: The photoluminescence spectra of the synthetic sample.

emission peak at 400–560 nm and a strong blue emission peak at 468 nm, attributed to the charge on the hybrid orbital formed by Bi^{6s} and O^{2p} on the Bi_2WO_6 valence band that transferred to the W^{5d} empty conduction band in the WO_6^{2-} complex [27]. It indicates that the recombination probability of photogenerated electron-hole pairs in pure Bi_2WO_6 is much higher, resulting in weaker photocatalytic activity. The PL intensity of silver-supported Bi_2WO_6 composites is much lower than that of pure Bi_2WO_6, which can effectively inhibit the recombination rate of excited electrons and holes and increase the photocatalytic activity.

Figure 7 shows the transient photocurrent test pattern of pure Bi_2WO_6 and AB-20 samples tested in 1 M NaOH solution at 300 W xenon lamp illumination. From Figure 7, it can be seen that the photocurrent densities of pure Bi_2WO_6 and AB-20 electrodes are 0.7 and 2.2 $\mu A/cm^2$, respectively. The photocurrent density of the Ag/Bi_2WO_6 electrode is higher than that of the pure Bi_2WO_6 electrode. This also shows that

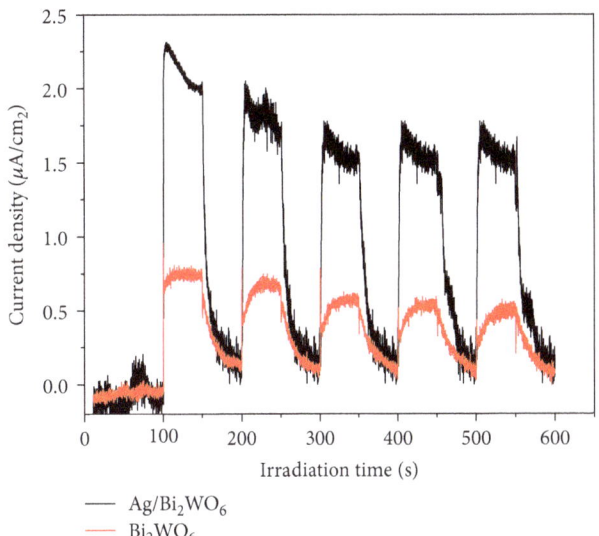

FIGURE 7: Transient photocurrent responses of the synthetic sample.

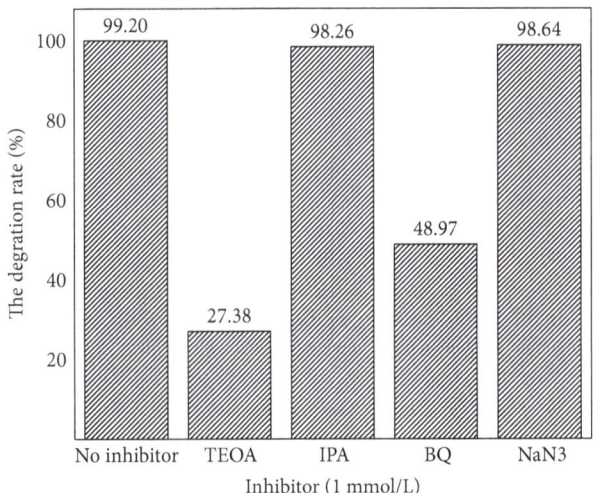

FIGURE 8: Photocatalytic degradation of RhB by Ag/Bi$_2$WO$_6$ with the presence of a capturing agent.

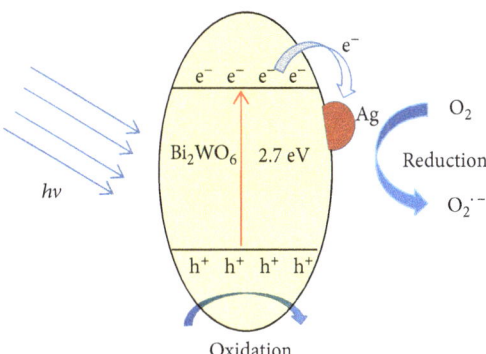

FIGURE 9: Schematic diagram of the photocatalytic mechanism for the Ag/Bi$_2$WO$_6$ composite.

rhodamine B, while the photocatalytic activity of the catalyst was reduced to 48.97% after the addition of BQ. In summary, h$_{VB}^+$ plays a key role in the photocatalytic degradation process, O$_2^{\bullet-}$ plays a part role in the degradation process, $^{\bullet}$OH and ^{1}O$_2$ have no effect, and the order of the active species is h$_{VB}^+$ > O2$^{\bullet-}$ ≫ $^{\bullet}$OH.

Combined with the radical trapping experiments and relevant literature, the possible photocatalytic degradation mechanism of rhodamine B with the Ag/Bi$_2$WO$_6$ composite photocatalyst is speculated as shown in Figure 8. The photocatalysis of Ag/Bi$_2$WO$_6$ composites is caused by the effective absorption of visible photons, the electrons (e$^-$) in the valence band of Bi$_2$WO$_6$ (VB) are excited to transition to the conduction band (CB), and holes (h$_{VB}$) are left on the valence band of Bi$_2$WO$_6$. Because the conduction band energy level of Bi$_2$WO$_6$ is higher than the conduction band energy level of Ag, the electrons can be quickly transferred to the Ag nanoparticles through the Schottky barrier at the metal-semiconductor interface, which is similar to other researchers' studies on the electron transfer from semiconductors to metals [28, 29]. Thus, the recombination probability of the photogenerated electrons in the conduction band (CB) and the photogenerated holes in the valence band (VB) was lessened, as revealed by photoluminescence experiment results and transient photocurrent test. Subsequently, the h$^+$ stored in the VB of Bi$_2$WO$_6$ can directly oxidize RhB, and the photogenerated electrons are quickly captured by molecular oxygen absorbed on the surface of the photocatalyst and generate superoxide radical (O$_2^{\bullet-}$) to further degrade RhB. Therefore, the enhancement of photocatalytic activity of Ag/Bi$_2$WO$_6$ is owed to photogenerated holes (h$_{VB}^+$) and superoxide radical (O$_2^{\bullet-)}$, which are consistent with the result of radical trapping experiments (Figure 9).

4. Conclusions

Three-dimensional flower-like Ag/Bi$_2$WO$_6$ composites were successfully prepared by a simple photoreduction method. The results of photocatalytic experiments show that Ag/Bi$_2$WO$_6$ composites have higher photocatalytic activity than pure Bi$_2$WO$_6$. The photocatalytic reaction rate constant of Ag/Bi$_2$WO$_6$ composites prepared at a photoreduction time of 20 min was 3.60 times bigger than that of pure Bi$_2$WO$_6$.

loading a certain amount of Ag can promote photogenerated electron-hole transfer and inhibit rapid recombination of photogenerated charges.

To analyze the photocatalytic reaction mechanism, it is of vital importance to identify the primary active species during the degradation process. Thus, the radical trapping experiments were performed (Figure 8). The capturing agents such as benzoquinone (BQ), triethanolammonium (TEOA), isopropanol (IPA), and sodium azide (NaN$_3$) with a concentration of 1 mmol/L are commonly used to capture superoxide radical (O$_2^{\bullet-}$), photogenerated hole (h$_{VB}^+$), hydroxyl radical ($^{\bullet}$OH), and singlet oxygen (^{1}O$_2$), respectively. When the photocatalytic reaction time is 40 min, the effect of adding TEOA is very significant, the degradation rate of rhodamine B is reduced to 27.38%, and the addition of IPA and NaN$_3$ had little effect on the degradation of

The excellent photocatalytic performance of Ag/Bi_2WO_6 composites is due to the introduction of silver which results in a synergistic effect of an increase of the light absorption range and an effective separation of photogenerated carriers. The synthesis of this material will help design new high-performance photocatalysts.

Conflicts of Interest

The authors declare no conflict of interest.

Acknowledgments

This work was supported by the National Natural Science Foundation of China (21772035, 21401088) and Scientific Research Foundation of Hunan Institute of Engineering (18RC009).

References

[1] M. R. Hoffmann, S. T. Martin, W. Choi, and D. W. Bahnemann, "Environmental applications of semiconductor photocatalysis," *Chemical Reviews*, vol. 95, no. 1, pp. 69–96, 1995.

[2] X. Chen, S. Shen, L. Guo, and S. S. Mao, "Semiconductor-based photocatalytic hydrogen generation," *Chemical Reviews*, vol. 110, no. 11, pp. 6503–6570, 2010.

[3] X. Chen and S. S. Mao, "Titanium dioxide nanomaterials: synthesis, properties, modifications, and applications," *Chemical Reviews*, vol. 107, no. 7, pp. 2891–2959, 2007.

[4] T. Saison, P. Gras, N. Chemin et al., "New insights into Bi_2WO_6 properties as a visible-light photocatalyst," *Journal of Physical Chemistry C*, vol. 117, no. 44, pp. 22656–22666, 2013.

[5] M. D. Hernández-Alonso, F. Fresno, S. Suárez, and J. M. Coronado, "Development of alternative photocatalysts to TiO_2: challenges and opportunities," *Energy & Environmental Science*, vol. 2, no. 12, p. 1231, 2009.

[6] Y. Qu and X. Duan, "Progress, challenge and perspective of heterogeneous photocatalysts," *Chemical Society Reviews*, vol. 42, no. 7, pp. 2568–2580, 2013.

[7] H. G. Kim, P. H. Borse, J. S. Jang et al., "Fabrication of $CaFe_2O_4$/$MgFe_2O_4$ bulk heterojunction for enhanced visible light photocatalysis," *Chemical Communications*, no. 39, pp. 5889–5891, 2009.

[8] X. Zhang, L. Zhang, T. Xie, and D. Wang, "Low-temperature synthesis and high visible-light-induced photocatalytic activity of BiOI/TiO2 heterostructures," *Journal of Physical Chemistry C*, vol. 113, no. 17, pp. 7371–7378, 2009.

[9] L. W. Zhang, Y. J. Wang, H. Y. Cheng, W. Q. Yao, and Y. F. Zhu, "Synthesis of porous Bi_2WO_6 thin films as efficient visible-light-active photocatalysts," *Advanced Materials*, vol. 21, no. 12, pp. 1286–1290, 2009.

[10] L. Zhang and Y. Zhu, "A review of controllable synthesis and enhancement of performances of bismuth tungstate visible-light-driven photocatalysts," *Catalysis Science & Technology*, vol. 2, no. 4, pp. 694–706, 2012.

[11] R. Shi, G. Huang, J. Lin, and Y. Zhu, "Photocatalytic activity enhancement for Bi_2WO_6 by fluorine substitution," *Journal of Physical Chemistry C*, vol. 113, no. 45, pp. 19633–19638, 2009.

[12] C. Bhattacharya, H. C. Lee, and A. J. Bard, "Rapid screening by scanning electrochemical microscopy (SECM) of dopants for Bi_2WO_6 improved photocatalytic water oxidation with Zn doping," *Journal of Physical Chemistry C*, vol. 117, no. 19, pp. 9633–9640, 2013.

[13] M. Shang, W. Wang, L. Zhang, and H. Xu, "Bi_2WO_6 with significantly enhanced photocatalytic activities by nitrogen doping," *Materials Chemistry and Physics*, vol. 120, no. 1, pp. 155–159, 2010.

[14] K. Awazu, M. Fujimaki, C. Rockstuhl et al., "A plasmonic photocatalyst consisting of silver nanoparticles embedded in titanium dioxide," *Journal of the American Chemical Society*, vol. 130, no. 5, pp. 1676–1680, 2008.

[15] J. Li, Z. Guo, and Z. Zhu, "Ag/Bi_2WO_6 plasmonic composites with enhanced visible photocatalytic activity," *Ceramics International*, vol. 40, no. 5, pp. 6495–6501, 2014.

[16] H. Fu, C. Pan, W. Yao, and Y. Zhu, "Visible-light-induced degradation of rhodamine B by nanosized Bi_2WO_6," *The Journal of Physical Chemistry B*, vol. 109, no. 47, pp. 22432–22439, 2005.

[17] S. Sakthivel, M. Janczarek, and H. Kisch, "Visible light activity and photoelectrochemical properties of nitrogen-doped TiO_2," *The Journal of Physical Chemistry. B*, vol. 108, no. 50, pp. 19384–19387, 2004.

[18] C. Hu, Y. Lan, J. Qu, X. Hu, and A. Wang, "Ag/AgBr/TiO_2 visible light photocatalyst for destruction of azodyes and bacteria," *The Journal of Physical Chemistry B*, vol. 110, no. 9, pp. 4066–4072, 2006.

[19] J. Ren, W. Wang, S. Sun, L. Zhang, and J. Chang, "Enhanced photocatalytic activity of Bi_2WO_6 loaded with Ag nanoparticles under visible light irradiation," *Applied Catalysis B: Environmental*, vol. 92, no. 1-2, pp. 50–55, 2009.

[20] D. Wang, G. Xue, Y. Zhen, F. Fu, and D. Li, "Monodispersed Ag nanoparticles loaded on the surface of spherical Bi_2WO_6 nanoarchitectures with enhanced photocatalytic activities," *Journal of Materials Chemistry*, vol. 22, no. 11, p. 4751, 2012.

[21] Y. Yu, M. Yu, Y. Zhang, W. Sun, and Y. Liu, "Preparation and mechanism of N-doped TiO_2 powders," *Chinese Journal of Inorganic Chemistry*, vol. 29, no. 8, pp. 1657–1662, 2013.

[22] A. Phuruangrat, A. Maneechote, P. Dumrongrojthanath, N. Ekthammathat, S. Thongtem, and T. Thongtem, "Visible-light driven photocatalytic degradation of rhodamine B by Ag/Bi_2WO_6 heterostructures," *Materials Letters*, vol. 159, pp. 289–292, 2015.

[23] P. Dumrongrojthanath, A. Phuruangrat, P. Junploy, S. Thongtem, and T. Thongtem, "Visible-light-driven photocatalysis of heterostructure Ag/Bi_2WO_6 nanocomposites and their photocatalytic degradation of dye under visible light irradiation," *Research on Chemical Intermediates*, vol. 42, no. 3, pp. 1651–1662, 2016.

[24] M.-A. Lavergne, C. Chanéac, D. Portehault, S. Cassaignon, and O. Durupthy, "Optimized design of Pt-doped Bi_2WO_6 nanoparticle synthesis for enhanced photocatalytic properties," *European Journal of Inorganic Chemistry*, vol. 2016, no. 13-14, pp. 2159–2165, 2016.

[25] J. Tang, Z. Zou, and J. Ye, "Photophysical and photocatalytic properties of $AgInW_2O_8$," *The Journal of Physical Chemistry B*, vol. 107, no. 51, pp. 14265–14269, 2003.

[26] K. Fujihara, S. Izumi, T. Ohno, and M. Matsumura, "Time-resolved photoluminescence of particulate TiO_2 photocatalysts suspended in aqueous solutions," *Journal of Photochemistry and Photobiology A: Chemistry*, vol. 132, no. 1-2, pp. 99–104, 2000.

[27] Q. Xiao, J. Zhang, C. Xiao, and X. Tan, "Photocatalytic degradation of methylene blue over Co_3O_4/Bi_2WO_6 composite under visible light irradiation," *Catalysis Communications*, vol. 9, no. 6, pp. 1247–1253, 2008.

[28] H. Tada, T. Mitsui, T. Kiyonaga, T. Akita, and K. Tanaka, "All-solid-state Z-scheme in CdS-Au-TiO_2 three-component nanojunction system," *Nature Materials*, vol. 5, no. 10, pp. 782–786, 2006.

[29] T. Hirakawa and P. V. Kamat, "Charge separation and catalytic activity of $Ag@TiO_2$ core-shell composite clusters under UV-irradiation," *Journal of the American Chemical Society*, vol. 127, no. 11, pp. 3928–3934, 2005.

10

Photoactivity of Titanium Dioxide Foams

Maryam Jami[ID],[1] **Ralf Dillert,**[1] **Yanpeng Suo**[ID],[1] **Detlef W. Bahnemann,**[1,2] and **Michael Wark**[3]

[1]*Institut für Technische Chemie, Leibniz Universität Hannover, Callinstraße 3, 30167 Hannover, Germany*
[2]*Laboratory "Photoactive Nanocomposite Materials", Saint Petersburg State University, Ulyanovskaya Str. 1, Peterhof, Saint Petersburg 198504, Russia*
[3]*Technische Chemie, Universität Oldenburg, Carl-von-Ossietzky Str. 9-11, 26111 Oldenburg, Germany*

Correspondence should be addressed to Maryam Jami; jamimaryam@ymail.com

Academic Editor: Leonardo Palmisano

TiO_2 foams have been prepared by a simple mechanical stirring method. Short-chain amphiphilic molecules have been used to stabilize colloidal suspensions of TiO_2 nanoparticles. TiO_2 foams were characterized by X-ray diffraction (XRD), X-ray photoelectron spectroscopy (XPS), UV-vis absorption spectroscopy, and scanning electron microscopy (SEM). The photoassisted oxidation of NO in the gas phase according to ISO 22197-1 has been used to compare the photoactivity of the newly prepared TiO_2 foams to that of the original powders. The results showed that the photoactivity is increased up to about 135%. Foam structures seem to be a good means of improving the photoactivity of semiconductor materials and can readily be used for applications such as air purification devices.

1. Introduction

Heterogeneous photocatalysis has been used as an effective technique for the remediation of chemical wastes. Titanium dioxide (TiO_2) is one of the most frequently investigated heterogeneous semiconductor photocatalysts and has been shown to be a relatively cheap and effective material to decompose various kinds of organic and inorganic wastes in both, the liquid and the gas phase [1–4]. As one of the most promising applications, the decomposition of nitrogen oxides (NO_x) in ambient air has been studied intensively. Nitrogen oxides (mainly NO and NO_2), which are emitted from sources such as automobiles and boilers, have become a serious environmental problem in urban areas and can also cause ozone depletion, photochemical smog, and acid deposition. Substantial efforts have been undertaken to develop methods to reduce the concentration of NO and of NO_2. In particular, the photoassisted oxidation of NO_x to nitric acid (HNO_3) using TiO_2 seems to be an effective, economical, and energy saving process for the treatment of diluted NO_x [5–10].

The photocatalytic properties of TiO_2 materials do not just depend on the chemical composition but also on the geometrical microstructure [11, 12]. During the last few decades, there have been numerous reports focusing on methods to enhance the photoactivity of TiO_2 materials [13, 14]. Several research groups have been working on the doping of TiO_2 with noble metals (Au, Pd, Ag, and Pt) or nonmetal atoms (N, F, S, C). All these investigations have developed our knowledge concerning TiO_2-based photocatalysts and have generated very interesting and important results in this research area. However, in general, the TiO_2-based photocatalysts suffer from rather low quantum efficiencies because of poor photoabsorption efficiencies and weak molecular transport capabilities [5, 15–22].

Another strategy to enhance the photoactivity of TiO_2 materials is the creation of new geometrical microstructures. It has, for example, been reported that the photoactivity of TiO_2 with structural, that is, ordered porosity is higher than that of nanoparticular TiO_2 with only interparticular nanovoids. Materials with structural porosity have high surface areas, and all pores are well interconnected. The macro-

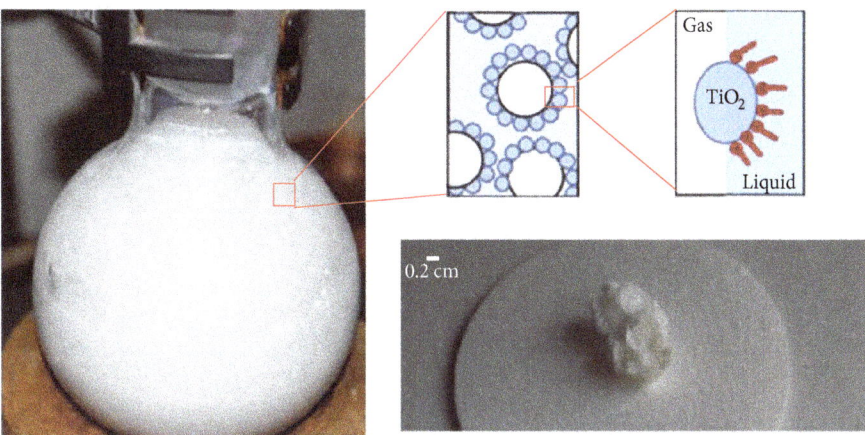

FIGURE 1: Photographs of wet particle-stabilized foams and a schematic illustration of the stabilization of gas bubbles with partially hydrophobized colloidal particles to finally obtain the TiO$_2$ foam.

and mesoporosity improves the adsorption and diffusion of reactants and products. The existence of cavities furthermore increases the light-capturing efficiency because of their strong light-scattering properties [21, 23–27].

Titanium dioxide foams have shown to have a higher photoactivity as the powder one by Zhao et al. [24]. However, it is not clear how the foam structure affected the photoactivity. One reason could be changes in the surface properties of TiO$_2$. Further investigation is necessary to confirm this theory. Photocatalytic deposition velocity is known to be an input value that predicts the possible effect of photocatalytically active surfaces on air pollution in urban areas.

In this article, it is reported that TiO$_2$ foams have been prepared by a simple mechanical stirring method employing short-chain amphiphilic molecules to surface-lyphobize commercially available TiO$_2$ nanoparticles. The aims of this work were to study the photoactivity and the changes in the surface properties of TiO$_2$ foams prepared in this way and compare it to that of the starting TiO$_2$ powder. In 2007, ISO published a standard test to evaluate the photoassisted oxidation efficiency of the air-cleaning products. The photoassisted degradation of NO in the gas phase according to ISO 22197-1 was used as the test reaction to compare the photoactivity of the TiO$_2$ foams to that of the original powders. Photocatalytic deposition velocities were determined to describe the changes in the surface properties. The morphology and the structural properties of TiO$_2$ foams have also been studied.

2. Experimental

2.1. Materials. Commercial titanium dioxide Evonik AEROXIDE® P25 (30% rutile, 70% anatase, 20–30 nm) has been used as received. Hexylic acid (hexanoic acid) and hexylamine from Sigma-Aldrich have been used as short-chain amphiphilic molecules. Other chemicals used in the experiments were deionized water and sodium hydroxide (Sigma-Aldrich).

2.2. Foam Preparation. Titanium dioxide foams were prepared according to the method described by Zhao et al.

[24]. In short, 2 g of P25 was added stepwise to water under vigorous stirring for 20 minutes. The pH value of the suspension was around 3.6. The suspension was homogenized for about 30 min using an Ultrasonic Vibra-Cell. Meanwhile, a solution containing short-chain amphiphilic molecules (0.3 g hexylic acid in water, the ratio of TiO$_2$ to amphiphilic molecules was adapted from [24]), was prepared. This solution was then added dropwise to the TiO$_2$ suspension under vigorous stirring for 30 min.

The pH was adjusted to pH 3.5 with 0.1 M NaOH if necessary (in case of using hexylic acid as amphiphilic agent). Finally, the wet TiO$_2$ foam was produced and this wet foam was air dried for one day at room temperature. Figure 1 shows photographs of freshly prepared wet TiO$_2$ foam and of a dry one.

2.3. Characterization. The crystal structures of the samples were determined by X-ray powder diffraction at room temperature using Cu Kα radiation on a Bruker D8 Advance diffractometer in the 2θ range of 10–80°. The particle size distributions were calculated from the peak broadening of the XRD patterns using the Scherrer equation [28]. X-ray photoelectron spectroscopy (XPS) measurements were carried out using a Thermo Fisher ESCALAB 250Xi equipped with monochromatic Al-Kα X-ray source (1486.6 eV). The binding energies (BE) were referenced to the adventitious carbon contamination C 1s peak (284.8 eV). High-resolution XPS spectra of Ti 2p, O 1s, and C 1s were recorded with 10 eV pass energy, and 0.02 eV step size with 5 spectra was averaged.

A scanning electron microscope (SEM) (JEOL JSM-6700F field-emission) was used to characterize the morphology of the TiO$_2$ foams, and the Varian Cary 4000 UV-vis spectrophotometer was used to measure the diffuse reflectance spectra of the samples. Band gap energies were calculated by using Tauc plots resulting from the Kubelka-Munk transformation of the respective diffuse reflectance spectra. The nitrogen sorption isotherms were measured at 77 K using a Quantachrome Autosorb 3. The foam samples were outgassed in vacuum at 25°C and P25-TiO$_2$ at 100°C for 24 hours prior the sorption measurement. The sorption data were analyzed with the quantachrome software ASiQwin 2.0.

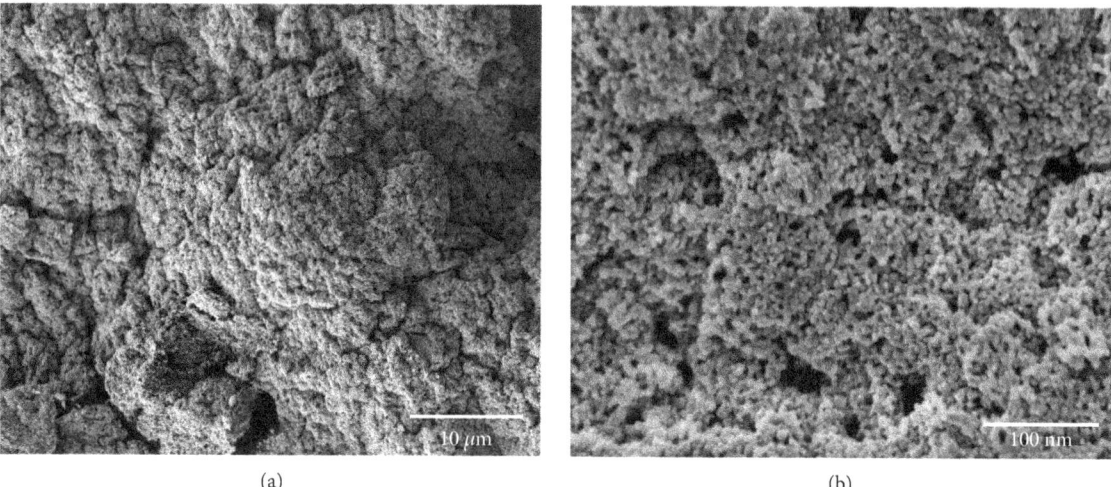

(a)   (b)

FIGURE 2: SEM pictures of the prepared TiO$_2$ foam with P25 as TiO$_2$ nanoparticle. The bar is 10 μm for the low-magnification image (a) and 100 nm for the high-magnification image (b).

2.4. Photoassisted Oxidation of NO. The photoassisted oxidation of NO was measured according to ISO 22197-1 [29]. The experimental details were described previously [9, 10, 30]. Briefly, the test equipment included the test gas supply, a humidifier, three mass flow controllers (Brooks Instrument), a photoreactor, a light source (Philips, Cleo Compact, l_{max} = 355 nm, 15 W), and a chemiluminescent NO–NO$_x$ analyzer (Horiba APNA 360). The concentration of the probe gas was followed by a NO/NO$_2$ analyzer in the dark until equilibrium was reached. The probe gas mixture had a concentration of 1 ppm$_v$ NO (Linde, 50 ppm$_v$ NO in N$_2$), a relative humidity of 50%, and a laminar volume flow $\dot{V}$ of 5×10^{-5} m^3/s. For a single degradation test, TiO$_2$ foam or the P25 sample was placed into the photoreactor with a geometric surface area of 18.49 cm^2 and covered with a UV(A) transparent borosilicate glass. The NO concentration was adjusted in the dark via a bypass mode followed by a dark adsorption of the pollutant on the sample surface by switching from bypass to reaction mode. After equilibration, the degradation of NO was followed under irradiation (light intensity 10 W/m^2) for 2 h until a steady state was reached. Subsequently, the system was kept in the dark until the initial concentration of NO was achieved again. For the determination of the photocatalytic deposition velocity use, a photoreactor with a geometric surface area of 39.6 cm^2 and the amount of the probe gas (NO) have been changed during the photoassisted oxidation. The NO concentration varied between 0.1 ppm$_v$ and 1.0 ppm$_v$. The flow rate of the probe gas (5×10^{-5} m^3/s), the temperature (25°C), and the relative humidity (50%) inside the photoreactor were kept constant.

3. Results and Discussions

3.1. TiO$_2$ Foam Characterization. The scanning electron microscopy (SEM) images of the TiO$_2$ foam show a porous, spongy, and rough morphology. The photograph and the low magnification image of the foam reveal macroporosity (Figure 2). The pore size distribution of the foams depends on the type of amphiphilic agent that has been used. For

example, TiO$_2$-foam-1 (with hexylic acid) has an average pore size of approx. 0.24 ± 0.09 cm and TiO$_2$-foam-2 (with hexylamine) has an average pore size of approx. 0.11 ± 0.04 cm. These large pores with extended networks provide many canals for diffusion of reactants. With increasing SEM magnification, the primary TiO$_2$ nanoparticles can be imaged (particle size distribution: 20–30 nm). Supplementary 1 presents the SEM photographs of TiO$_2$-P25, TiO$_2$-foam-1, and TiO$_2$-foam-2.

The UV-vis diffuse reflectance spectra (Supplementary Material, Fig. S2) from P25 and from the TiO$_2$-foam-1 have been measured, and the energy band gaps of the materials were obtained using the Kubelka-Munk function. The band gap energy of the materials was not changed ($E_g \sim 3.2$ eV) indicating that the semiconductor properties of the TiO$_2$ foam are same as those of the TiO$_2$ powder.

BET surface areas have been determined from the N$_2$ adsorption/desorption measurements (see Supplementary Material, Fig. S3). There is no big difference between BET surface areas of the P25 and to those of TiO$_2$-foam-2 (with hexylamine). P25 shows a surface area of 58.9 ± 1.8 m$^2 \cdot$g^{-1} whereas TiO$_2$-foam-1 (with hexylic acid) has a surface area of 51.7 ± 1.6 m$^2 \cdot$g^{-1} and TiO$_2$-foam-2 (with hexylamine) has a surface area of 58.1 ± 2.5 m$^2 \cdot$g^{-1}. Based on the experimental data, we recognize that the nitrogen sorption is not a suitable method to determine the pore size distribution of our foams. It is obvious that the TiO$_2$ foams have large pore size which this method could not determine.

Figure 3 shows the XRD pattern of the TiO$_2$ foam. This XRD pattern is, as expected, consistent with the anatase and rutile crystal phases of TiO$_2$ (JCPDS file number anatase: 00-0210-1272; rutile: 01-070-7347). The three characteristic sharp peaks at $2\theta = 25.2°$, $37.7°$, and $48.0°$ correspond to the (101), (044), and (200) planes of the anatase phase, respectively, and the peaks at $2\theta = 27.4°$, $54.3°$, and $36.0°$ fit to the (110), (211), and (101) planes of the rutile phase, respectively. The average particle sizes of the TiO$_2$ nanoparticles in the foam were calculated using the Scherrer equation yielding approx. 17 nm for anatase and 20 nm for rutile, which is in

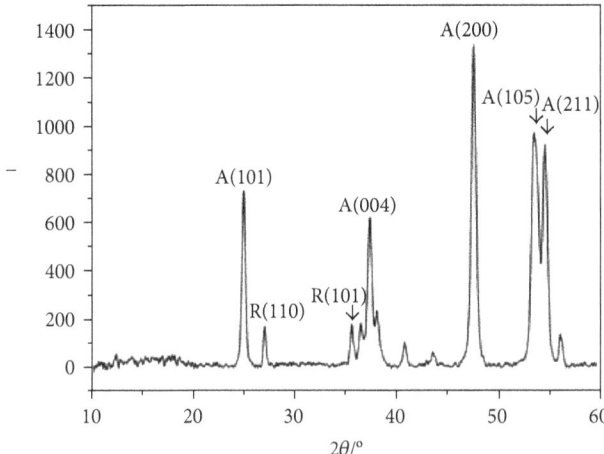

FIGURE 3: XRD pattern of the as-prepared TiO_2 foam. It is a mixture of anatase (A) and rutile (R) crystal phases.

agreement with the SEM observations. Supplementary 4 presents a comparison between the XRD pattern of TiO_2-foam-1 with original used P25 (Fig. S4) and the XRD pattern of all three samples (TiO_2-P25, TiO_2-foam-1, and TiO_2-foam-2; Fig. S5). All three samples have the same crystal structure, which is a mixture of anatase and rutile phases.

The surface composition and chemical states of the TiO_2-foam-1 and the starting TiO_2 powder (P25) were investigated by XPS. The survey spectra for both samples show exclusively the presence of C, O, and Ti (see Figure 4(a)). The Ti 2p photoelectron line (Figure 4(b)) consists of the Ti $2p_{3/2}$ and Ti $2p_{1/2}$ doublets at binding energies of 458.9 (459 eV) and 464.6 (464.7 eV), respectively. These values are closely related to the reported Ti 2p lines ranging 458.5–459.5 eV and 464.2–464.7 eV, respectively [31, 32].

The peak separation (~5.7 eV) and the strong satellite peak 14 eV above the main line indicate the Ti^4 oxidation state. The main O 1s line (Figure 4(c)) is located at 530.1 (530.2 eV) assigned to the lattice oxygen with additional broad peak centered at 531.1 (530.9 eV) related to low coordinated surface O atoms (with low electron density than the lattice O^{2-} ions), respectively (PCCP 2000, 2, 1319–1324). The Ti:O ratios were found to be 0.48 and 0.52 for the TiO_2 foam and the P25, respectively.

3.2. Photoactivity of TiO_2 Foams. The photoassisted degradation of NO in the gas phase was used to compare the photoactivity of the TiO_2 foams to that of the original powders (P25). The mechanism and kinetics of this reaction have been studied intensively before [9, 10]. It has been reported that using Degussa P25 as a catalyst, the determination of the reaction rate by the Langmuir–Hinshelwood model has some limitations. Wang et al. reported about the effect of variations of the inlet NO concentrations. Their observations showed that the reaction is first-order at low concentrations and zero-order at high concentrations [31, 32]. Studies conducted by Dillert et al. have shown that the NO oxidation at TiO_2 surfaces also depends on the light intensity. According to their analysis, the rates of the photocatalytic NO oxidation on UV(A)-irradiated TiO_2 samples at varying nitrogen(II)

oxide concentrations and photon fluxes and possibly at varying humidity can be predicted by a mathematical model [10].

Figure 5 shows the concentration changes of NO, NO_2, and NO_x as a function of UV illumination time during the photoassisted oxidation of NO with employing AERO-XIDE P25 and TiO_2 foam, respectively, under the same conditions. The typical experimental data for NO oxidation can be divided into four zones. Zone A is characterized by the adjustment of the NO concentration (1 ppm) before UV light illumination. Zone B is characterized by the adsorption of NO in the dark followed by the saturation of the surface of the photocatalyst sample. Once there is no further change in the NO concentration, the light source can be switched on to start the photocatalytic reaction (Zone C). After 2 h of irradiation with UV light, the light source is switched off (Zone D).

The photoassisted oxidation of NO is assumed to be a surface reaction mediated between NO and photogenerated hydroxyl radicals. And its mechanism has been already investigated. The three major steps during the photoassisted oxidation are oxidation of NO to HNO_2, oxidation of HNO_2 to NO_2 (in the transient state), and finally oxidation of NO_2 to HNO_3. If the catalyst is saturated with HNO_3, a steady state is attained and the oxidation reaction can go only as far as to NO_2 [2, 9, 10, 33, 34].

Considering the reaction dynamics during the entire photocatalytic process, the photocatalytic efficiency is determined by the number of photogenerated charge carriers which can avoid the recombination reaction. The photonic efficiency (ζ) of the NO degradation is calculated from the degradation rate of NO and the incident photon flux according to the following equation [33]:

$$\text{Photonic efficiency } (\zeta) = \frac{\text{degradation rate}}{\text{incident photon flux}}. \quad (1)$$

The incident photon flux (J_0) is calculated based on (2) with its value being $5 \times 10-8$ mol·s-1 (UV-A light intensity 1 mW·cm^{-2}, average illumination wavelength $\lambda = 350$ nm, irradiated surface area 18.49 cm^2, N_A Avogadro number, h Planck constant, and c light velocity).

$$J_0 = \frac{IA_r\lambda}{N_A hc}, \quad (2)$$

where I is the UV-A light intensity (1 mW·cm^{-2}), A_r is the illuminated surface area (18.49 cm^2), λ is the average illumination wavelength (350 nm), N_A denotes Avogadro number; h denotes Planck constant; and c denotes light velocity.

The degradation rate of NO is calculated using the following equation (where q_V is the volumetric flow rate):

$$\frac{\Delta n_{NO}}{\Delta t} = (c_{NO,in} - c_{NO,out})q_V \cdot \frac{P}{RT}. \quad (3)$$

The results are summarized and tabulated in Table 1. In this table, the results are presented as degraded concentration (C_{gas} in ppm), calculated degradation rate ($\Delta n_{NO}/\Delta t$ in mol·s^{-1}), and as photonic efficiency (ζ in %). The photonic efficiency (ζ %) for the complete degradation of the incoming

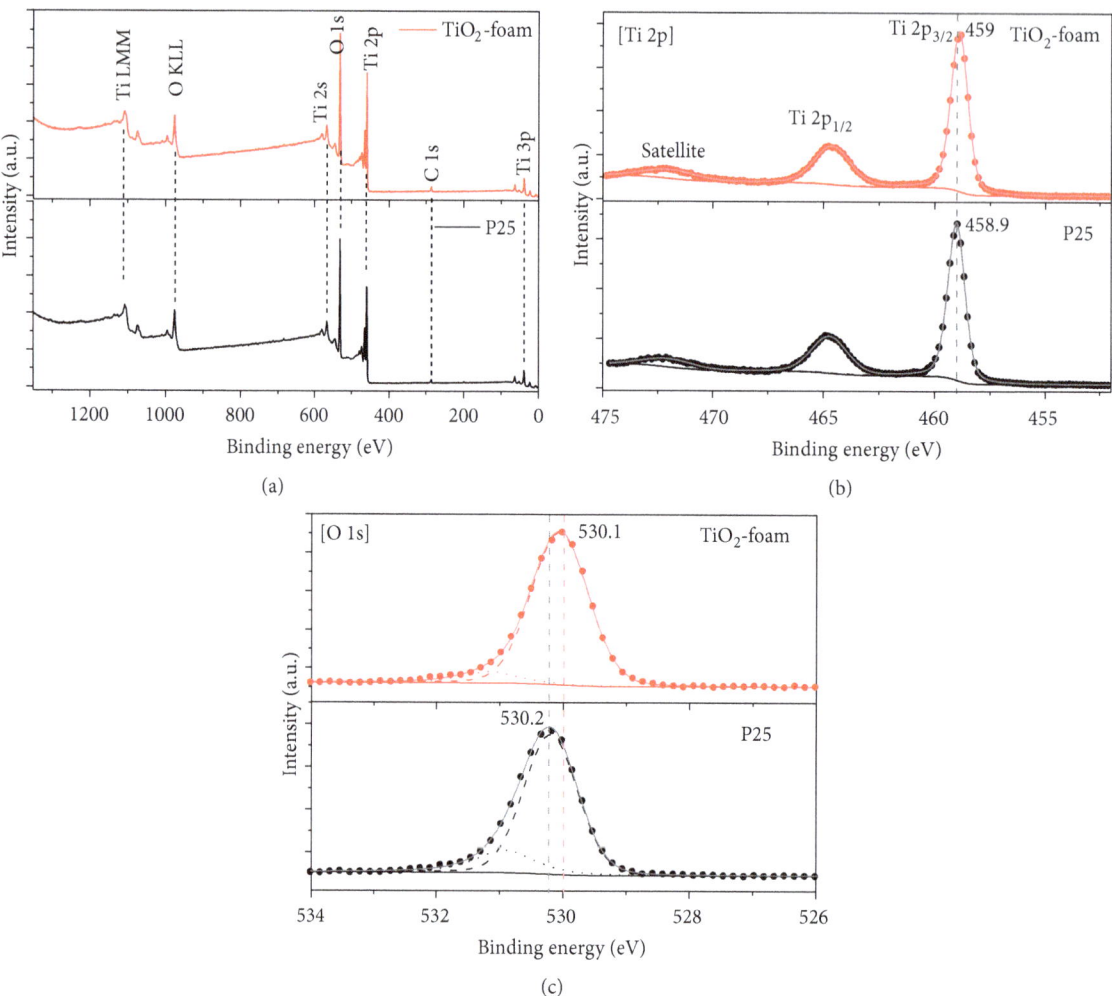

FIGURE 4: (a) XPS survey spectra of TiO$_2$-foam-1 and TiO$_2$-P25. (b) Ti 2p photoelectron line. (c) O 1s line.

NO gas under ISO 22197-1 conditions for a sample with a surface area of 18.49 cm^2 is approx. 4% (for a sample with a surface area of 50 cm^2, it is 1.43%).

From the results presented in Table 1, the following conclusions can be drawn: the TiO$_2$-foam-1 (using hexylic acid as amphiphilic molecule) adsorbs more NO than P25 powder and produces more NO$_2$ and NO$_x$ (HNO$_3$). This could be an indication of the activity of TiO$_2$ foams under the UV illumination. And also it shows that the foam structures are photocatalytically more active than the original powders.

Engel et al. [30] published an experimental method for the determination of the photocatalytic deposition velocity. To gain more information about our samples and their differences, we used this method. For this investigation, we prepared two different TiO$_2$ foams (using different amphiphilic molecule) and change the amount of the probe gas (NO) during the photoassisted oxidation. The following equation has been used to calculate photocatalytic deposition velocity:

$$\frac{\ln c_{NO,out}/c_{NO,in}}{c_{NO,out} - c_{NO,in}} = AkiKi \cdot \frac{A_r}{q_V(c_{NO,out} - c_{NO,out})} - Ki. \quad (4)$$

This equation is a linear equation of the two variables $\ln(c_{NO,out}/c_{NO,in})/(c_{NO,out} - c_{NO,in})$ and $A_r/(q_V(c_{NO,out} - c_{NO,out}))$ having the slope $AkiKi$ and the intercept with the ordinate K_i. With the calculation of the slope, the photocatalytic deposition velocity will be determined.

Figure 6 presents the change of the NO concentration at the reactor outlet during irradiation of AEROXIDE P25, TiO$_2$-foam-1(with hexylic acid), and TiO$_2$-foam-2 (with hexylamine) with varying NO inlet concentrations. All other experimental parameters, that is, temperature, relative humidity, and UV (A) photon flux, were kept constant during these experimental runs. With the value of illumination surface area A_r (3.96·10^{-3} m^2), the volumetric flow rate q_V (3.0 L min^{-1}), and the measured values of the NO inlet and outlet concentration, the values of the two variables of (4) have been calculated and plotted for each individual experimental run. These plots clearly show that the TiO$_2$-foam-2 (with hexylamine) has the more active surface for the NO oxidation and both foams are photocatalytically more active than the original powders.

The photoactivity depends on various factors such as the employed catalyst, the light collection efficiency, and the rate

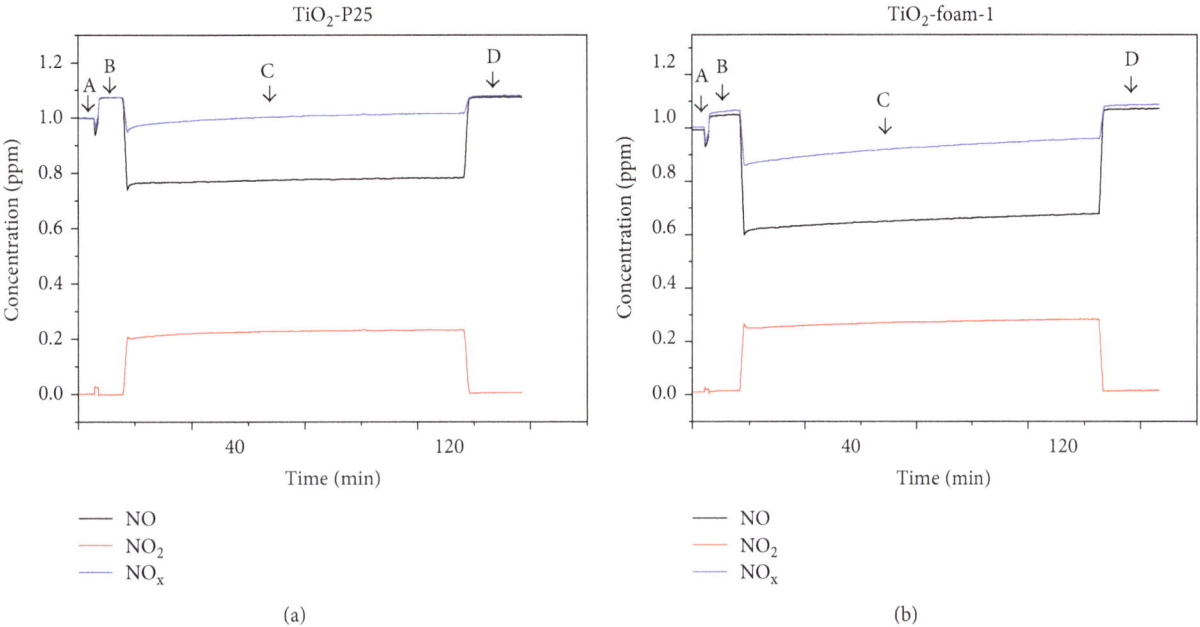

FIGURE 5: Concentration changes of NO, NO$_2$, and NO$_x$ as a function of UV illumination time in a typical experimental run for a AEROXIDE P25 (a) and TiO$_2$-foam-1 (with hexylic acid) sample (b).

TABLE 1: Concentration changes and calculated photonic efficiencies (ζ%).

Sample	ΔC_{NO} (ppm)	ΔC_{NO_x} (ppm)	ΔC_{NO2} (ppm)	$\Delta n_{NO}/\Delta t$ (10^{-10} mol·S^{-1})	ζ%
TiO$_2$ P25	0.3	0.1	0.2	6.26	1.4
TiO2-foam-1*	0.5	0.2	0.3	10.4	1.9

*Hexylic acid as amphiphilic molecule.

of molecular diffusion [7, 33, 35]. The activities of catalysts with the same chemical compositions (P25) have been compared (Table 1 and Figure 5). The light intensity for all experiments was the same although the light collection by the catalyst could be influenced by its structure. The interesting point from our results is that due to the macroporous/mesoporous structure of the foam, the photocatalyst did not own large surface area. The N$_2$ sorption results could not show big difference between the surface area of P25 and that of the foams. Although there is no big difference between the surface areas, the foam structures show more photoactivity.

TiO$_2$ foam shows high photoactivity due to the structure of macroporous/mesoporous channels that provide fast intraparticle molecular transfers, which improves the light harvesting and the adsorption of reactant molecules. This macroporous/mesoporous structure also provides proper interfaces for a facile interparticle charge transfer while the reactants can freely diffuse through the pores. At the same time, the light absorption and the energy transfer through the foam structure into the inner surface of the macroporous/mesoporous TiO$_2$ are apparently as high as expected.

It has been shown that mass transfer and surface reactions control heterogeneous catalytic reactions [10, 36]. Yang et al. [35] reported that the mass transfer coefficient is affected by the flow velocity. Beyond a certain flow velocity, the photocatalytic reaction process was founded to be independent of the mass transfer and was only controlled by the surface reactions. In this work, the value of the flow velocity is rather high and identical for all reactions. Therefore, the photocatalytic process is apparently independent of any mass transfer limitations and will only be influenced by surface properties.

The surface properties of the photocatalyst play an important role as heterogeneous photocatalytic reactions usually take place at the interface of solid/liquid or solid/gas phases. The large surface of the foam structure is apparently very effective due to the fact that the photoassisted oxidation of NO in the gas phase requires the adsorption of the reactants at the TiO$_2$ surface. The amount of adsorbed substances increases in the TiO$_2$ foam and thus enhancing the photocatalytic process.

4. Conclusions

Titanium dioxide foam was prepared by a simple mechanical stirring method. The obtained TiO$_2$ foam has been characterized using X-ray powder diffraction and SEM. The crystal structure of the TiO$_2$ foam remains unchanged as compared with the initial TiO$_2$ particles. The SEM investigations showed very rough and porous surface structures of the TiO$_2$ foam. These macroporous/mesoporous foams were used to measure the photoassisted degradation of NO according to ISO 22197-1. The photocatalytic properties of the TiO$_2$ foam, compared to those

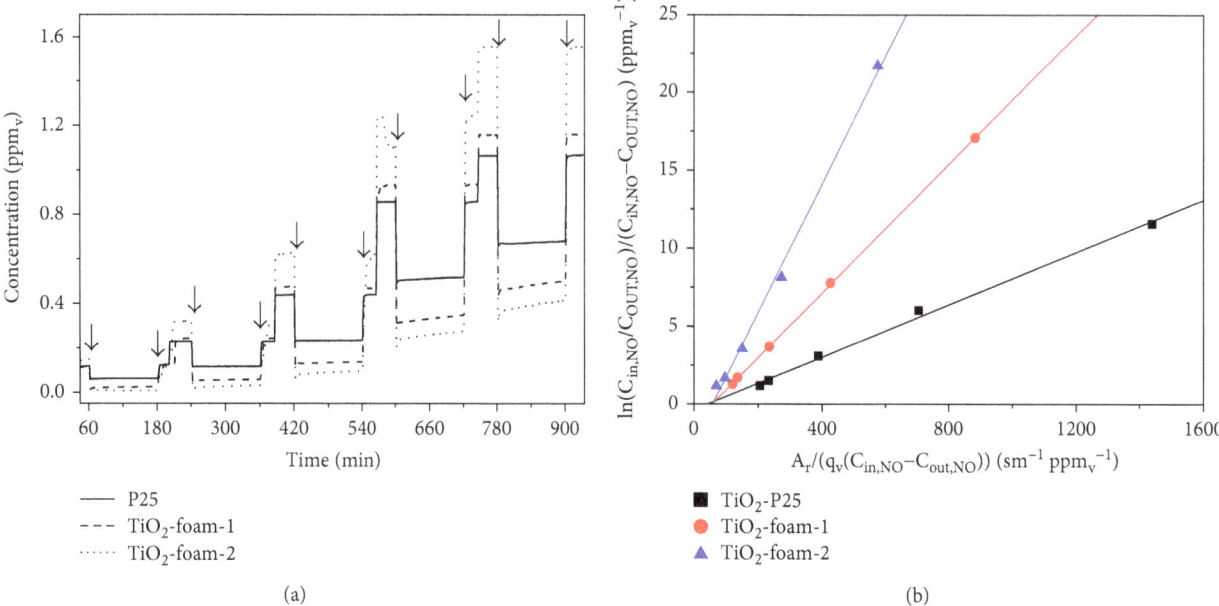

(a)   (b)

FIGURE 6: (a) Concentration changes of NO at the reactor outlet observed during three typical experimental run for a AEROXIDE P25, TiO$_2$-foam-1 (with hexylic acid), and TiO$_2$-foam-2 (with hexylamine) as a function of UV illumination. The arrows indicate switching on and off the radiation source. (b) The data analysis according to (4).

of the original photocatalyst powder, improved significantly and increased by about 135%. The photocatalytic properties of TiO$_2$ materials do not only depend on their chemical composition but also on their geometrical microstructure. The porous structure is an attractive way to achieve high photocatalytic activities.

Conflicts of Interest

The authors declare that they have no conflicts of interest.

Acknowledgments

The authors thank F. Steinbach (PCI) Leibniz Universität Hannover for his support during the SEM measurements and M. Jahns and A. Mohmeyer (ACI) Leibniz Universität Hannover for their support during the N$_2$ sorption measurements. The authors also thank Dr. D. H. Taffa from Carl von Ossietzky Universität Oldenburg for the XPS measurements. Maryam Jami would like to thank Hannover School for Nanotechnology for her doctoral scholarship.

Supplementary Materials

Supplementary 1. Fig. S1: (a) SEM photographs of TiO$_2$-P25 (Evonik AEROXIDE P25) with different magnifications. (b) SEM photographs of TiO$_2$-foam-1 (hexylic acid as amphiphilic molecule) with different magnifications. (c) SEM photographs of TiO$_2$-foam-2 (hexylamine as amphiphilic molecule) with different magnifications. Supplementary 2. Fig. S2: comparison between the diffuse reflectance spectra of TiO$_2$-P25 and prepared TiO$_2$ foam. Inset shows the modified Kubelka-Munk function versus the photon energy of these two samples. Supplementary 3. Fig. S3: left: nitrogen

adsorption-desorption isotherm of TiO$_2$-foam-1 (hexylic acid as amphiphilic molecule), TiO$_2$-foam-2 (hexylamine as amphiphilic molecule), and TiO$_2$-P25; right: the BJH graphs indicated corresponding pore size distributions of samples. Supplementary 4. Fig. S4: comparison between the XRD pattern of TiO$_2$ foam (with hexylic acid) and commercial used TiO$_2$ (Evonik AEROXIDE P25). Fig. S5: XRD patterns of TiO$_2$-P25, TiO$_2$-foam-1, and TiO$_2$-foam-2, the respective reflection peaks of anatase (A) and rutile (R) phases are labeled with their Miller indices. (*Supplementary Materials*)

References

[1] M. R. Hoffmann, S. T. Martin, W. Choi, and D. W. Bahnemann, "Environmental applications of semiconductor photocatalysis," *Chemical Reviews*, vol. 95, no. 1, pp. 69–96, 1995.

[2] X. Chen and S. S. Mao, "Titanium dioxide nanomaterials: synthesis, properties, modifications, and applications," *Chemical Reviews*, vol. 107, no. 7, pp. 2891–2959, 2007.

[3] A. O. Ibhadon, G. M. Greenway, Y. Yue, P. Falaras, and D. Tsoukleris, "The photocatalytic activity of TiO$_2$ foam and surface modified binary oxide titania nanoparticles," *Journal of Photochemistry and Photobiology A: Chemistry*, vol. 197, no. 2-3, pp. 321–328, 2008.

[4] L. Chen, C. Huang, G. Xu, S. L. Hutton, and L. Miao, "Macroporous TiO$_2$ foam with mesoporous walls," *Materials Characterization*, vol. 75, pp. 8–12, 2013.

[5] K. Skalska, J. S. Miller, and S. Ledakowicz, "Trends in NO$_x$ abatement: a review," *Science of The Total Environment*, vol. 408, no. 19, pp. 3976–3989, 2010.

[6] Z. Wu, Z. Sheng, Y. Liu, H. Wang, N. Tang, and J. Wang, "Characterization and activity of Pd-modified TiO$_2$ catalysts

for photocatalytic oxidation of NO in gas phase," *Journal of Hazardous Materials*, vol. 164, no. 2-3, pp. 542–548, 2009.

[7] Y. Ohko, Y. Nakamura, A. Fukuda, S. Matsuzawa, and K. Takeuchi, "Photocatalytic oxidation of nitrogen dioxide with TiO_2 thin films under continuous UV-light illumination," *The Journal of Physical Chemistry C*, vol. 112, no. 28, pp. 10502–10508, 2008.

[8] T. Ibusuki and K. Takeuchi, "Removal of low concentration nitrogen oxides through photoassisted heterogeneous catalysis," *Journal of Molecular Catalysis*, vol. 88, no. 1, pp. 93–102, 1994.

[9] R. Dillert, J. Stötzner, A. Engel, and D. W. Bahnemann, "Influence of inlet concentration and light intensity on the photocatalytic oxidation of nitrogen(II) oxide at the surface of Aeroxide® TiO_2 P25," *Journal of Hazardous Materials*, vol. 211-212, pp. 240–246, 2012.

[10] R. Dillert, A. Engel, J. Große, P. Lindner, and D. W. Bahnemann, "Light intensity dependence of the kinetics of the photocatalytic oxidation of nitrogen(II) oxide at the surface of TiO_2," *Physical Chemistry Chemical Physics*, vol. 15, no. 48, pp. 20876–20886, 2013.

[11] V. Kalousek, J. Tschirch, D. Bahnemann, and J. Rathouský, "Mesoporous layers of TiO_2 as highly efficient photocatalysts for the purification of air," *Superlattices and Microstructures*, vol. 44, no. 4-5, pp. 506–513, 2008.

[12] U. Diebold, "The surface science of titanium dioxide," *Surface Science Reports*, vol. 48, no. 5-8, pp. 53–229, 2003.

[13] T. A. Kandiel, R. Dillert, and D. W. Bahnemann, "Enhanced photocatalytic production of molecular hydrogen on TiO_2 modified with Pt–polypyrrole nanocomposites," *Photochemical & Photobiological Sciences*, vol. 8, no. 5, pp. 683–690, 2009.

[14] S. Sakthivel, M. C. Hidalgo, D. W. Bahnemann, S.-U. Geissen, V. Murugesan, and a. Vogelpohl, "A fine route to tune the photocatalytic activity of TiO_2," *Applied Catalysis B: Environmental*, vol. 63, no. 1-2, pp. 31–40, 2006.

[15] A. L. Linsebigler, A. L. Linsebigler, J. T. Yates Jr., G. Lu, G. Lu, and J. T. Yates, "Photocatalysis on TiO2 surfaces: principles, mechanisms, and selected results," *Chemical Reviews*, vol. 95, no. 3, pp. 735–758, 1995.

[16] Y. Wang, J. Li, P. Peng, T. Lu, and L. Wang, "Preparation of S-TiO_2 photocatalyst and photodegradation of L-acid under visible light," *Applied Surface Science*, vol. 254, no. 16, pp. 5276–5280, 2008.

[17] T. Ohno, M. Akiyoshi, T. Umebayashi, K. Asai, T. Mitsui, and M. Matsumura, "Preparation of S-doped TiO_2 photocatalysts and their photocatalytic activities under visible light," *Applied Catalysis A: General*, vol. 265, no. 1, pp. 115–121, 2004.

[18] S. U. M. Khan, M. Al-Shahry, and W. B. Ingler, "Efficient photochemical water splitting by a chemically modified n-TiO_2," *Science*, vol. 297, no. 5590, pp. 2243–2245, 2002.

[19] R. Asahi, T. Morikawa, T. Ohwaki, K. Aoki, and Y. Taga, "Visible-light photocatalysis in nitrogen-doped titanium oxides," *Science*, vol. 293, no. 5528, pp. 269–271, 2001.

[20] H. Kisch, L. Zang, C. Lange, W. F. Maier, C. Antonius, and D. Meissner, "Modified, amorphous titania—a hybrid semiconductor for detoxification and current generation by visible light," *Angewandte Chemie International Edition*, vol. 37, no. 21, pp. 3034–3036, 1998.

[21] A. A. Ismail, A. Hakki, and D. W. Bahnemann, "Mesostructure Au/TiO_2 nanocomposites for highly efficient catalytic

[22] A. A. Ismail, D. W. Bahnemann, and S. A. Al-Sayari, "Synthesis and photocatalytic properties of nanocrystalline Au, Pd and Pt photodeposited onto mesoporous RuO_2-TiO_2 nanocomposites," *Applied Catalysis A: General*, vol. 431-432, pp. 62–68, 2012.

[23] J. Tschirch, D. Bahnemann, M. Wark, and J. Rathouský, "A comparative study into the photocatalytic properties of thin mesoporous layers of TiO_2 with controlled mesoporosity," *Journal of Photochemistry and Photobiology A: Chemistry*, vol. 194, no. 2-3, pp. 181–188, 2008.

[24] Y. Zhao, X. Zhang, J. Zhai et al., "Ultrastable TiO_2 foams derived macro-/meso-porous material and its photocatalytic activity," *Microporous and Mesoporous Materials*, vol. 116, no. 1-3, pp. 710–714, 2008.

[25] A. A. Ismail, D. W. Bahnemann, J. Rathousky, V. Yarovyi, and M. Wark, "Multilayered ordered mesoporous platinum/titania composite films: does the photocatalytic activity benefit from the film thickness?," *Journal of Materials Chemistry*, vol. 21, no. 21, pp. 7802–7810, 2011.

[26] T. A. Kandiel, A. A. Ismail, and D. W. Bahnemann, "Mesoporous TiO_2 nanostructures: a route to minimize Pt loading on titania photocatalysts for hydrogen production," *Physical Chemistry Chemical Physics*, vol. 13, no. 45, pp. 20155–20161, 2011.

[27] A. A. Ismail and D. W. Bahnemann, "One-step synthesis of mesoporous platinum/titania nanocomposites as photocatalyst with enhanced photocatalytic activity for methanol oxidation," *Green Chemistry*, vol. 13, no. 2, pp. 428–435, 2011.

[28] J. I. Langford and A. J. C. Wilson, "Scherrer after sixty years: a survey and some new results in the determination of crystallite size," *Journal of Applied Crystallography*, vol. 11, no. 2, pp. 102–113, 1978.

[29] ISO 22197-1, *Fine Ceramics (advanced ceramics, advanced technical ceramics) – Test Method for Air-Purification Performance of Semiconducting Photocatalytic Materials*, Part 1. Removal of Nitric Oxide, Draft International Organization for Standardization, 2007, https://www.iso.org/obp/ui/#iso:std:iso:22197:-1:ed-2:v1:en.

[30] A. Engel, A. Glyk, A. Hülsewig, J. Große, R. Dillert, and D. W. Bahnemann, "Determination of the photocatalytic deposition velocity," *Chemical Engineering Journal*, vol. 261, pp. 88–94, 2015.

[31] M. C. Biesinger, B. P. Payne, A. P. Grosvenor, L. W. M. Lau, A. R. Gerson, and R. S. C. Smart, "Resolving surface chemical states in XPS analysis of first row transition metals, oxides and hydroxides: Cr, Mn, Fe, Co and Ni," *Applied Surface Science*, vol. 257, no. 7, pp. 2717–2730, 2011.

[32] G. Ketteler, S. Yamamoto, H. Bluhm et al., "The nature of water nucleation sites on TiO_2(110) surfaces revealed by ambient pressure X-ray photoelectron spectroscopy," *The Journal of Physical Chemistry C*, vol. 111, no. 23, pp. 8278–8282, 2007.

[33] S. Devahasdin, C. Fan, K. Li, and D. H. Chen, "TiO_2 photocatalytic oxidation of nitric oxide: transient behavior and reaction kinetics," *Journal of Photochemistry and Photobiology A: Chemistry*, vol. 156, pp. 161–170, 2003.

[34] H. Wang, Z. Wu, W. Zhao, and B. Guan, "Photocatalytic oxidation of nitrogen oxides using TiO_2 loading on woven glass fabric," *Chemosphere*, vol. 66, no. 1, pp. 185–190, 2007.

[35] L. Yang, A. Cai, C. Luo, Z. Liu, W. Shangguan, and T. Xi, "Performance analysis of a novel TiO_2-coated foam-nickel PCO air purifier in HVAC systems," *Separation and Purification Technology*, vol. 68, no. 2, pp. 232–237, 2009.

[36] C. A. Bignozzi, "Topics in current chemistry," in *Photocatalysis*, Springer-Verlag, Berlin, Heidelberg, 2011.

11

Modelling and Simulation of the Radiant Field in an Annular Heterogeneous Photoreactor Using a Four-Flux Model

O. Alvarado-Rolon,[1] R. Natividad ⓘ,[2] R. Romero ⓘ,[2] L. Hurtado ⓘ,[2] and A. Ramírez-Serrano ⓘ[1]

[1]Facultad de Química, Universidad Autónoma del Estado de México, Paseo Colon esq. Paseo Tollocan s/n, 50120 Toluca, MEX, Mexico
[2]Facultad de Química, Centro Conjunto de Investigación en Química Sustentable UAEM-UNAM, Universidad Autónoma del Estado de México, Carretera Toluca–Atlacomulco, Km 14.5, Unidad San Cayetano, 50200 Toluca, MEX, Mexico

Correspondence should be addressed to R. Natividad; reynanr@gmail.com and A. Ramírez-Serrano; aramirezs@uaemex.mx

Academic Editor: Detlef W. Bahnemann

This work focuses on modeling and simulating the absorption and scattering of radiation in a photocatalytic annular reactor. To achieve so, a model based on four fluxes (FFM) of radiation in cylindrical coordinates to describe the radiant field is assessed. This model allows calculating the local volumetric rate energy absorption (LVREA) profiles when the reaction space of the reactors is not a thin film. The obtained results were compared to radiation experimental data from other authors and with the results obtained by discrete ordinate method (DOM) carried out with the Heat Transfer Module of Comsol Multiphysics® 4.4. The FFM showed a good agreement with the results of Monte Carlo method (MC) and the six-flux model (SFM). Through this model, the LVREA is obtained, which is an important parameter to establish the reaction rate equation. In this study, the photocatalytic oxidation of benzyl alcohol to benzaldehyde was carried out, and the kinetic equation for this process was obtained. To perform the simulation, the commercial software COMSOL Multiphysics v. 4.4 was employed.

1. Introduction

In the last decades, photocatalytic processes have been the subject of different studies such as wastewater treatment [1–6], air purification in polluted environments with volatile organic compounds [7–9], and synthesis of fine organic compounds such as benzaldehyde [10, 11]. According to literature [1, 3–6, 11–16], the following different variables are crucial in a photocatalytical process efficiency: (a) catalyst type and concentration, (b) reagent type and concentration, (c) geometry and type of reactor, and (d) characteristics of the radiation inside the photoreactor. Because of the number of variables and the interaction among them, the modeling of this type of processes is expected to be rather useful not only for reactor design but also to achieve a better insight and understanding of the process.

The mathematical modeling and simulation of a photocatalytic reactor imply a great challenge due to the numerous involved variables; however, the computational analysis of these variables aids to accomplish such a task. Furthermore, the computational analysis allows evaluating hydrodynamic effects and kinetics without employing physical prototypes. The full modeling of photocatalytic reactors requires to include several submodels to simulate the physical phenomena occurring inside the reactor. Some of these necessary submodels are (a) radiation emission and incidence, (b) radiation absorption and scattering, (c) photoconversion kinetics, and (d) hydrodynamics [5, 13–15, 17, 18]. These are the result of mass, energy, and momentum balances, as well as radiation distribution and optical characterization of reaction space [6, 16, 19, 20]. These submodels are strongly interlinked. For example, the kinetics is a function of radiation absorption, which is in turn a function of catalyst characteristics and hydrodynamics. The conversion and performance of a photocatalytic reaction are a function of the local volumetric rate energy absorption (LVREA), which is defined as the energy due to photons absorbed per time and volume inside the photoreactor [21]. To evaluate the LVREA is necessary to solve the radiation transfer equation (RTE) [22–25].

$$\frac{dI_\lambda(x,\Omega)}{dx} = -\beta_\lambda I_\lambda(x,\Omega) + \frac{\sigma_\lambda}{2}\int_{4\pi} I_\lambda(x,\Omega)p\left(\Omega \to \Omega'\right)d\Omega,$$

(1)

where $I_\lambda(x,\Omega)$ is the spectral radiation intensity, λ represents the wavelength, β_λ is the extinction coefficient, which is the sum of the absorption coefficient, κ_λ, and σ_λ is the scattering coefficient. The ratio $\omega = \sigma_\lambda/\beta_\lambda$ is the scattering albedo coefficient which is inherent to each photocatalyst since it represents its photon absorption capacity. Ω is the solid angle, and $p(\Omega \to \Omega')$ is the phase function representing the redistribution of radiation after the scattering event. According to the first term in the right side of (1), the intensity is diminished by the effect of mainly two phenomena, scattering and absorption. This decrease is characterized by the extinction coefficient. There is also an increase in the intensity due to the scattering from other directions, and it is represented by the second term in the right-hand side of (1) [24, 26, 27].

The analytical solution of the RTE is a rather complex task, unless it is limited to simple reactor geometries with specific assumptions. Even when using specialized software, the radiation field simulation is a task that requires a high computational effort. Comsol Multiphysics v. 4.4 contains the physics of radiation in participating media (rpm), in the Heat Transfer Module, which is designed to solve 3D radiation transfer problems, taking into account the phenomena of emission, dispersion, and absorption of radiation. The Comsol Multiphysics v. 4.4 Heat Transfer Module employs the discrete ordinate method (DOM). This method consists the transformation of the integral-differential RTE into a system of algebraic equations to describe the transport of photons in such way that can be solved following the direction of propagation, starting from the values provided by the boundary conditions. However, RTE is solved by discretizing the solid angle at every discrete position in the 3D domain, which is computationally very demanding and may result in unrealistic results when the discretization of the solid angle is not refined enough.

A viable alternative is to employ numerical computational methods as the statistical method Monte Carlo (MC), which is known as highly accurate but requires a great computational effort [21, 28, 29]. Also, it is possible to employ analytical simplified methods like the two-flux model (TFM) and the six-flux model (SFM). These models consist of several algebraic equations developed for flat slab geometries [15–17, 30], which were obtained by solving a system of differential equations with specific boundary conditions, for example, the outer wall of the reactor is opaque. SFM is very accurate for cylindrical geometries [14] in which the space where the reaction occurs, δ, is much smaller than the radius of the reactor, R_R.

$$\delta \ll R_R,$$
$$\frac{R_R}{R_R + \delta} \sim 1.$$

(2)

However, in this investigation, a reactor in which the lamp is immersed in the reaction medium was used, so (2) is not satisfied. The geometry used in this work is shown in Figure 1. This paper aims to evaluate the effectiveness of a modified model based on four flux of radiation (FFM), whose equations are based on a cylindrical geometry, to mathematically represent the radiation field in a stirred annular photoreactor. This model is coupled to a reaction rate model representing the benzyl alcohol oxidation. The FFM evaluates the incident radiation in each point of the reaction space. This model considers that the incident radiation is the sum of radiation fluxes traveling from the light source towards this point and the fluxes due from both axial and radial scattering. As this model is developed from cylindrical geometries, its solution is expected to better represent the radiant field inside an annular photocatalytic reactor than the models developed from slab plane geometries where the reaction space is only a thin film.

The main objective of this work was to validate the proposed four-flux model, which is specifically designed for annular photocatalytic reactors with a relationship, that is, the reactor is not thin-walled. FFM is tested against the results with experimental data of the photocatalytic and selective oxidation of benzyl alcohol towards benzaldehyde. Moreover, the radiation profiles were compared to those calculated by MC, DOM, and SFM. The FFM and DOM were carried out with commercial software Comsol Multiphysics 4.4, which is a powerful differential equation solver.

2. Methodology

The main objective of this work was to test a proposed FFM to efficiently represent the radiant field inside an annular reactor when the reaction space is not a thin film. In order to validate the proposed model, the profiles obtained with FFM were compared to those previously reported in the literature. Also, the FFM was applied to describe the radiant field in a batch annular photoreactor employed to experimentally obtain benzyl alcohol oxidation data. Then, the kinetics of this reaction was established as function of LVREA.

2.1. Source Data

2.1.1. System 1. The profiles obtained in a thin-film slurry reactor of inner wall (TFSIW) reported by Li Puma et al. [5, 14, 17] and obtained by the six flow model were replicated for comparison purposes. In this case, the relation $R_R/(R_R + \delta) = 0.76$. The characteristics of the system are summarized in Table 1.

2.1.2. System 2. This photoreactor was previously reported [28, 29] and was named Photo-CREC Water II and employs TiO_2 (anatase) as a catalyst. In such a reaction system, the lamp is annulus centered. The relationship

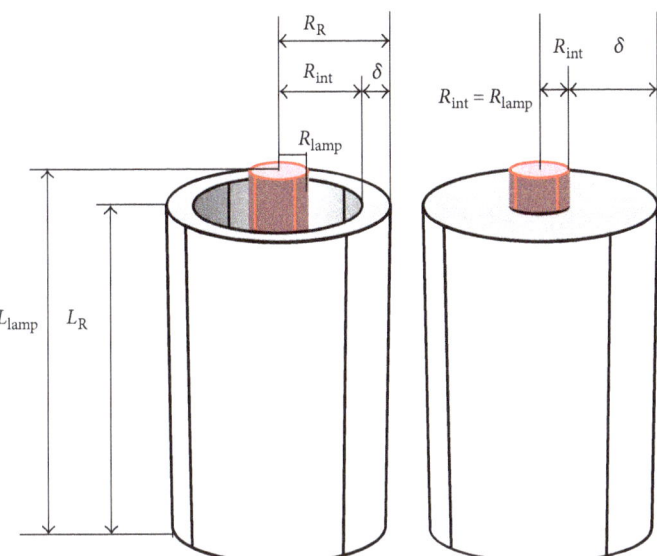

FIGURE 1: Schematic representation of the geometry of the assessed annular photocatalytic reactor.

TABLE 1: Characteristics of systems.

	Catalyst	Lamp characteristics	Reactor characteristics
System 1 Li Puma [14]	TiO$_2$ DP 25 (i) $\sigma_\lambda = 1.02\ C_{cat}$ (1/m) (ii) $\kappa_\lambda = 0.338\ C_{cat}$ (1/m)	Power: 4 W Wavelength: 300 nm Radius: 0.00775 m Length: 0.213 m	Length: 0.225 m Ext. radius: 0.019 m Int. radius: 0.013 m $R_R/(R_R + \delta) = 0.76$
System 2 Moreira et al. [28, 29]	TiO$_2$ anatase (i) $\sigma_\lambda = 3.1149\ C_{cat}$ (1/m) (ii) $\kappa_\lambda = 0.3957\ C_{cat}$ (1/m)	Power: 8 W Wavelength: 250 nm Radius: 0.0133 m Length: 0.413 m	Length: 0.445 m Ext. radius: 0.0444 m Int. radius: 0.01755 m $R_R/(R_R + \delta) = 0.62315$
System 3	LiVMoO$_6$ (i) $\sigma_\lambda = 0.24128\ C_{cat}$ (1/m) (ii) $\kappa_\lambda = 0.03092\ C_{cat}$ (1/m)	Power: 8 W Wavelength: 254 nm Radius: 0.005 m Length: 0.23 m	Length: 0.25 m Ext. radius: 0.025 m Int. radius: none $R_R/(R_R + \delta) = 0.5555$

$R_R/(R_R + \delta) = 0.62315$. The characteristics of the system are also summarized in Table 1.

2.1.3. System 3. Once the radiation model was validated with data reported in systems 1 and 2, this model was applied to simulate the radiation field in system 3 during benzyl alcohol selective oxidation towards benzaldehyde. Experimental data of benzyl alcohol oxidation were obtained in an annular cylindrical photocatalytic reactor. It is worth pointing out that in this reaction system, the lamp was placed at the center of the reactor without any additional physical protection (e.g., quartz sleeve). For this reason, the relationship $R_R/(R_R + \delta) = 0.5555$.

The employed catalyst was LiVMoO$_6$, and a detailed characterization has been previously reported [31]. The characteristics of the system are shown in Table 1. The FFM was used to describe the radiant field in this reactor and to obtain the kinetics of benzyl alcohol oxidation as a function of LVREA.

2.2. Mathematical Modeling of Radiation Emission. The emission of radiation from the cylindrical lamp is modeled using the linear source spherical emission (LSSE). This model considers that the lamp is a linear source, and each point on the line emits radiation isotropically and in every direction. It is assumed that the radiation emitted by each point of the lamp is constant along the axial length of the lamp [5]. According to the literature, the intensity of the incident radiation entering the inner wall of the annulus can be calculated as

$$I_{R_{int},z} = \frac{S_1}{4\pi R_{int}} \left[a \tan\left(\frac{2z - L_R + L_{lamp}}{2R_{int}}\right) - a \tan\left(\frac{2z - L_R - L_{lamp}}{2R_{int}}\right) \right], \tag{3}$$

where

$$S_1 = 2\pi R_{lamp} I_w. \tag{4}$$

The experimental emitted radiation was measured by a UVX radiometer equipped with a sensor of 254 nm placed at the lamp wall and 0.01 m from the lamp.

2.3. Mathematical Modeling of Absorption and Scattering Radiation. To establish the mathematical FFM, the following assumptions were made: (a) reactor with slurry catalyst, (b) heterogeneous model, (c) isothermal process, (d) perfect mixing and therefore the catalyst concentration is homogeneous at all reaction space, (e) photons are absorbed only by catalyst particles, (f) the flux of photons occurs only in four directions, two radial, and two axial directions, (g) the emission of photons by the lamp is isocratic, (h) oxygen bubbles do not affect the radiation fluxes, and (i) the scattering of photons by the catalyst is isotropic.

FFM was employed to evaluate the incident radiation on a given point inside the reaction space. In this model, the total radiation flux is taken as the sum of the flux of photons traveling from the light source towards that point and flux of photons from scattering in both two axial directions and two both radial directions. In concordance, a photon balance was performed in a differential volume element shell shaped in cylindrical coordinates (Figure 2).

The flux of incident radiation (g_f), the flux entering the differential element due to backscattering (g_b), and the fluxes entering from bottom and upper walls $(g_a$ and $g_c)$ are the four fluxes that this model accounts for. The parameters p_f, p_b, p_a, and p_c represent the probabilities of occurring backscattering in the corresponding directions. These parameters were calculated by MC method employing an isotropic phase function, and their values are $p_f = 0.405$, $p_b = 0.303$, $p_a = 0.146$, and $p_c = 0.146$ for LiVMoO$_6$ and $p_f = 0.357$, $p_b = 0.351$, $p_a = 0.146$, and $p_c = 0.146$ for TiO$_2$. The number and external area of the catalytic particles are n_p and a_p, respectively, so in order to establish that the FFM is necessary to perform a balance of incident radiation (g_f), in the four considered directions. For example, the following radiation balance in the radial direction can be written as

$$\{\text{Input photons}\} - \{\text{output photons}\} = \{\text{absorbed photons}\}. \tag{5}$$

So the balance is

$$\begin{aligned}
&g_f(2\pi r\Delta z)|_r - g_f(2\pi r\Delta z)|_{r+\Delta r} + g_b\omega(2\pi r\Delta r\Delta z)(n_p a_p p_b) \\
&\quad + g_a\omega(2\pi r\Delta r\Delta z)(n_p a_p p_a) + g_c\omega(2\pi r\Delta r\Delta z) \\
&\quad \cdot (n_p a_p p_c) - g_f(n_p a_p(\omega p_a + \omega p_b + \omega p_c)) \\
&\quad \cdot (2\pi r\Delta r\Delta z) \\
&= g_f(1-\omega)(n_p a_p)(2\pi r\Delta r\Delta z).
\end{aligned} \tag{6}$$

By reordering and applying $\lim_{\Delta r \to 0}$,

$$\frac{d(rg_f)}{dr} = \left(\frac{r}{\beta \cdot C_{\text{cat}}}\right)(\omega(g_b p_b + g_a p_a + g_c p_c) - g_f(1-\omega p_f)), \tag{7}$$

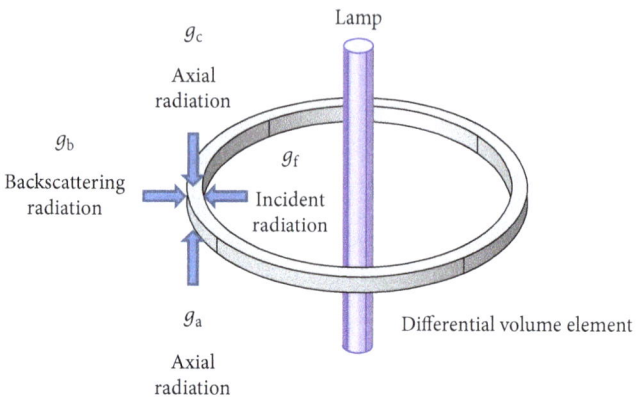

FIGURE 2: Directions of the fluxes of photons in the four-flux model.

where

$$\omega p_b + \omega p_a + \omega p_c + \omega p_f + (1-\omega) = 1, \tag{8}$$

$$\left(\frac{1}{(1/n_p a_p)}\right) = \left(\frac{1}{\beta C_{\text{cat}}}\right). \tag{9}$$

The term $(1/(1/n_p a_p))$ is the extinction characteristic length. It has been suggested [5] that the extinction characteristic length can be replaced by the inverse of the extinction volumetric coefficient $(1/\beta C_{\text{cat}})$. Physically, this represents the mean free path of the photons in the slurry. Doing a similar balance in the backscattering directions, the following equations are obtained:

$$\frac{d(rg_b)}{dr} = \left(\frac{r}{\beta \cdot C_{\text{cat}}}\right)(\omega(g_f p_b + g_a p_c = g_c p_a) - g_b(1-\omega p_f)), \tag{10}$$

$$\frac{d(rg_a)}{dz} = \left(\frac{r}{\beta C_{\text{cat}}}\right)(\omega(g_f p_c + g_b p_a + g_c p_b) - g_a(1-\omega p_f)), \tag{11}$$

$$\frac{d(rg_c)}{dz} = \left(\frac{r}{\beta \cdot C_{\text{cat}}}\right)(\omega(g_f p_a + g_a p_b + g_b p_c) - g_c(1-\omega p_f)). \tag{12}$$

Equations (7), (10), (11), and (12) are simultaneously solved by applying the following boundary conditions.

(1) BC 1: at wall lamp or inner wall:

$$\begin{aligned}
g_f(r = R_{\text{lamp}}) &= I_{r,z}A_{\text{lamp}}p_f \\
g_b(r = R_{\text{lamp}}) &= (g_f + g_a + g_c)p_b \\
g_a(r = R_{\text{lamp}}) &= (g_f + g_b + g_c)p_a \\
g_c(r = R_{\text{lamp}}) &= (g_f + g_a + g_b)p_c.
\end{aligned} \tag{13a}$$

(2) BC 2: at external reactor wall (opaque wall):

$$g_f(r = R_R) = (I_{r,z}p_f)\exp(-\beta\delta)$$

$$g_b(r = R_R) = 0$$

$$g_a(r = R_R) = (g_f + g_b + g_c)p_a \qquad (13b)$$

$$g_c(r = R_{lamp}) = (g_f + g_a + g_b)p_c.$$

(3) BC 3: at upper and bottom wall:

$$g_a(z = L_R) = g_c(z = 0) = 0. \qquad (13c)$$

These boundary conditions are shown in Figure 3. Furthermore, considering an infinitely long reactor, the following condition can be established, along the axial axis:

$$\frac{\partial g_a}{\partial z} = \frac{\partial g_c}{\partial z} = 0. \qquad (14)$$

The LVREA using the four-flux model can be calculated by the following expression:

$$\text{LVREA} = \frac{g_{total}\kappa_\lambda}{V}\left(\frac{R_{lamp}\delta}{4r^2}\right). \qquad (15)$$

2.4. Simulation of Radiant Field. The software COMSOL Multiphysics version 4.4 and subroutines performed in Matlab® were employed to solve the FFM and kinetic models, respectively. To carry out the simulation, the geometric domain of both, reaction space and lamp, was established. The model is two-dimensional and symmetric with respect to the axial axis. A nonuniform mesh was used, with a size of element calibrated to plasma, giving major emphasis on the inner wall of the annulus, using a fine mesh at this boundary and coarser in the outer wall of the reactor to accurately assess each border (Figure 4(a)). As a result, color maps are obtained, which represent the distribution of LVREA within the photocatalytic reactor. The red zone represents the highest values, and the colors are decreasing towards blue which represents low values of LVREA. The modeling instructions for FFM can be found in the complementary content (Appendix A).

The results obtained by FFM were compared with the following.

(A) Discrete ordinate method (DOM) carried out with the physics of radiation in participating media of the Heat Transfer Module of Comsol Multiphysics 4.4. To do so, the geometric domain of reaction space was established as 3D model. Several preliminary simulations were run using this method. In these trials, the mesh in all domains was refined incrementally until the physical ram limit of the workstation (8 Gb) was reached. Geometry and mesh employed are shown in Figure 4(b). The modeling instructions for DOM can be found in the complementary content (Appendix B).

(B) Six-flux model (SFM) was implemented in programming language Matlab according to the methodology reported by Li Puma [14, 17].

(C) Monte Carlo Method (MC) was also implemented in programming language Matlab based on Moreira

et al. [28, 29]. In this case, the number of used photons was 1×10^7. In addition, subroutines were programmed to generate random numbers.

The codes to solve the applied models, SFM and MC, are rather lengthy. However, they can be provided upon request.

2.5. Kinetic Model. To determine the radiation effect on reaction rate, a kinetic expression as function of LVREA can be obtained.

$$\frac{dC_{AB}}{dt} = k_r f(C_{AB})g(\text{LVREA}), \qquad (16)$$

where $f(C_{AB})$ is a function of reagent concentration (benzyl alcohol) and the dependence of reaction rate with LVREA is given by $g(\text{LVREA})$. To describe $f(C_{AB})$ is possible to employ a power law model. This is accepted when the reagent absorption on the catalytic surface is negligible and therefore the LHHW model becomes a pseudo first-order equation. Although this kind of equation does not include the effect of reactive intermediaries, it still provides reasonable results [6, 11]. Several authors have studied the kinetics of photocatalytic oxidation of aromatic alcohols to corresponding aldehydes, and they claim a first-order kinetics regarding alcohol concentration [10, 11].

$$-\frac{dC_{AB}}{dt} = K_{Ap}C_{AB}, \qquad (17)$$

where K_{Ap} is the apparent kinetic coefficient that includes the effect of catalyst concentration, temperature, oxidant concentration, and so forth. Furthermore, since there is reaction due to photolysis only (without catalyst), this can be considered within the reaction rate expression.

$$\left(\frac{-dC_{AB}}{dt}\right)_{Total} = \left(\frac{-dC_{AB}}{dt}\right)_{Without\ catalyst} + \left(\frac{-dC_{AB}}{dt}\right)_{With\ catalyst}, \qquad (18)$$

$$\left(\frac{-dC_{AB}}{dt}\right)_{Total} = k_{r1}C_{AB} + k_{r2}C_{AB} = (k_{r1} + k_{r2})C_{AB} = K_{Ap}C_{AB}, \qquad (19)$$

where k_{r1} is the intrinsic constant of reaction rate without catalyst and k_{r2} is the reaction rate constant with catalyst, which is a function of LVREA. Therefore, K_{Ap} can be expressed as

$$K_{Ap} = k_{r1} + k_{r2} = k_{r3}(\text{LVREA})^m, \qquad (20)$$

by linear regression, both the order of LVREA and k_{r3} were calculated. Employing the FFM method, the values of averaged LVREA corresponding to each catalyst concentration were calculated. The contribution due to photolysis is negligible; for this reason, the LVREA due to the reactive species was not added in the photolysis term. It is worth noticing that in other cases, when the reactant molecule has a strong absorption of photons, this contribution must also be taken into account. This also applies for intermediaries. In the present case, however, the reaction kinetics was established with

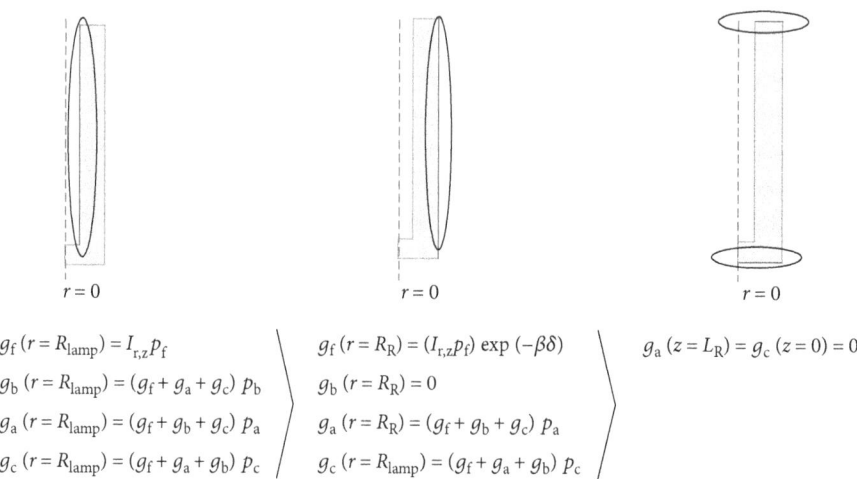

$$g_f\,(r=R_{lamp})=I_{r,z}p_f$$
$$g_b\,(r=R_{lamp})=(g_f+g_a+g_c)\,p_b$$
$$g_a\,(r=R_{lamp})=(g_f+g_b+g_c)\,p_a$$
$$g_c\,(r=R_{lamp})=(g_f+g_a+g_b)\,p_c$$

$$g_f\,(r=R_R)=(I_{r,z}p_f)\exp(-\beta\delta)$$
$$g_b\,(r=R_R)=0$$
$$g_a\,(r=R_R)=(g_f+g_b+g_c)\,p_a$$
$$g_c\,(r=R_{lamp})=(g_f+g_a+g_b)\,p_c$$

$$g_a\,(z=L_R)=g_c\,(z=0)=0$$

FIGURE 3: Boundary conditions used in the four-flux model.

very low conversion data and therefore the presence of intermediaries was considered rather low as to contribute to LVREA.

3. Results

3.1. Emission Model. Figure 5 shows the emitted radiation profiles calculated by both, MC and LSSE methods. In addition, the radiation values experimentally measured by a UVx radiometer equipped with a detector 254 nm were plotted. Figure 5(a) shows the values of the emitted radiation, $I_{(R_{lamp},z)}$ on the wall of the lamp, and Figure 5(b) shows the values of $I_{(R_{lamp}+0.01m,z)}$ at 0.01 m from the wall of the lamp. It can be seen that both methods are in good agreement with experimentally obtained data, which justifies the use of both Monte Carlo method and LSSE model in this research.

3.2. Absorption of Radiation Model. The results of the proposed model ((7), (8), (9), (10), (11), (12), (13a), (13b), (13c), (14), and (15)) were compared with those obtained by MC, SFM, and DOM. It was assumed that MC is the method that best represents the radiant field in the photocatalytic reactor. Even though the DOM is robust, it requires a very refined mesh to give congruent results.

3.2.1. System 1: TFSIW. The first analyzed photocatalytic reactor was a TFSIW reported by Li Puma et al. [5, 14, 17]. This reactor has a radius ratio $R_R/(R_R+\delta)=0.76$. Figure 6 shows the radial profiles of LVREA at $z=L_R/2$ for this system, calculated by the four methods and parity diagram. It can be seen that SFM and DOM represent LVREA profiles better than the FFM with regard MC, especially when the catalyst concentration is low. However, FFM results can be considered to be adequate also if a rapid estimation of LVREA is required. Both the FFM and the SFM have small deviations in the inner wall when the catalyst loading is large. With the mesh used for the DOM, the computing time was approximately 40 minutes. Using a finer mesh could increase the computing time by several hours. The solving time for

FFM was about 2 minutes regardless the elements number in the mesh.

3.2.2. System 2 (Photo-CREC II). The results obtained by FFM method are in agreement with the data previously reported by Moreira et al., which were obtained from MC for Photo-CREC water II [28, 29]. Figure 7 shows the radial profiles for the LVREA at different photocatalyst concentrations for TiO$_2$ anatase and parity diagram obtained by SFM, DOM, and FFM versus MC. In this case, it is observed that the results obtained by FFM and DOM are quite congruent although they tend to deviate slightly from those obtained by MC. This reactor has a ratio of radius $R_R/(R_R+\delta)=0.6231$. The mesh used in DOM for this relationship can be considered as semicoarse and give good results in about 1 hour of computing time.

3.2.3. System 3. This system was theoretically and experimentally studied. Figure 8 shows the comparison of the LVREA profiles obtained from the three methods for the catalyst LiVMoO$_6$. It is worth noticing that the catalyst with the highest extinction coefficient values $(\beta=\sigma+\kappa)$ (TiO$_2$ DP 25) produces higher values of LVREA at the same catalyst concentration. The values of LVREA obtained by LiVMoO$_6$ catalyst are smaller than the values obtained by TiO$_2$ catalyst; however, the special interest on LiVMoO$_6$ catalyst resides on that it presents catalytic activity even in the visible spectrum [31].

In Figure 8, it can be observed that near the lamp wall (dimensionless radius = 0.4), LVREA is maximum and rapidly decreases as dimensionless radius increases. This effect is considered by both, MC and FFM; however, the SFM does not account for it. This can be attributed to SFM being explicitly developed for thin-walled annular reactors and presents significant deviations when $R_R/(R_R+\delta)\ll 1$. Also, DOM presents a great deviation with respect to MC. This is because the meshing is not fine enough. However, using a more refined mesh causes the available RAM to be exceeded.

Table 2 shows a comparison of correlation coefficients for SFM, FFM, and DOM considering that MC is the most

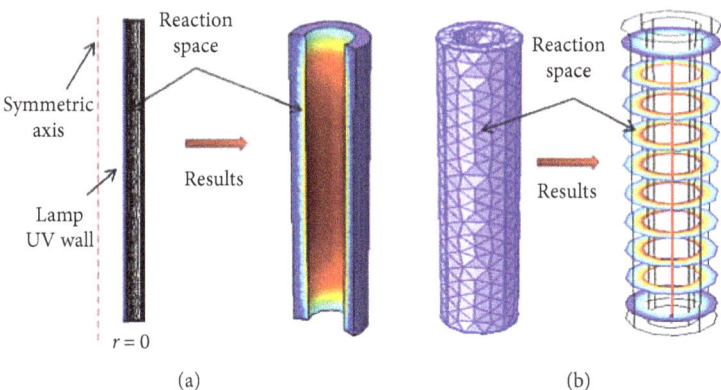

FIGURE 4: Graphical representation of photocatalytic reactor and geometry employed to solve (a) four-flux model and (b) discrete ordinate method, in COMSOL Multiphysics.

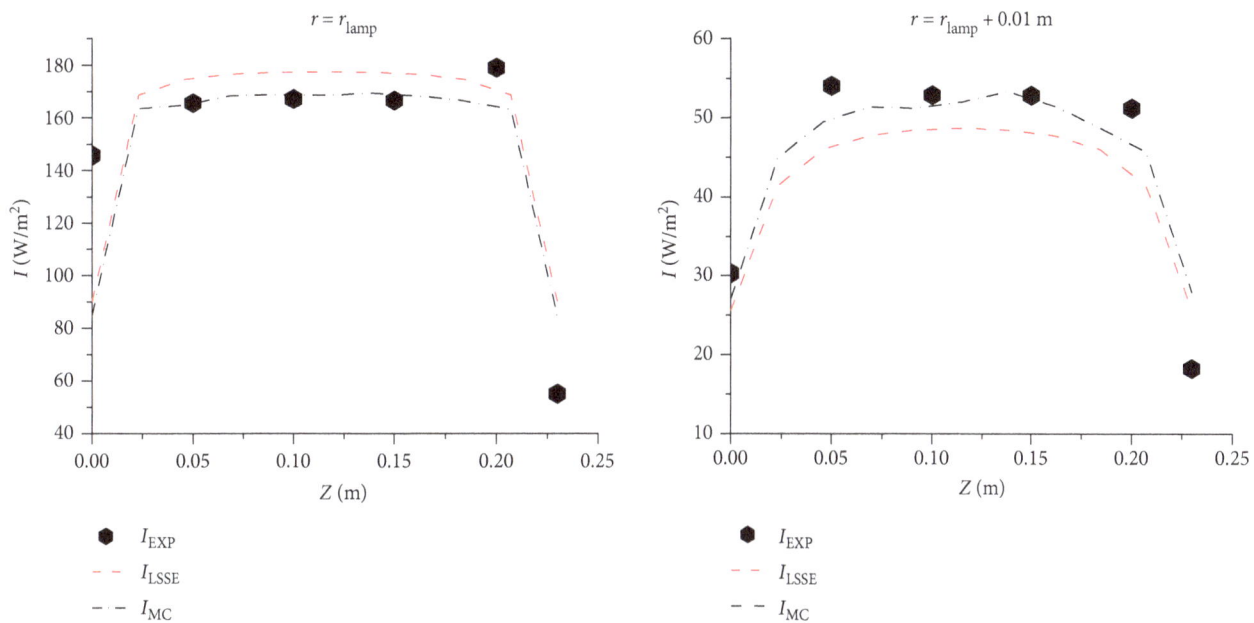

FIGURE 5: Emission model results. Calculated and experimental incident radiation profiles.

accurate one. The percentage of the area under the curve of the radiation profiles obtained by the different methods in relation to the area under the Monte Carlo method curve is also shown. It can be seen that the FFM better predicts the profiles of LVREA when $R_R/(R_R + \delta) \ll 1$ and the catalyst loading is relatively low, for example in system 3.

Through Figures 6, 7, and 8, it can be seen that near the inner wall of the reaction space, LVREA is maximum and rapidly decreases as dimensionless radius increases. This can be ascribed to an obstruction effect produced by catalyst particles. One can also notice that in cases where the photocatalyst concentration is relatively high, the particles closer to the inner radius absorb most of the radiation entering the reactor. According to the results, it may be seen that low values of LVREA are obtained at low catalyst concentrations; however, the effectively irradiated zone is greater. At high catalyst concentrations, high values of LVREA are

achieved near the wall of the lamp; however, the effectively irradiated zone is drastically diminished in the radial direction. It is important to note this effect since it is desirable to obtain high values of LVREA, but at the same time maximize the irradiated zone. In dark zones, absorption of photons does not occur, which provokes the effective volume of the reaction being smaller, that is, the reactor volume is being subutilized. This effect is shown in Figure 9. A sufficiently high photocatalyst concentration produces zones with dark areas towards the external radius. Therefore, there is an optimal catalyst concentration that provides an optimal irradiated reactor space. Photocatalyst concentrations above this maximum show an essentially negligible effect on LVREA. This optimal concentration can be seen in Figure 10, and it is in agreement with those reported [28, 29] for TiO_2 catalyst (system 1 and system 2). For system 3, the optimal concentration is achieved at $1 \, kg/m^3$.

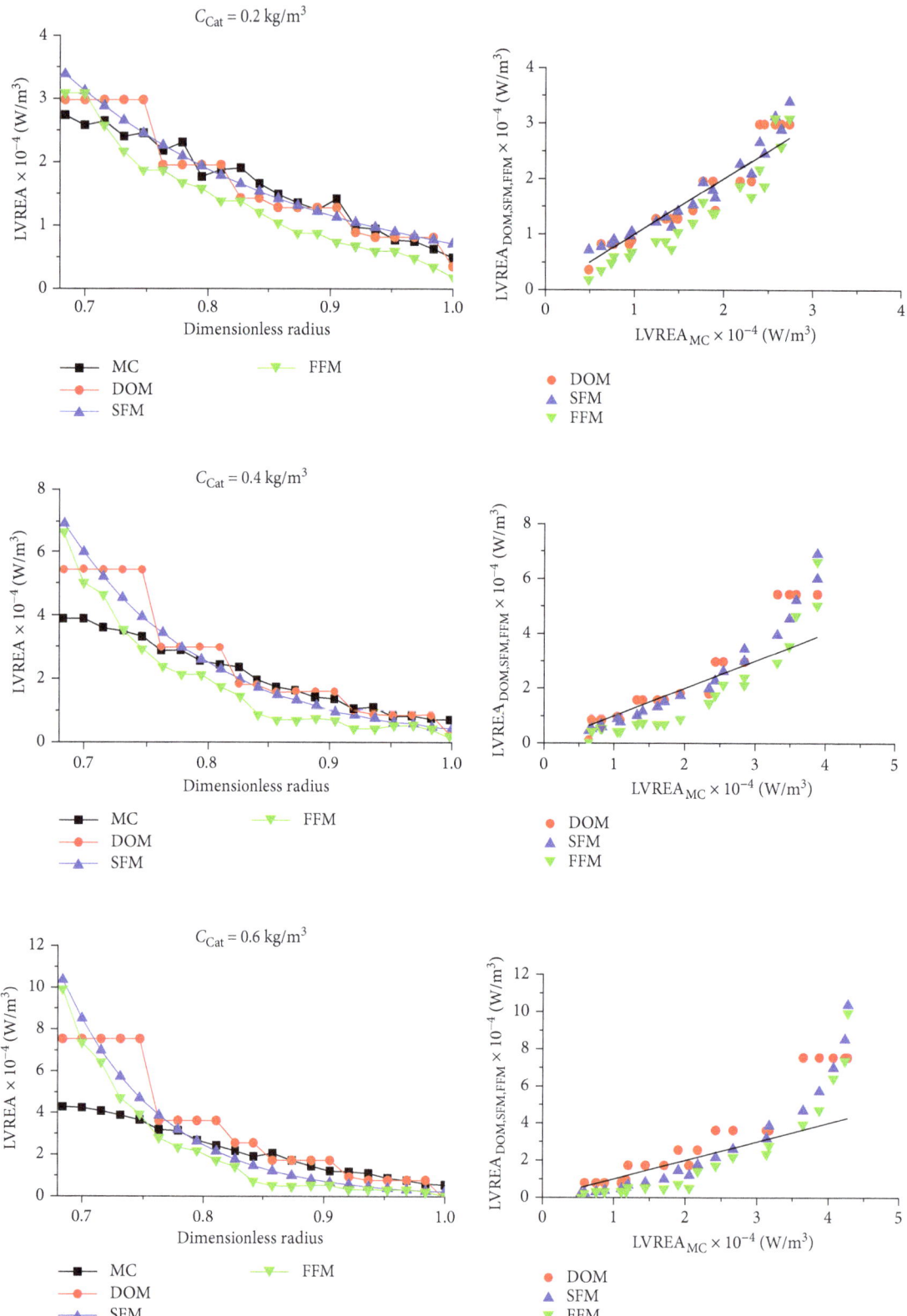

FIGURE 6: Radial profiles of local volumetric rate energy absorption (LVREA) obtained with four-flux model at different concentrations of catalyst for system 1 and its comparison with the other models.

3.3. Kinetic Model. To obtain a kinetic expression for photocatalytic oxidation of benzyl alcohol, the integral method was employed. An adjust by least squares was performed

for different models, including the LHHW model, and it was found that the better adjustment is at pseudo first order in respect of concentration of benzyl alcohol. This

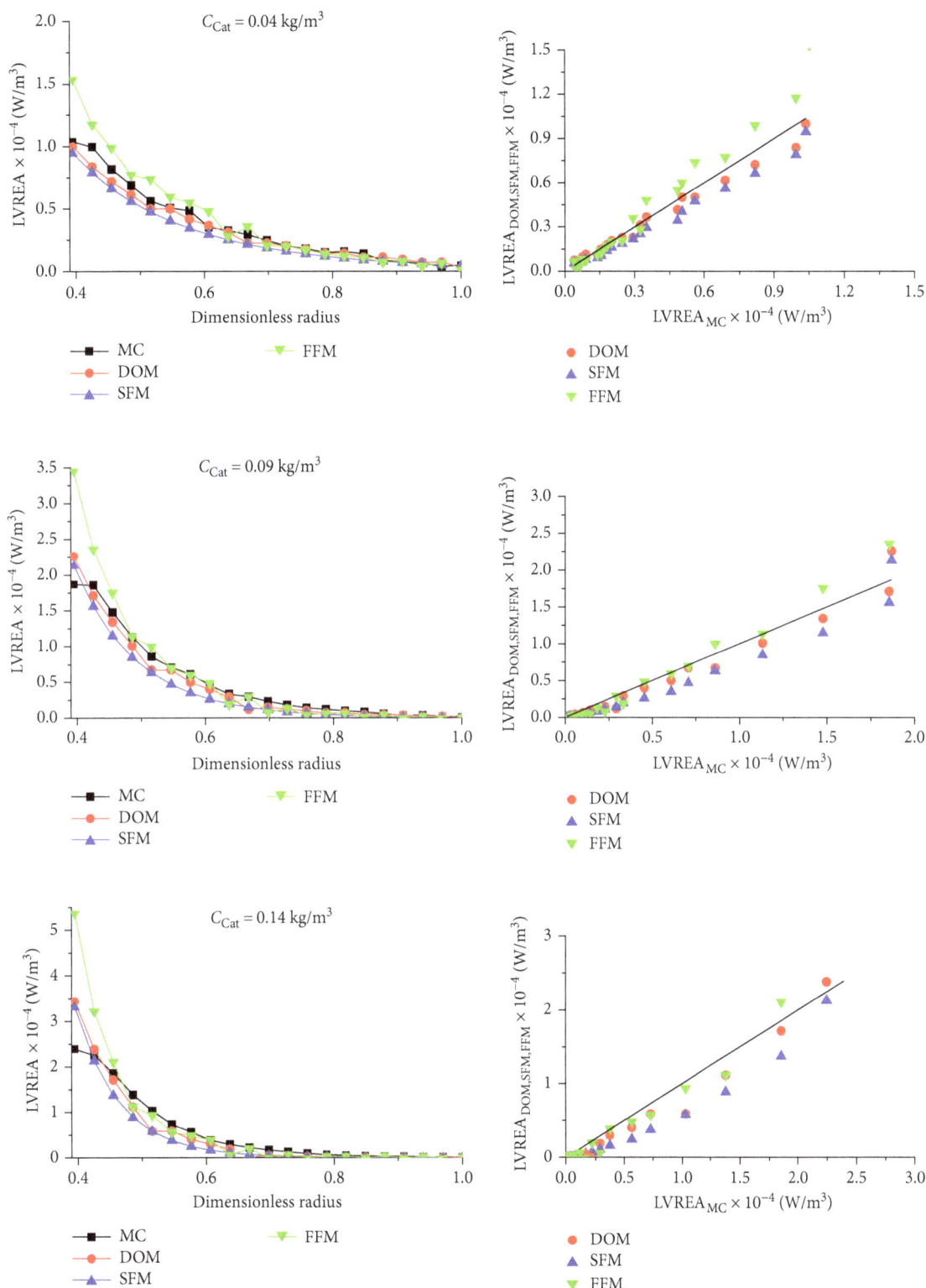

FIGURE 7: Radial profiles of local volumetric rate energy absorption (LVREA) obtained with four-flux model at different catalyst concentration for system 2 and its comparison with the other models.

result is in agreement with the results reported by [10, 11], albeit with other catalysts. Figure 11 shows the comparison of the results for the adjustment by least squares according to experimental data of benzyl alcohol oxidation, at

different catalyst loading, employing a pseudo first-order power model.

Taking into account the data of concentration—time obtained at each catalyst loading, an adjustment by least

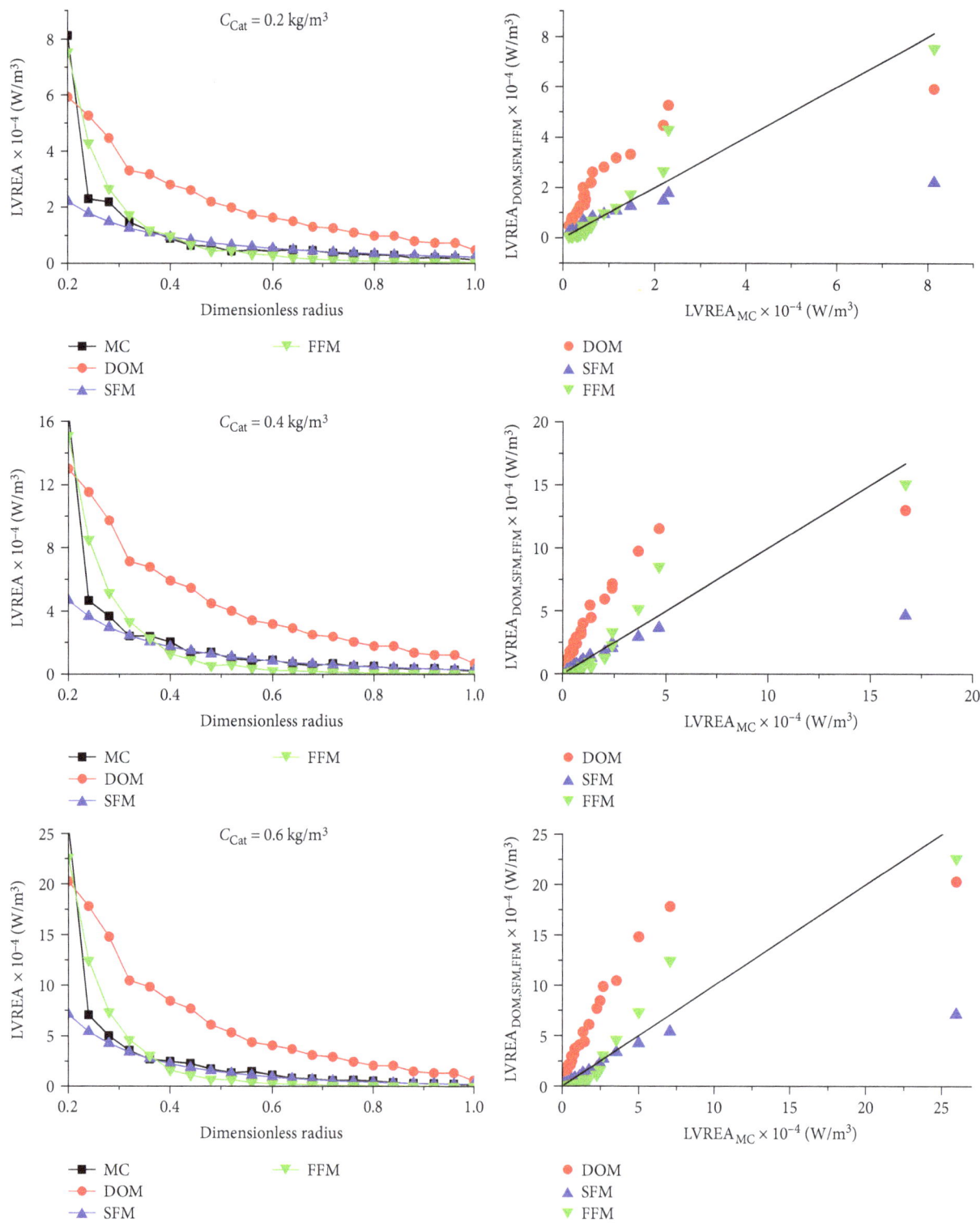

FIGURE 8: Radial profiles of local volumetric rate energy absorption (LVREA) obtained by four-flux model at different concentration of catalyst (LiVMoO$_6$) for system 2 and its comparison with the other models.

squares was performed to obtain the dependence of rate constants with the LVREA (Table 3).

Figure 12 shows the plot of $\ln(K_{Ap} - k_{r1})$ as a function of $\ln(\text{LVREA})$. The slope of the line represents the order of the reaction with respect to the LVREA, and the intercept provides

the $\ln(k_{r3})$. Hence $k_{r1} = 0.0101\,\text{h}^{-1}$, intrinsic reaction constant without catalyst; $k_{r3} = 0.01887\,\text{h}^{-1}\left(\text{W/m}^3\right)^{-0.1464}$; and $m = 0.1467$ is the power of LVREA. This fractional exponent of LVREA was expected. It has even been reported that the exponent is equal to 0.5 in the presence of TiO$_2$ for system 1 [18]. It

TABLE 2: Comparison of correlation coefficients of different radiation absorption models for studied systems.

System	Catalyst	C_{CAT} (mg/L)	DOM		SFM		FFM	
			R^2	%A_{MC}	R^2	%A_{MC}	R^2	%A_{MC}
1 $R_R/(R_R + \delta) = 0.76$	TiO$_2$	0.20	0.9530	101.69	0.9611	103.21	0.9458	80.78
		0.40	0.9602	120.54	0.9583	110.80	0.9167	84.22
		0.60	0.9595	143.91	0.9242	116.33	0.8906	94.20
2 $R_R/(R_R + \delta) = 0.62315$	TiO$_2$ anatase	0.04	0.9943	91.83	0.9945	80.09	0.9868	114.16
		0.09	0.9833	88.33	0.9744	73.70	0.9642	110.29
		0.14	0.9629	87.47	0.9421	73.52	0.9285	113.02
3 $R_R/(R_R + \delta) = 0.5555$	LiVMoO$_6$	0.20	0.8089	237.11	0.8468	81.79	0.9605	100.88
		0.40	0.8004	257.06	0.8518	76.16	0.9553	95.18
		0.60	0.7984	258.14	0.8571	70.79	0.9564	95.73

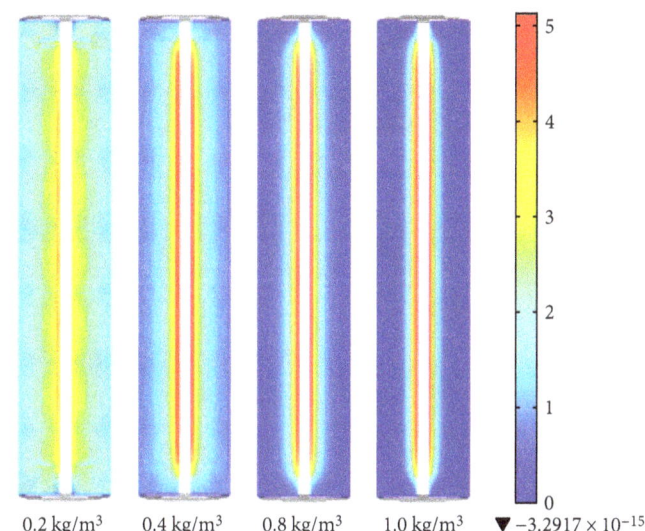

FIGURE 9: Effect of catalyst concentration on the reactor radial section where photon absorption occurs (LVREA map) in system 3.

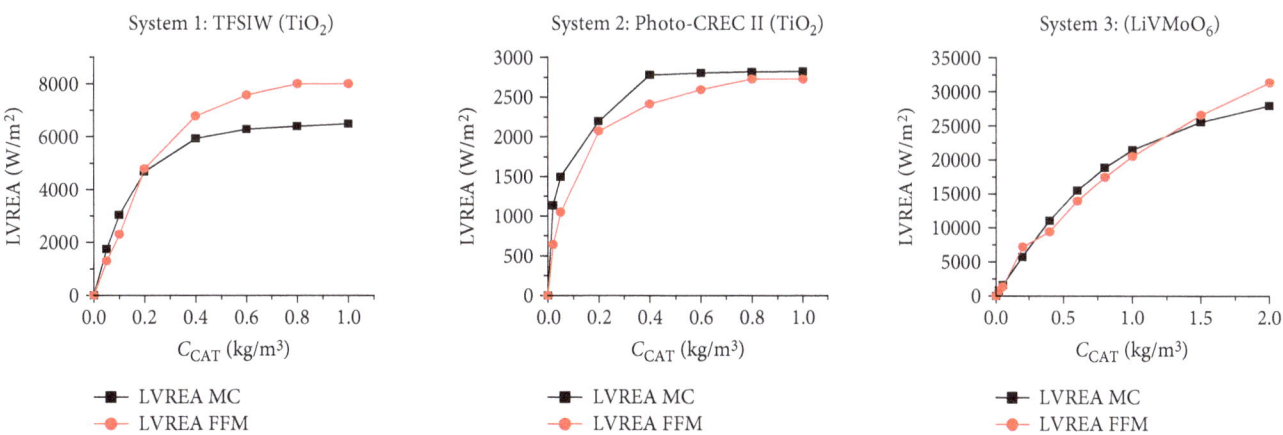

FIGURE 10: Simulated results of incident radiation as function catalyst loading.

should be noted that the value of m is relatively independent on the type of substrate. Instead, it should be dependent on the radiation intensity level over the catalyst. A fractional order dependence of photocatalytic reaction rate from the LVRPA is obtained when the rate of electron-hole recombination in the catalyst particles becomes predominant [18].

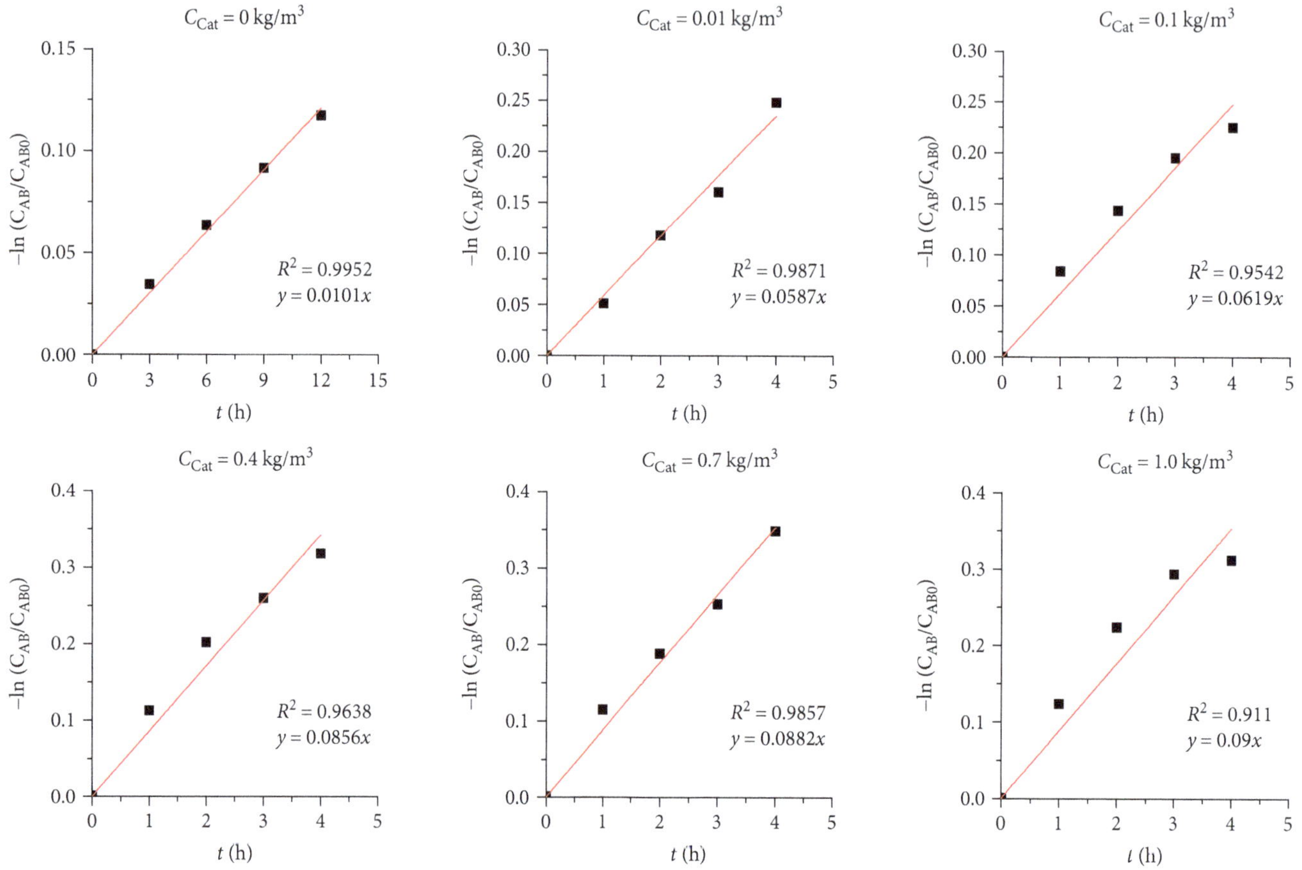

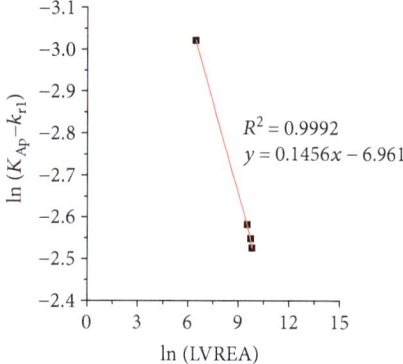

FIGURE 11: Test for a pseudo first-order kinetics as a function of concentration.

TABLE 3: First-order kinetic constants and values of LVREA at different catalyst loadings.

C_{cat} (kg/m^3)	K_{Ap} (h^{-1})	LVREA (W/m^3)
0.00	0.0101	0.00
0.01	0.0587	630.00
0.10	0.0750	5300.00
0.40	0.0856	13600.00
0.70	0.0882	16700.00
1.00	0.0900	18000.00

FIGURE 12: Adjustment for dependence of K_{Ap} with the LVREA.

Therefore, the kinetic equation that describes the photocatalytic oxidation of benzyl alcohol to benzaldehyde is

$$-\frac{dC_{AB}}{dt} = \left(0.0101\,\text{h}^{-1} + 0.01887\,\text{h}^{-1} \left(\frac{\text{W}}{\text{m}^3}\right)^{-0.1464} (\text{LVREA})^{0.1464} \right) C_{AB}. \tag{21}$$

Equation (20) shows that the reaction rate depends on the LVREA values and the amount of irradiated catalyst. Figure 13 shows the concordance of the proposed mathematical model with experimental data of benzyl alcohol oxidation at different catalyst loadings.

In Figure 13, a linear decrease of benzyl alcohol concentration is observed. This is in agreement with that previously

reported [10, 11]. On the other hand, the conversion increases with the catalyst loading up to a point where a further increase on catalyst loading does not produce a significant improvement on conversion, due to LVREA reaches a maximum at this point, establishing that the optimal catalyst loading is 1.0 kg/m^3 for LiVMoO$_6$.

4. Conclusions

The proposed mathematical model (FFM) describes the radiant field in a photocatalytic annular reactor. Its

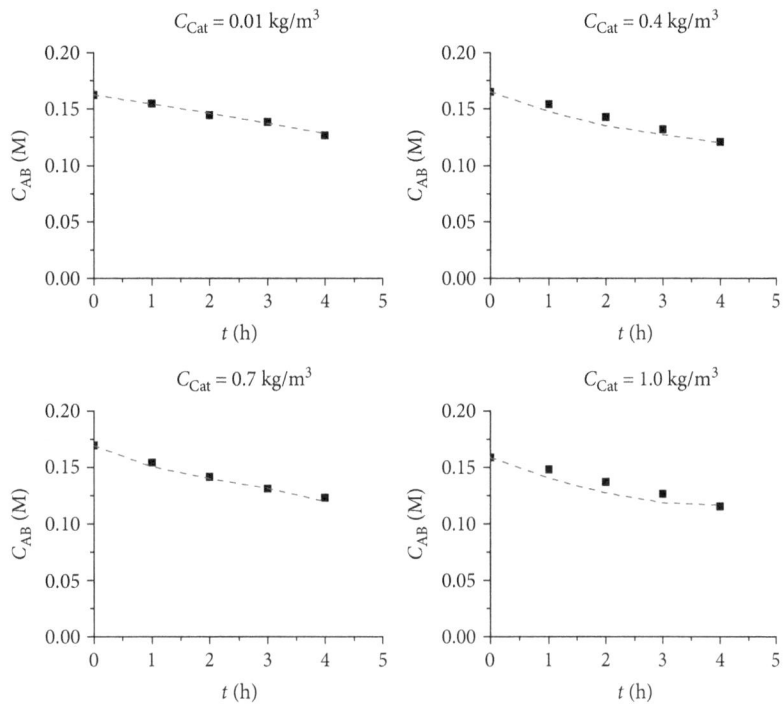

FIGURE 13: Comparison of experimental concentration profiles (dots) with those obtained by the proposed model.

numerical solution corresponds appropriately with experimental and numerical data, and it requires a minor computational effort than the other models, such as DOM, which is very robust and accurate but requires a high RAM capacity. The FFM was specifically designed for cylindrical geometries with the lamp located at the axial axis of the reactor submerged in reaction medium. The FFM predicts the LVREA profiles better than the other models when $R_R/(R_R + \delta) \ll 1$, and the catalyst loading is low.

The obtained kinetic equation describes the reaction rate in the photocatalytic reactor for selective oxidation of benzyl alcohol as function of the LVREA. The FFM allows the evaluation of LVREA at different catalyst loadings, power lamp, or reactor dimensions. Therefore, it allows the calculation of reaction rates at different experimental setups.

Within the range of studied variables, the reaction rate of the selective oxidation of benzyl alcohol adequately fits a first-order kinetics, where the kinetic coefficient is a function of LVREA, and this depends on catalyst loading, power lamp, and annulus width.

Appendix

A. Simulation of Radiant Field Employing the Four-Flux Model (FFM) in Comsol Multiphysics v. 4.4

(1) Model 2D: from the File menu, choose New. In the New window, click Model Wizard and select a 2D axisymmetric model.

(2) Interface for ordinary differential equations, ODE: in the Select Physics tree, select Mathematics > ODEs Interface > ODEs in general form (g). Click Add. The Study is Stationary.

(3) Parameters: go to Global definitions section and insert Parameters. In the Settings window for Parameters, locate the Parameters section and add the necessary parameters, such as reactor dimensions (reactor length, internal radius, and external radius), dispersion probabilities (p_f, p_b, p_a, and p_c), characteristics of the lamp (power, dimensions, and wavelength), catalyst charge, and optical properties of catalyst (absorption, kappa_s; dispersion, sigma_s; and extinction coefficients, beta_s), from Table 1.

(4) Global variable: go to Global definitions section and insert Variables. In the Settings windows for Variable, write the expression for radiation intensity (Irz), according to (3) and (4).

(5) Geometry: set reactor geometry as rectangular section that represents the 2D axial section of the reactor. It can be drawn as a simple rectangle.

(6) Model definitions: in model definitions, insert section for variables. In the Settings windows for Variable, insert the LVREA expression. It can be introduced with (15).

(7) Set of differential equations: in the ODE interfaces, select all domains. Go to the ODE general form window settings and introduce the differential equations system defined for (7), (10), (11), and (12).

(8) Incident intensity boundary: in ODE interfaces, insert a Dirichlet boundary condition. In this setting section, select the internal boundary (inner radius of reaction section). Locate boundary condition field and introduce (13a).

(9) External wall boundary condition: in ODE interfaces, insert a Dirichlet boundary condition. In this setting section, select the external boundary (external radius of reaction section). Locate boundary condition field and introduce (13b).

(10) Upper and bottom wall boundary condition: in ODE interfaces, insert a Dirichlet boundary condition. In this setting section, select the upper and bottom boundary. Locate boundary condition field and introduce (13c).

(11) Weak form of ODE: in ODE interfaces, insert a Weak Form for ODE condition. In this setting section, select all domains. In Weak expression field, introduce the following expressions:

WEAK = 0; 0; −test(gaz)*gaz + test(ga); −test(gcz)*gcz + test(gc).

(12) Mesh: in the mesh section, introduce a mesh using the option of free quadratic mesh. You can try different mesh sizes. In free quadratic mesh, add distribution, select inner wall and locate the input section, and introduce 500 in number of elements.

(13) Go to the study section and Run model.

B. Simulation of Radiant Field Employing the Discrete Ordinate Method (DOM) in Comsol Multiphysics v. 4.4

(1) Model 3D: from the File menu, choose New. In the New window, click Model Wizard and select a 3D model.

(2) Radiation in participating media: in the Select Physics tree, select Heat Transfer > Radiation > Radiation in Participating Media (rpm). Click Add. The Study is Stationary.

(3) Parameters: go to Global definitions section and insert Parameters. In the Settings window for Parameters, locate the Parameters section and add the necessary parameters, such as reactor dimensions (reactor length, internal radius, and external radius), characteristics of the lamp (power, dimensions, and wavelength), catalyst charge, and optical properties of catalyst (absorption, kappa_s; dispersion, sigma_s; and extinction coefficients, beta_s)

(4) Global variable: go to Global definitions section and insert variables. In the Settings windows for Variable, write the expression for radiation intensity (Irz), according to (3) and (4).

(5) Geometry: set reactor geometry as annular section. It can be drawn as a Boolean difference from two cylinders.

(6) Model definitions: in model definitions, insert section for variables. In the Setting windows for Variable, insert the LVREA expression. It can be introduced as LVREA = rpm.G*kappa.

(7) Radiation in participating media (rpm): in the physics for radiation in participating media go to Radiation with participating media window settings and introduce dispersion and absorption coefficients in the model input sections.

(8) In radiation in participating media, insert Incident intensity section. In this setting section, select the internal boundaries. Locate the incident intensity field and introduce Irz variable.

(9) Opaque surface: in radiation in participating media, insert opaque surface. In the setting section, select external boundaries of contours of the domain. In wall adjust, select Black Wall.

(10) Mesh: in the mesh section, introduce a mesh using the option of free tetrahedral mesh. You can try different mesh sizes. A too fine mesh can cause the available RAM to be exceeded.

(11) Go to the study section and Run model.

Nomenclature

a_{lamp}:	Area of lamp (m^2)
a_p:	Catalytic particle area (m^2)
C_{AB}:	Benzylic acid concentration ($mol \cdot dm^{-3}$)
C_{cat}:	Catalyst concentration ($kg \cdot m^{-3}$)
g_f:	Flux incident radiation ($Watts \cdot m^{-2}$)
g_a:	Upwards flux scattering radiation ($Watts \cdot m^{-2}$)
g_b:	Flux backscattering radiation ($Watts \cdot m^{-2}$)
g_c:	Downwards flux scattering radiation ($Watts \cdot m^{-2}$)
I_λ:	Spectral radiation intensity ($Watts \cdot m^{-2} \cdot sr^{-1}$)
K_{Ap}:	Apparent reaction constant (s^{-1})
k_{r1}:	Intrinsic reaction constant without catalyst (s^{-1})
k_{r2}:	Reaction constant with catalyst (s^{-1})
k_{r3}:	Intrinsic reaction constant with catalyst ($Watts^{(-m)} \cdot m^{(3m)}$)
L:	Length (m)
L_{lamp}:	Length of lamp (m)
L_{Reac}:	Length of the reactor (m)
LVREA:	Local volumetric rate of absorption of energy ($Watts \cdot m^{-3}$)
m:	Reaction order with respect to LVREA
n_p:	Number of catalyst particles (m^{-3})
p_a:	Probabilities of scattering toward up
p_b:	Probabilities of backscattering
p_f:	Probabilities of forward scattering
$p(\Omega \rightarrow \Omega')$:	Phase function

r: Radial coordinate (m)

R_{int}:   Inner radius of the annulus (m)

R_{lamp}:   Radius of lamp (m)

R_R: Radius of reactor (m)

t: Time (h)

V: Volume (m^3)

X: Conversion

z: Coordinate axial (m).

Greek Letters

β_λ: Extinction coefficient (m^{-1})

Δ: Thickness of the annulus (m)

κ_λ: Absorption coefficient (m^{-1})

λ: Wavelength

σ_λ: Scattering coefficient (m^{-1})

ω: Albedo coefficient

Ω: Solid angle (sr).

Acronyms

DOM: Discrete ordinate method

FFM: Four-flux model

LVREA: Local volumetric rate of energy absorption

LHHW: Langmuir–Hinselwwod–Hougen–Watson   kinetic model

MC: Monte Carlo model

RTE: Radiation transfer equation

SFM: Six-flux model

TFM: Two-flux model

TFSIW: Thin-film slurry reactor of inner wall.

Conflicts of Interest

The authors declare that they have no conflicts of interest.

Acknowledgments

The authors are grateful to PRODEP for the financial support through Project 103.5/13/5257 and CONACYT through Project 269093. Mr. O. Alvarado-Rolon is grateful to CONACYT for the financial support (Scholarship 401273) to conduct postgraduate studies. Citlalit Martínez Soto is acknowledged for the technical support.

References

[1] H. I. De Lasa, B. Serrano, and M. Salaices, *Photocatalytic Reaction Engineering*, Springer, 2005, April 2016 http://link.springer.com/content/pdf/10.1007/0-387-27591-6.pdf.

[2] L. F. Garcés Giraldo, E. A. Mejía Franco, and J. J. Santamaría Arango, "La fotocatálisis como alternativa para el tratamiento de aguas residuales," *Revista Lasallista de investigación*, vol. 1, pp. 83–92, 2004.

[3] G. Li Puma and P. L. Yue, "A novel fountain photocatalytic reactor for water treatment and purification: modeling and design," *Industrial and Engineering Chemistry Research*, vol. 40, no. 23, pp. 5162–5169, 2001.

[4] L. Zhang, W. Anderson, and Z. Zhang, "Development and modeling of a rotating disc photocatalytic reactor for wastewater treatment," *Chemical Engineering Journal*, vol. 121, no. 2-3, pp. 125–134, 2006.

[5] G. Li Puma, J. N. Khor, and A. Brucato, "Modeling of an annular photocatalytic reactor for water purification: oxidation of pesticides," *Environmental Science & Technology*, vol. 38, no. 13, pp. 3737–3745, 2004.

[6] G. Li Puma and P. L. Yue, "Modelling and design of thin-film slurry photocatalytic reactors for water purification," *Chemical Engineering Science*, vol. 58, no. 11, pp. 2269–2281, 2003.

[7] G. E. Imoberdorf, A. E. Cassano, H. A. Irazoqui, and O. M. Alfano, "Simulation of a multi-annular photocatalytic reactor for degradation of perchloroethylene in air: parametric analysis of radiative energy efficiencies," *Chemical Engineering Science*, vol. 62, no. 4, pp. 1138–1154, 2007.

[8] F. Shiraishi, T. Nomura, S. Yamaguchi, and Y. Ohbuchi, "Rapid removal of trace HCHO from indoor air by an air purifier consisting of a continuous concentrator and photocatalytic reactor and its computer simulation," *Chemical Engineering Journal*, vol. 127, no. 1-3, pp. 157–165, 2007.

[9] S. Romero-Vargas Castrillón and H. I. de Lasa, "Performance evaluation of photocatalytic reactors for air purification using computational fluid dynamics (CFD)," *Industrial and Engineering Chemistry Research*, vol. 46, no. 18, pp. 5867–5880, 2007.

[10] V. Augugliaro, H. Kisch, V. Loddo et al., "Photocatalytic oxidation of aromatic alcohols to aldehydes in aqueous suspension of home prepared titanium dioxide," *Applied Catalysis A: General*, vol. 349, no. 1-2, pp. 189–197, 2008.

[11] S. Higashimoto, N. Kitao, N. Yoshida et al., "Selective photocatalytic oxidation of benzyl alcohol and its derivatives into corresponding aldehydes by molecular oxygen on titanium dioxide under visible light irradiation," *Journal of Catalysis*, vol. 266, no. 2, pp. 279–285, 2009.

[12] N. Qi, H. Zhang, B. Jin, and K. Zhang, "CFD modelling of hydrodynamics and degradation kinetics in an annular slurry photocatalytic reactor for wastewater treatment," *Chemical Engineering Journal*, vol. 172, no. 1, pp. 84–95, 2011.

[13] A. Gora, B. Toepfer, V. Puddu, and G. Li Puma, "Photocatalytic oxidation of herbicides in single-component and multicomponent systems: reaction kinetics analysis," *Applied Catalysis B: Environmental*, vol. 65, no. 1-2, pp. 1–10, 2006.

[14] G. Li Puma, "Modeling of thin-film slurry photocatalytic reactors affected by radiation scattering," *Environmental Science & Technology*, vol. 37, no. 24, pp. 5783–5791, 2003.

[15] G. Li puma, "Dimensionless analysis of photocatalytic reactors using suspended solid photocatalysts," *Chemical Engineering Research and Design*, vol. 83, no. 7, pp. 820–826, 2005.

[16] A. Brucato, A. E. Cassano, F. Grisafi, G. Montante, L. Rizzuti, and G. Vella, "Estimating radiant fields in flat heterogeneous photoreactors by the six-flux model," *AICHE Journal*, vol. 52, no. 11, pp. 3882–3890, 2006.

[17] G. Li Puma and A. Brucato, "Dimensionless analysis of slurry photocatalytic reactors using two-flux and six-flux radiation absorption–scattering models," *Catalysis Today*, vol. 122, no. 1-2, pp. 78–90, 2007.

[18] G. Li Puma, V. Puddu, H. K. Tsang, A. Gora, and B. Toepfer, "Photocatalytic oxidation of multicomponent mixtures of estrogens (estrone (E1), 17β-estradiol (E2), 17α-ethynylestradiol (EE2) and estriol (E3)) under UVA and UVC

radiation: photon absorption, quantum yields and rate constants independent of photon absorption," *Applied Catalysis B: Environmental*, vol. 99, no. 3-4, pp. 388–397, 2010.

[19] G. E. Imoberdorf, A. E. Cassano, H. A. Irazoqui, and O. M. Alfano, "Optimal design and modeling of annular photocatalytic wall reactors," *Catalysis Today*, vol. 129, no. 1-2, pp. 118–126, 2007.

[20] V. K. Pareek and A. A. Adesina, "Light intensity distribution in a photocatalytic reactor using finite volume," *AICHE Journal*, vol. 50, no. 6, pp. 1273–1288, 2004.

[21] M. e. M. Zekri and C. Colbeau-Justin, "A mathematical model to describe the photocatalytic reality: what is the probability that a photon does its job?," *Chemical Engineering Journal*, vol. 225, pp. 547–557, 2013.

[22] M. L. Satuf, R. J. Brandi, A. E. Cassano, and O. M. Alfano, "Scaling-up of slurry reactors for the photocatalytic degradation of 4-chlorophenol," *Catalysis Today*, vol. 129, no. 1-2, pp. 110–117, 2007.

[23] M. L. Satuf, R. J. Brandi, A. E. Cassano, and O. M. Alfano, "Modeling of a flat plate, slurry reactor for the photocatalytic degradation of 4-chlorophenol," *International Journal of Chemical Reactor Engineering*, vol. 5, no. 1, 2007.

[24] S. L. Orozco, C. A. Arancibia-Bulnes, and R. Suárez-Parra, "Radiation absorption and degradation of an azo dye in a hybrid photocatalytic reactor," *Chemical Engineering Science*, vol. 64, no. 9, pp. 2173–2185, 2009.

[25] Y. Boyjoo, M. Ang, and V. Pareek, "Light intensity distribution in multi-lamp photocatalytic reactors," *Chemical Engineering Science*, vol. 93, pp. 11–21, 2013.

[26] M. L. Satuf, R. J. Brandi, A. E. Cassano, and O. M. Alfano, "Experimental method to evaluate the optical properties of aqueous titanium dioxide suspensions," *Industrial and Engineering Chemistry Research*, vol. 44, no. 17, pp. 6643–6649, 2005.

[27] G. Sagawe, M. L. Satuf, R. J. Brandi et al., "Analysis of photocatalytic reactors employing the photonic efficiency and the removal efficiency parameters: degradation of radiation absorbing and nonabsorbing pollutants," *Industrial and Engineering Chemistry Research*, vol. 49, no. 15, pp. 6898–6908, 2010.

[28] J. Moreira, B. Serrano, A. Ortiz, and H. de Lasa, "Evaluation of photon absorption in an aqueous TiO_2 slurry reactor using Monte Carlo simulations and macroscopic balance," *Industrial and Engineering Chemistry Research*, vol. 49, no. 21, pp. 10524–10534, 2010.

[29] J. Moreira, B. Serrano, A. Ortiz, and H. de Lasa, "TiO_2 absorption and scattering coefficients using Monte Carlo method and macroscopic balances in a photo-CREC unit," *Chemical Engineering Science*, vol. 66, no. 23, pp. 5813–5821, 2011.

[30] G. L. Puma and P. L. Yue, "A laminar falling film slurry photocatalytic reactor. Part I—model development," *Chemical Engineering Science*, vol. 53, no. 16, pp. 2993–3006, 1998.

[31] L. Hurtado, R. Natividad, E. Torres-García, J. Farias, and G. Li Puma, "Correlating the photocatalytic activity and the optical properties of LiVMoO6 photocatalyst under the UV and the visible region of the solar radiation spectrum," *Chemical Engineering Journal*, vol. 262, pp. 1284–1291, 2015.

12

Highly Efficient Photocatalytic Hydrogen on CoS/TiO$_2$ Photocatalysts from Aqueous Methanol Solution

Yu Niu ⓘ,[1] Fuying Li ⓘ,[1] Kai Yang,[2] Qiyou Wu,[1] Peijing Xu,[1] and Renzhang Wang[1]

[1]*Fujian Provincial Collaborative Innovation Center for Clean Coal Gasification, Technology College of Resources and Chemical Engineering, Sanming University, Sanming 365004, China*
[2]*School of Metallurgy and Chemical Engineering, Jiangxi University of Science and Technology, Ganzhou 341000, China*

Correspondence should be addressed to Yu Niu; niuyu200704@163.com

Academic Editor: Chunling Wang

The photocatalyzed water splitting reaction in aqueous methanol solution is an efficient preparation method for hydrogen and methanal under mild conditions. In this work, metal sulfide-loaded TiO$_2$ photocatalysts for hydrogen and methanol production were synthesized by hydrothermal method (180°C/12 h) and characterized by X-ray diffraction (XRD), UV-visible diffuse reflectance spectroscopy (DRS), scanning electron microscopy (SEM), and energy-dispersive X-ray spectroscopy (EDX). The crystal structures of the samples are the typical anatase phase of TiO$_2$ and exhibit a spherical morphology. When TiO$_2$ was loaded with CoS, ZnS, and Bi$_2$S$_3$, respectively, the resulting catalysts showed photocatalytic activities for water decomposition to hydrogen in aqueous methanol solution under 300 W Xe lamp irradiation. Among the photocatalysts with various compositions, the 20 wt% CoS/TiO$_2$ sample with a 2.1 eV band gap showed the maximum photocatalytic activity for the photocatalytic reaction, which indicated that CoS improved the separation ratio of photoexcited electrons and holes. The enhanced activity can be attributed to the intimate junctions that are formed between CoS and TiO$_2$, which can reduce the electron-hole recombination. The production rate of hydrogen with 20 wt% CoS/TiO$_2$ photocatalyst was about 5.6 mmol/g/h, which was 67 times higher than that of pure TiO$_2$. The formation rate of HCHO was 1.9 mmol/g/h with 98.7% selectivity. Moreover, the CoS/TiO$_2$ photocatalyst demonstrated good reusability and stability. In the present study, it is demonstrated that CoS can act as an effective cocatalyst to enhance the photocatalytic hydrogen and methanal production activity of TiO$_2$. The highly improved performance of the CoS/TiO$_2$ composite was mainly ascribed to the efficient charge separation.

1. Introduction

Photocatalytic water splitting into hydrogen, a renewable, clean-burning, and environmental-friendly fuel for future energy sources, is considered as one of the most significant and attractive solutions to solve the global energy and environmental problems [1–3]. A previous study found that adding methanol (CH$_3$OH) to pure water can dramatically enhance H$_2$ production, suggesting that CH$_3$OH plays a crucial role in H$_2$ production [4]. Methanol is used as a raw material for the industrial production of methanol through an oxidation reaction using Ag, Cu, or V$_2$O$_5$ as catalysts. However, this process requires high temperatures of 700–900 K and expensive catalysts. Photocatalytic production of

both hydrogen and methanol from aqueous methanol solution using photocatalysts is an efficient approach to address the above problems. Moreover, the photocatalytic reaction conditions are mild compared to industrial methods. Fujishima and Honda first observed the splitting of water at a TiO$_2$ electrode under the irradiation of ultraviolet (UV) light in 1972 [5]. Since then, TiO$_2$ is considered one of the most promising semiconductor photocatalysts due to its superior photo reactivity, nontoxicity, long-term stability, and low cost [1, 6]. TiO$_2$ has also received a lot of attention as a photocatalyst for hydrogen production [7, 8]. However, the photocatalytic decomposition of water on pure TiO$_2$ photocatalyst is ineffective. One reason is that the production of hydrogen is limited by the rapid recombination of photoexcited holes and electrons. To improve the photocatalytic efficiency, one of the effective strategies is to develop cocatalyst-modified photocatalysts [9–13]. According to previous research, CdS photocatalysts facilitate the production of H_2 by promoting the separation of photoexcited electrons and holes [14–16]. However, CdS is noxious, environmentally hazardous, and costly [17]. Therefore, developing suitable photocatalysts for H_2 production is important and extremely urgent. Our research is focused on the development of nontoxic, environmentally friendly, and inexpensive promoters, such as CoS, ZnS, and Bi_2S_3. Although metal sulfides have demonstrated high activity in H_2 involving reactions in heterogeneous catalysis, CoS has rarely been used as a cocatalyst in photocatalytic H_2 production. In our research, different photocatalysts were successfully prepared through the hydrothermal method and were characterized using XRD, UV-visible DRS, SEM, and EDX analyses. CoS, ZnS, and Bi_2S_3 were investigated as cocatalysts for photocatalytic H_2 and methanal production from methanol solution under 300 W Xe lamp irradiation. The stability and reusability of the catalyst were also evaluated.

2. Experimental

2.1. Catalyst Preparation. All the chemicals were of reagent grade and used as received without any further purification.

The metal sulfide samples were prepared by the hydrothermal method [18]. In a typical procedure, 5 mL deionized water and 20 mL ethyl alcohol were stirred at room temperature for 0.5 h. Different metal salts ($Co(NO_3)_3 \cdot 6H_2O$, 1.46 g; $Zn(CH_3COO)_2$, 0.92 g; $Bi(NO_3)_3 \cdot 5H_2O$, 1.62 g) and thiourea ($(NH_4)_2S$, 0.38 g), which were used as the Co, Zn, Bi, and S precursors, were added to the above solution. The mixture was continuously stirred for 0.5 h and ultrasonicated for 0.5 h to obtain a well-mixed solution. The mixture was then transferred to a Teflon-lined autoclave and heated at 180°C for 12 h. The resulting precipitate was collected by centrifugation and washed successively with distilled water and ethanol three times to remove unbound impurities. It was then dried at 60°C in air for 12 h and ground for 1 h.

The metal sulfide-loaded TiO_2 samples were prepared by the hydrothermal method [19]. In a typical procedure, different amounts of CoS powder were dissolved in 1 mL tetrabutyl titanate and 5 mL ethyl alcohol, and the mixture was stirred at room temperature for 0.5 h. Then, 20 mL deionized water and ammonia water (to adjust pH = 10) were added to the above solution. The solution was continuously stirred for 0.5 h and ultrasonicated for 0.5 h to achieve a well-mixed solution. The mixture was then transferred to a Teflon-lined autoclave and heated at 180°C for 12 h. The resulting precipitate was collected by centrifugation and washed successively with distilled water and ethanol three times to remove unbound impurities. The product was then dried at 60°C in air for 12 h and ground for 1 h and labeled as CoS/TiO_2.

The procedure for the preparation of TiO_2, ZnS/TiO_2, and Bi_2S_3/TiO_2 was the same as that for CoS/TiO_2, except for the different precursors.

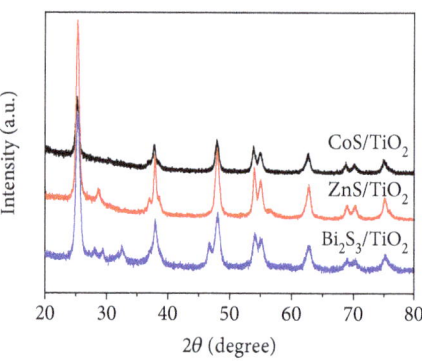

FIGURE 1: XRD patterns of CoS/TiO_2, ZnS/TiO_2, and Bi_2S_3/TiO_2.

2.2. Characterization of Catalysts. The phase compositions of the samples were determined from their XRD patterns, which were obtained using an X'Pert X-ray diffractometer (PANalytical, Netherlands) using Cu Kα radiation ($\lambda = 0.15406$ nm) at a scan rate of 2°/min from 20° to 80° (2θ). The accelerating voltage and applied current were 40 kV and 30 mA, respectively [20].

The micro structures of the samples were determined using SEM images obtained at an accelerating voltage of 20 kV using a ZEISS SIGMA instrument.

The UV–vis diffuse reflection spectra (DRS) were recorded using a Varian Cary 500 Scan UV–vis–NIR spectrometer with $BaSO_4$ as the reference sample. The reflectance spectra were transformed into absorption intensity by using Kubelka-Munk method.

2.3. Catalytic Performance. The photocatalytic reactions were carried out in a sealed quartz tube reactor (volume, 25 mL). The light source was a 300 W Xe lamp. The solid catalyst powder (25 mg) was ultrasonically dispersed in 5.0 mL of mixed solution containing 76 wt% CH_3OH and 24 wt% H_2O. Then, the reactor was evacuated and filled with high-purity (99.999%) nitrogen. The photocatalytic reaction was carried out at room temperature for 12 h. After the reaction, the liquid products were analyzed by high-performance liquid chromatography (HPLC, Shimadzu LC-20A) with both refractive index and UV detectors. The stationary phase was a Shodex SUGARSH-1011 column (8×300 mm) and the mobile phase was a dilute H_2SO_4 aqueous solution. H_2 contents were analyzed by an Agilent Micro GC3000 equipped with a molecular sieve 5A column and a high-sensitivity thermal conductivity detector [21].

3. Results and Discussion

3.1. Characterization of the Samples. The crystalline phases of the samples were characterized by their XRD patterns. Figure 1 shows the XRD patterns of CoS/TiO_2, ZnS/TiO_2, and Bi_2S_3/TiO_2 nanoparticles. For all the samples, the peaks at $2\theta = 25.1°$, 37.6°, 48.0°, 53.8°, 55.0°, and 62.7° can be attributed to the typical anatase phase of TiO_2 (JCPDS: 21–1272) [22]. The XRD patterns show that the loading of metal sulfide nanoparticles did not change the crystal structure of TiO_2. The peaks at $2\theta = 28.7°$ and 29.9° with low intensity can be attributed to CoS [23]. The peak at $2\theta = 28.6°$ with a low intensity is due to ZnS [24]. The peaks at $2\theta = 28.0°$, 29.3°, and 32.5° with low intensity can be attributed to Bi_2S_3 [25].

3.2. UV–Vis Diffuse Reflection Spectra (DRS). Figure 2 shows the DRS of TiO_2, CoS/TiO_2, ZnS/TiO_2, and Bi_2S_3/TiO_2 samples. It can be observed from the spectra that the metal sulfide-loaded TiO_2 samples have enhanced absorption in the visible-light region compared to pure TiO_2. Specifically, CoS/TiO_2 showed stronger absorption than ZnS/TiO_2 and Bi_2S_3/TiO_2. Compared to pure TiO_2 and metal sulfide-loaded TiO_2, the broader absorption bands can be attributed to the type of loaded metal sulfide nanoparticles. The $(ahv)^{1/2}$ vs (hv) spectra were obtained from the corresponding diffuse reflectance spectra by means of the Kubelka-Munk function [26]. Figure 3 shows the curves of $(ahv)^{1/2}$ vs (hv) for the samples. By extrapolating the linear portion of the curves to $(ahv)^{1/2} = 0$, the E_g values of TiO_2, CoS/TiO_2, ZnS/TiO_2, and Bi_2S_3/TiO_2 were determined to be 3.4 eV, 2.1 eV, 3.2 eV, and 2.4 eV, respectively. As a result, CoS/TiO_2 has the largest visible light absorption capacity and the smallest band gap energy. This result is consistent with the fact that the increase in wavelength range of absorption edge in semiconductors is related to the decrease in optical absorption edge energy.

3.3. Morphologies of Samples. SEM and EDX analyses of the samples were carried out to determine the morphologies, polycrystalline structure, and elemental composition of the samples. The SEM images of 20 wt% CoS/TiO_2, 40 wt% CoS/TiO_2, and 60 wt% CoS/TiO_2 samples are presented in Figures 4(a)–4(c), respectively. It can be seen in Figure 4(a) that most of the crystallites are spherical, and their morphologies are almost the same. It can be seen in Figures 4(b) and 4(c) that the crystallite shape transforms from particles to platelets with increase in content of CoS. The SEM images indicate that 20 wt% CoS/TiO_2 nanoparticles showed the best dispersion among all the samples. The EDX spectrum in Figure 4(d) for 20 wt% CoS/TiO_2 sample shows the signals of Ti, O, Co, and S elements.

3.4. Photocatalytic Performance of Samples. The photocatalytic activities of TiO_2, CoS/TiO_2, ZnS/TiO_2, and Bi_2S_3/TiO_2 samples were evaluated using the photocatalytic hydrogen generation reaction in aqueous methanol solution. The results are shown in Figure 5. As can be seen from the figure, the loaded metal sulfides have a significant influence on the photocatalytic activity of TiO_2. When there was no metal sulfide, pure TiO_2 showed low photocatalytic activity because of the rapid recombination between Conduction Band (CB) electrons and Valence Band (VB) holes [27]. Moreover, we found that CoS is a better cocatalyst for H_2 production than ZnS and Bi_2S_3. The photocatalytic activity of the samples decreased in the following order: $CoS/TiO_2 > Bi_2S_3/TiO_2 > ZnS/TiO_2 > TiO_2$. In the liquid phase reaction, HCHO was the major product along with H_2.

Figure 6 shows a comparison of the photocatalytic H_2 production activities of the 5 wt%, 10 wt%, 20 wt%, 40 wt%, and 60 wt% CoS/TiO_2 samples in aqueous methanol

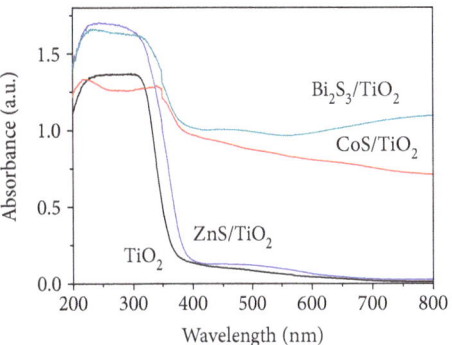

FIGURE 2: UV–vis DRS of the samples.

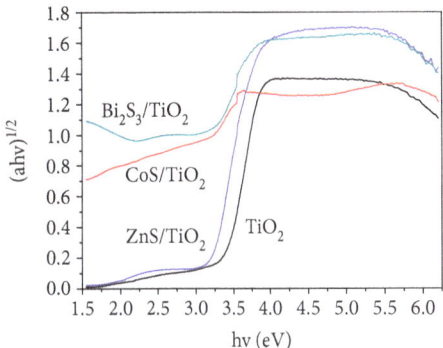

FIGURE 3: Plots of $(ahv)^{1/2}$ vs (hv) for estimating optical band gaps of the samples.

solution. As can be seen from the figure, the content of CoS has a significant influence on the photocatalytic activity of TiO_2. The photocatalytic activity of the samples increased as the content of CoS increased from 5% to 20%. The highest hydrogen and methanal production rates were obtained for the 20 wt% CoS/TiO_2 sample. The H_2 formation rate was 5.6 mmol/g/h, which is 67 times higher than that of pure TiO_2. The formation rate of HCHO was 1.9 mmol/g/h with 98.7% selectivity. As shown in Figure 6, further increase in CoS content resulted in reduced photocatalytic activity. Based on the Debye-Scherrer equation, the calculated crystalline lattice sizes are summarized in Table 1. It is clear that the lattice size increased with the content of the CoS composite, indicating that the introduction of CoS can accelerate the aggregation and growth of TiO_2 nanocrystals. As a result, although an appropriate CoS content plays a role in increasing the photocatalytic activity, the larger TiO_2 nanocrystals lead to decreased photocatalytic activity. Moreover, we speculate that the reaction mechanism involves the activation of C–H bond and O–H bond in methanol by photoexcited holes on CoS/TiO_2 surface. The photogenerated electrons will transfer to the surface of CoS/TiO_2 and reduce protons to H_2.

The capability for reuse is one of the most important factors for an ideal photocatalyst. Hence, the reusability and stability of the 20 wt% CoS/TiO_2 sample were investigated. The sample was collected after each photocatalytic H_2 production experiment and reused for five times. Figure 7 shows

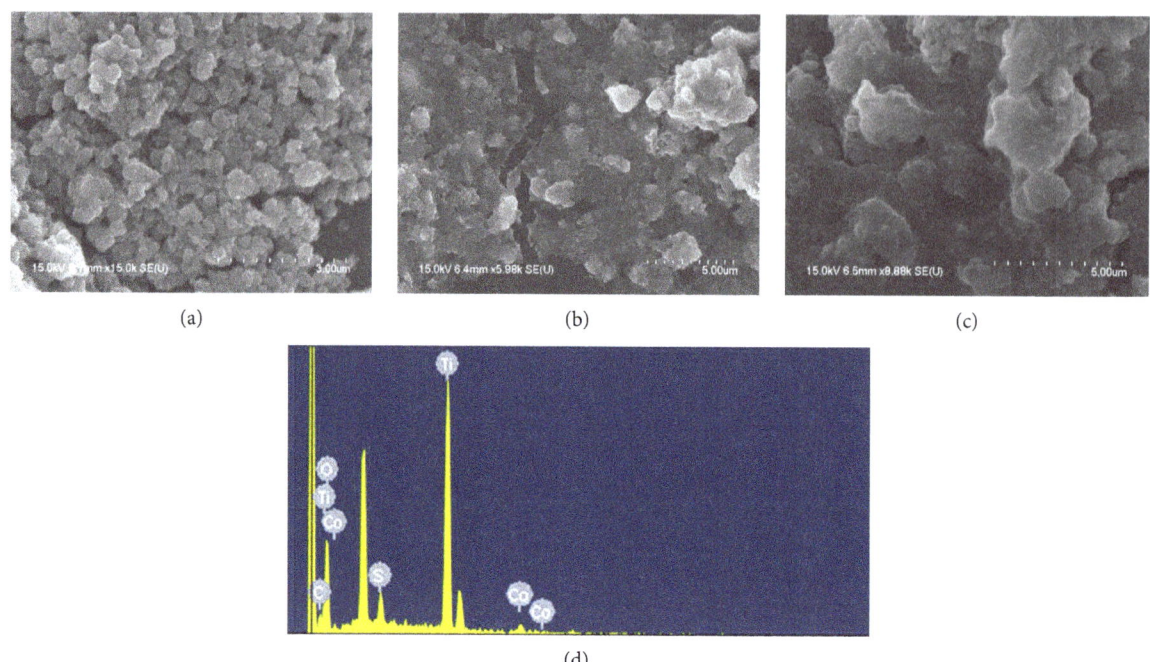

(a)   (b)   (c)

(d)

FIGURE 4: (a) SEM image of 20 wt% CoS/TiO$_2$, (b) SEM image of 40 wt% CoS/TiO$_2$, (c) SEM image of 60 wt% CoS/TiO$_2$, and (d) EDX spectrum of 20 wt% CoS/TiO$_2$.

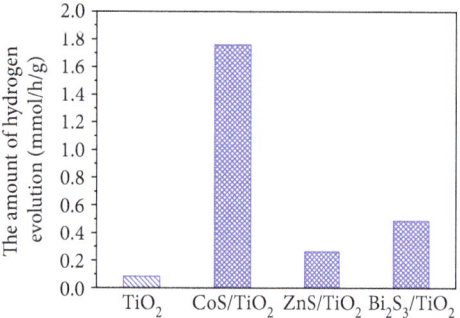

FIGURE 5: Photocatalytic H$_2$ production activity of the samples.

TABLE 1: Lattice size of CoS/TiO$_2$.

Sample	FWHM (rad)	2Theta (°)	Lattice size (nm)
5 wt% CoS/TiO$_2$	0.90	25.24	8.85
10 wt% CoS/TiO$_2$	0.88	25.24	9.37
20 wt% CoS/TiO$_2$	0.87	25.24	9.98
40 wt% CoS/TiO$_2$	0.55	25.24	15.79
60 wt% CoS/TiO$_2$	0.51	25.25	17.03

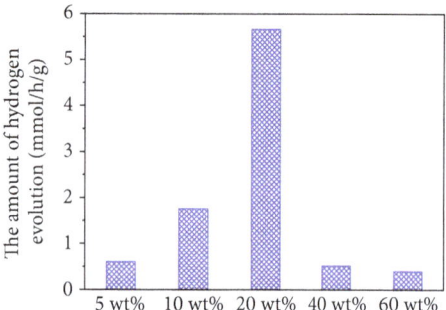

FIGURE 6: Photocatalytic H$_2$ production activity of CoS/TiO$_2$.

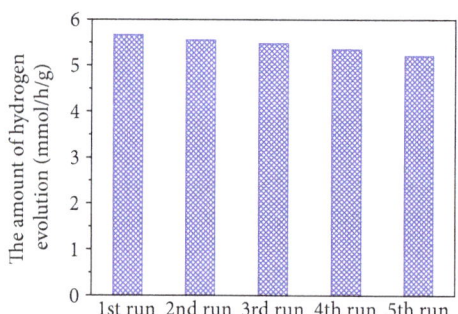

FIGURE 7: Recycling of 20 wt% CoS/TiO$_2$ for photocatalytic H$_2$ production.

the results of five successive H$_2$ production runs under the same experimental conditions. It can be seen that 20 wt% CoS/TiO$_2$ does not exhibit a significant loss in photocatalytic activity in the five recycles.

3.5. Reaction Mechanism. A possible mechanism for the H$_2$ and HCHO production over the CoS/TiO$_2$ photocatalyst proposed is shown in Figure 8. Obviously, the CoS/TiO$_2$ sample as an oxidation and reduction semiconductor can be excited under simulated solar light irradiation. Subsequently, the photogenerated holes will migrate to the host photocatalyst surface, react with methanol, and drive the

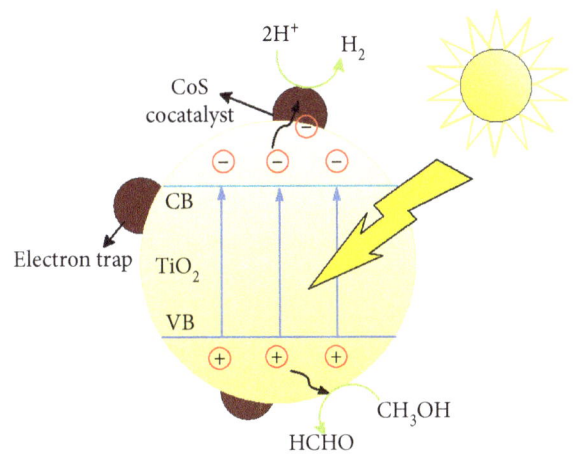

FIGURE 8: Schematic diagram of charge transfer process.

generation of HCHO. Photogenerated electrons in the CB of TiO_2 could quickly transfer to CoS and recombine with holes in the VB of CoS. Then, the electrons in the CB of CoS with stronger reduction ability could drive the generation of H_2. Clearly, TiO_2 lacks the active sites for H_2 evolution, so the rate of H_2 evolution on pure TiO_2 is extremely low. However, when the photogenerated electrons transfer from TiO_2 to the CoS particles, protons can be efficiently reduced to produce H_2 because CoS is a good cocatalyst for the reduction of protons. Moreover, intimate junctions can be formed between CoS and TiO_2, which can facilitate the electron transfer from TiO_2 to CoS and reduce the electron and hole recombination.

4. Conclusion

Metal sulfide-modified TiO_2 catalysts were synthesized using the hydrothermal method. We found that H_2 formation on CoS/TiO_2 is considerably more efficient than on ZnS/TiO_2 and Bi_2S_3/TiO_2. The results showed that CoS/TiO_2 had the best photocatalytic activity in the H_2 and HCHO production reactions under 300 W Xe lamp irradiation. The enhanced activity can be attributed to the intimate junctions that are formed between CoS and TiO_2, which can facilitate the electron transfer from TiO_2 to CoS and reduce the electron-hole recombination. The experimental results showed that a suitable amount of CoS could significantly enhance the photocatalytic activity of TiO_2 for H_2 and methanal production. This result is consistent with SEM analysis of the samples, which showed the highly dispersed nature of the 20 wt% CoS/TiO_2 sample. The maximum photocatalytic activity was obtained for 20 wt% CoS/TiO_2, with hydrogen formation rate of 5.6 mmol/g/h and HCHO formation rate of 1.9 mmol/g/h with selectivity of 98.7%. Moreover, the CoS/TiO_2 photocatalyst showed good reusability and stability.

Conflicts of Interest

The authors declare that there is no conflict of interests regarding the publication of this paper.

Acknowledgments

This work was financially supported by the Fujian Provincial Collaborative Innovation Center for Clean Coal Gasification Technology, the Key Project Young Natural Science Foundation of Fujian Provincial University (JZ160478), Outstanding Youth Scientific Research Talent Incubation plan in Universities of Fujian Province ([2017]52), and the National Natural Science Foundation of China (21707055).

References

[1] M. Ni, M. K. H. Leung, D. Y. C. Leung, and K. Sumathy, "A review and recent developments in photocatalytic water-splitting using TiO₂ for hydrogen production," *Renewable and Sustainable Energy Reviews*, vol. 11, no. 3, pp. 401–425, 2007.

[2] A. Kudo and Y. Miseki, "Heterogeneous photocatalyst materials for water splitting," *Chemical Society Reviews*, vol. 38, no. 1, pp. 253–278, 2009.

[3] Z. Zou, J. Ye, K. Sayama, and H. Arakawa, "Direct splitting of water under visible light irradiation with an oxide semiconductor photocatalyst," *Nature*, vol. 414, no. 6864, pp. 625–627, 2001.

[4] C. Xu, W. Yang, Q. Guo, D. Dai, M. Chen, and X. Yang, "Molecular hydrogen formation from photocatalysis of methanol on anatase-TiO₂ (101)," *Journal of the American Chemical Society*, vol. 136, no. 2, pp. 602–605, 2014.

[5] A. Fujishima and K. Honda, "Electrochemical photolysis of water at a semiconductor electrode," *Nature*, vol. 238, no. 5358, pp. 37-38, 1972.

[6] K. Hashimoto, H. Irie, and A. Fujishima, "TiO₂ photocatalysis: a historical overview and future prospects," *Japanese Journal of Applied Physics*, vol. 44, no. 12, pp. 8269–8285, 2005.

[7] J. Zhang, Q. Xu, Z. Feng, M. Li, and C. Li, "Importance of the relationship between surface phases and photocatalytic activity of TiO₂," *Angewandte Chemie International Edition*, vol. 47, no. 9, pp. 1766–1769, 2008.

[8] J. Yu, W. Wang, B. Cheng, and B.-L. Su, "Enhancement of photocatalytic activity of mesoporous TiO₂ powders by hydrothermal surface fluorination treatment," *Journal of Physical Chemistry C*, vol. 113, no. 16, pp. 6743–6750, 2009.

[9] S. G. Kumar and L. G. Devi, "Review on modified TiO₂ photocatalysis under UV/visible light: selected results and related mechanisms on interfacial charge carrier transfer dynamics," *The Journal of Physical Chemistry A*, vol. 115, no. 46, pp. 13211–13241, 2011.

[10] R. Daghrir, P. Drogui, and D. Robert, "Modified TiO₂ for environmental photocatalytic applications: a review," *Industrial & Engineering Chemistry Research*, vol. 52, no. 10, pp. 3581–3599, 2013.

[11] H. R. Liang and L. J. Guo, "Photocatalytic H₂ evolution under visible-light irradiation on modified TiO₂ catalysts," *Advanced Materials Research*, vol. 512-515, pp. 1426–1431, 2012.

[12] X. Ren, X. Qi, Y. Shen et al., "2D co-catalytic MoS_2 nanosheets embedded with 1D TiO_2 nanoparticles for enhancing photocatalytic activity," *Journal of Physics D: Applied Physics*, vol. 49, no. 31, article 315304, 2016.

[13] P. Ganesan, A. Sivanantham, and S. Shanmugam, "CoS_2–TiO_2 hybrid nanostructures: efficient and durable bifunctional electrocatalysts for alkaline electrolyte membrane water electrolyzers," *Journal of Materials Chemistry A*, vol. 6, no. 3, pp. 1075–1085, 2018.

[14] L. Cheng, Q. Xiang, Y. Liao, and H. Zhang, "CdS-based photocatalysts," *Energy & Environmental Science*, vol. 11, no. 6, pp. 1362–1391, 2018.

[15] Z. Hu, H. Quan, Z. Chen, Y. Shao, and D. Li, "New insight into an efficient visible light-driven photocatalytic organic transformation over CdS/TiO_2 photocatalysts," *Photochemical & Photobiological Sciences*, vol. 17, no. 1, pp. 51–59, 2018.

[16] N. L. Reddy, V. N. Rao, M. M. Kumari, P. Ravi, M. Sathish, and M. V. Shankar, "Effective shuttling of photoexcitons on CdS/NiO core/shell photocatalysts for enhanced photocatalytic hydrogen production," *Materials Research Bulletin*, vol. 101, pp. 223–231, 2018.

[17] H. Yan, J. Yang, G. Ma et al., "Visible-light-driven hydrogen production with extremely high quantum efficiency on Pt–PdS/CdS photocatalyst," *Journal of Catalysis*, vol. 266, no. 2, pp. 165–168, 2009.

[18] Q. Xiang, J. Yu, and M. Jaroniec, "Synergetic effect of MoS_2 and graphene as cocatalysts for enhanced photocatalytic H_2 production activity of TiO_2 nanoparticles," *Journal of the American Chemical Society*, vol. 134, no. 15, pp. 6575–6578, 2012.

[19] Q. Wang, N. An, Y. Bai et al., "High photocatalytic hydrogen production from methanol aqueous solution using the photocatalysts CuS/TiO_2," *International Journal of Hydrogen Energy*, vol. 38, no. 25, pp. 10739–10745, 2013.

[20] Y. Niu, P. Huang, F. Li et al., "Noble metal decoration and presulfation on TiO_2: increased photocatalytic activity and efficient esterification of *n*-butanol with citric acid," *International Journal of Photoenergy*, vol. 2016, Article ID 4618924, 12 pages, 2016.

[21] S. Xie, Z. Shen, J. Deng et al., "Visible light-driven C–H activation and C–C coupling of methanol into ethylene glycol," *Nature Communications*, vol. 9, no. 1, article 1181, 2018.

[22] Y. Sui, Q. Liu, T. Jiang, and Y. Guo, "Synthesis of nano-TiO_2 photocatalysts with tunable Fe doping concentration from Ti-bearing tailings," *Applied Surface Science*, vol. 428, pp. 1149–1158, 2018.

[23] X. Fang, J. Song, T. Pu et al., "Graphitic carbon nitride-stabilized CdS@CoS nanorods: an efficient visible-light-driven photocatalyst for hydrogen evolution with enhanced photo-corrosion resistance," *International Journal of Hydrogen Energy*, vol. 42, no. 47, pp. 28183–28192, 2017.

[24] Y. Zhang, N. Zhang, R. Tang Z, and Y.-J. Xu, "Graphene transforms wide band gap ZnS to a visible light photocatalyst. The new role of graphene as a macromolecular photosensitizer," *ACS Nano*, vol. 6, no. 11, pp. 9777–9789, 2012.

[25] J. Cao, B. Xu, H. Lin, B. Luo, and S. Chen, "Novel heterostructured $Bi_2S_3/BiOI$ photocatalyst: facile preparation, characterization and visible light photocatalytic performance," *Dalton Transactions*, vol. 41, no. 37, pp. 11482–11490, 2012.

[26] X. Gao, G. Huang, H. Gao et al., "Facile fabrication of Bi_2S_3/SnS_2 heterojunction photocatalysts with efficient photocata-

lytic activity under visible light," *Journal of Alloys and Compounds*, vol. 674, pp. 98–108, 2016.

[27] S. Yu, "Ultrasonic aerosol spray-assisted preparation of TiO_2/In_2O_3 composite for visible-light-driven photocatalysis," *Journal of Catalysis*, vol. 310, pp. 84–90, 2014.

13

Inspired Preparation of Zinc Oxide Nanocatalyst and the Photocatalytic Activity in the Treatment of Methyl Orange Dye and Paraquat Herbicide

Ghaida H. Munshi [1,2] **Amal M. Ibrahim** [2,3] **and Laila M. Al-Harbi** [1]

[1]*Chemistry Department, Faculty of Science, King Abdulaziz University, P.O. Box 80203, Jeddah 21589, Saudi Arabia*
[2]*Chemistry Department, Faculty of Science, University of Jeddah, P.O. Box 80327, Jeddah 23218, Saudi Arabia*
[3]*Physical Chemistry Department, National Research Centre, P.O. Box 12311, Cairo 12622, Egypt*

Correspondence should be addressed to Laila M. Al-Harbi; lalhrbi@kau.edu.sa

Academic Editor: Jinn Kong Sheu

As the need to use green chemistry routes increases, environmentally friendly catalytic processes are a demand. One of the most important and abundant naturally occurring catalysts is chlorophyll. Chlorophyll is the first recognized catalyst; it is a reducing agent due to its electron-rich structure. The effects of spinach on the preparation of zinc oxide nanoparticles and the photocatalytic degradation of methyl orange and paraquat in sunlight and under a UV lamp and photocatalytic degradation in sunlight were studied. Different parameters of the catalytic preparation process and photocatalytic degradation process were studied. Characterization of differently prepared samples was carried out using different analytical techniques such as XRD, SEM, and EDX and finally the photocatalytic activity towards decomposition of methyl orange and paraquat.

1. Introduction

Nanostructure materials have attracted increasing attention during the past few decades due to their marvelous properties and a wide range of applications such as catalysis, electronics, optics, and environmental and biotechnology applications [1–6]. The increase in world demand to manipulate nanostructure materials trying to solve technological and environmental situations should be accomplished by green chemistry processes to minimize the hazards of the traditional chemical processes. Among the most widely used nanostructured metal oxides, ZnO is considered as one of the most important metal oxides with unique properties as high-surface energy, high-electron mobility, cheap, and environmentally nontoxic, and the most significant property is at wide bandgap (3.3 ev). This wide bandgap makes ZnO a field of many studies in using and improving the applications in electronics light emitters, chemical sensors, and photocatalysts [7, 8].

In general, many types of research discuss the synthesis of nanostructured materials and zinc oxide nanoparticles especially that these methods include chemical precipitation, sol-gel synthesis, hydrothermal reaction, and microwave. Many other chemical and physical attempts were carried out to improve and control the synthesis of metal oxide nanoparticles [9–12].

Green route synthesis is one of the most promising techniques for the synthesis of nanostructured materials being simple, environmentally safe (mild reaction conditions and no need for toxic chemicals) and inexpensive and can be produced in a large-scale process [13–15]. In the present work, a traditional and microwave-assisted green synthesis of ZnO using spinach extract was studied and the resulting ZnO nanoparticles were characterized using scanning electron microscopy (SEM) energy, dispersive X-ray spectroscopy (EDX), and X-ray diffraction spectroscopy (XRD); also, the catalytic activity of the obtained ZnO nanoparticles is evaluated.

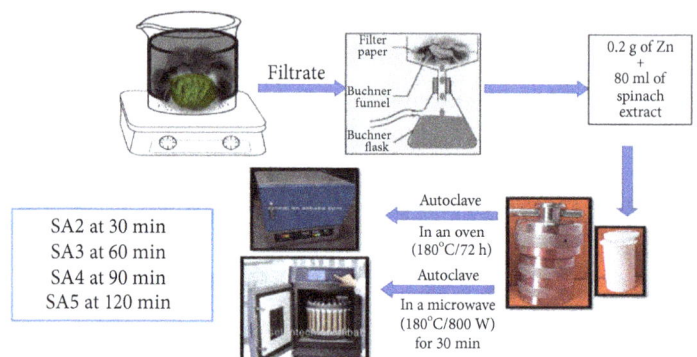

SCHEME 1: Preparation equipment of ZnO catalyst both with conventional heating technique and microwave-assisted technique.

2. Materials and Methods

2.1. Materials. Zinc powder (Ranbaxy Chemicals, >5 μm) has been used without any preheated producer or any further purification, and deionized water has been prepared in the laboratory. For the synthesis of nanomaterials, a closed cylindrical Teflon-lined stainless steel chamber of 100 ml capacity (autoclave, Latech Scientific Supply Pte. Ltd. Company) was used. 4, 4$'$-Bipyridine and methyl orange indicator were purchased from Sigma-Aldrich Chemical Co., and they were used without further purification.

2.2. Preparation of the Fresh Spinach Extract. Spinach (100 g) bought from a local market was washed with deionized water and patted dry with wipes. The fresh spinach leaves were heated using 200 ml of deionized water and filtered using a filter paper. The filtrate was then used to prepare zinc oxide nanoparticles.

2.3. Synthesis of Zinc Oxide Nanoparticles

2.3.1. Synthesis of Zinc Oxide Nanoparticles by Traditional Technique. Zinc metal (powder) (0.1 g) was added to 40 ml deionized (DI) water, transferred into a stainless steel Teflon-lined metallic bomb of 100 ml capacity, and sealed under inert conditions. Then, it was placed in the microwave (Milestone company (high-performance microwave digestion system) ETHOS One, SN: 1301 0243) at 800 W/180°C for 30 minutes. The furnace allowed to cool after the desired time, and the resulting suspension was centrifuged to retrieve the product (S1), washed, and then finally vacuum dried for few hours. Zinc metal (powder) (0.2 g) was added to 80 ml aqueous fresh spinach extract transferred into a stainless steel Teflon-lined metallic bomb of 100 ml capacity and sealed under inert conditions. The closed chamber was then placed in a preheated box furnace, and the mixture was heated slowly (2°C/min) to 180°C and maintained at this temperature for 72 hours. The furnace allowed to cool after the desired time, and the resulting suspension was centrifuged to retrieve the product (SA1), washed, and then finally vacuum dried for few hours.

2.3.2. Microwave-Assisted Heating Technique. Zinc metal (0.2 gram) was added to 40 ml aqueous fresh spinach extract transferred into a stainless steel Teflon-lined metallic bomb

of 100 ml capacity and sealed under inert conditions. Then, it was placed in the microwave (Milestone company (high-performance microwave digestion system) ETHOS One, SN: 1301 0243) at 800 W/180°C for 30 minutes (SA2), 60 minutes (SA3), 90 minutes (SA4), and 120 minutes (SA5). The furnace was allowed to cool after the desired time, and the resulting suspension was centrifuged to retrieve the product, washed, and then finally vacuum dried for few hours Scheme 1.

2.4. The Photocatalytic Degradation of the Methyl Orange (MO). The photocatalytic degradation of methyl orange (MO) was performed in a Pyrex beaker using as-synthesized ZnO nanoparticles as a photocatalyst under UV illumination for various time intervals and sunlight at 48°C. MO dye (10 ppm) solution was prepared in 100 ml DI water and mixed with 0.3 g of different kinds of synthesized ZnO, in which nanoparticles powder was added to it. The resulting suspension was equilibrated by stirring for 30 min to stabilize the absorption of MO dye over the surface of the photocatalyst, that is, ZnO nanoparticles, before exposing to the light. The photocatalytic decomposition of MO was examined by measuring the absorbance at regular time intervals by using the ultraviolet and visible (UV-Vis) spectrophotometer wavelength at 465 nm. Analytical samples were taken from the reaction suspension at regular time intervals for 10 minutes and were then analyzed for their absorption using UV-Vis spectrophotometer.

2.5. The Photocatalytic Degradation of the Herbicide "Paraquat." Photocatalytic degradation experiments were carried in a reactor system that has been already described. It has a Pyrex glass tube reactor (200 ml) that can be irradiated with sunlight at 48°C temperature. Since the concentration of organic pollutants is a very important parameter in the wastewater treatment processes, we have studied the effect of paraquat initial concentration on the reaction rate to develop a kinetic model for the photocatalytic degradation of paraquat. Several experiments were carried out with paraquat aqueous solutions with different initial concentration (10 and 50–100 ppm), and different kinds of synthesized ZnO nanoparticles powder was added to it. For each of the experiments, 100 ml of a paraquat solution was placed inside the glass reactor and mixed with 0.3 g of ZnO. This slurry was

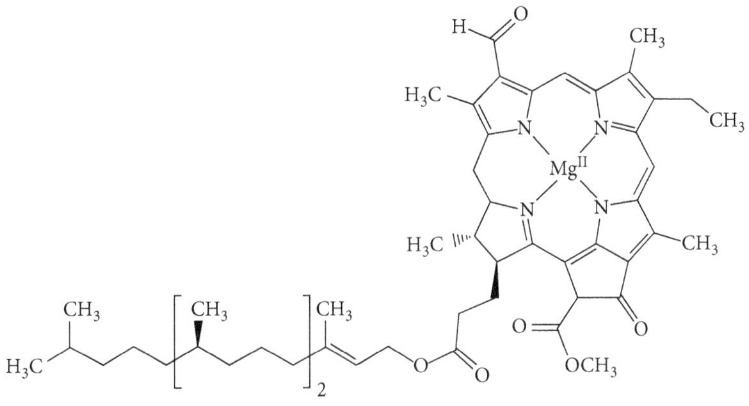

FIGURE 1: Chlorophyll a (Chl a).

SCHEME 2: A mechanism proposed for the catalytic preparation of ZnO catalyst.

agitated with a magnetic stirrer. Samples for analysis were taken at different times to monitor the reaction. Analytical samples were taken from the reaction suspension at regular time intervals for 60 min and measuring the absorbance at regular time intervals by using the UV-Vis spectrophotometer wavelength at 230 nm wavelength [16, 17].

3. Results and Discussion

Spinach is considered as an amazing green chemistry candidate, which indeed contains chlorophyll. Chlorophyll is considered as the most important naturally occurring photocatalyst on earth. The supposed mechanism of the preparation of zinc oxide could be started with the reducing effect of chlorophyll a (Chl a) (Figure 1), which can lose an electron from its aromatic л-electron system of porphyrin [18]. The reducing properties of Chl a and the produced electrons from its aromatic л-electron attack the water molecule to produce hydroxyl group (OH⁻) which can attack the zinc metal to give zinc hydroxide which is eventually converted into zinc oxide (ZnO) (Scheme 2) [19].

3.1. Scanning Electron Microscope Spectroscopy (SEM). The SEM images were shown in Figure 2, representing differently prepared samples with and without the presence of spinach extract using the microwave-assisted technique as well as the conventional method. It was observed that the presence of spinach extract affected the morphology of the produced ZnO nanoparticles. Sample S1 represents the blank sample without the presence of spinach extract showing nonhomogeneous morphology with a short nanorod shape. Meanwhile, SA1 is representing ZnO nanoparticles produced in the presence of spinach extract using the traditional technique. The image showed the homogeneous morphology of an elongated rod shape with an average diameter of 150 nm. This well-crystalline morphology could be attributed to a long time of the experiment. On the other hand, samples SA2, SA3, SA4, and SA5 represent the morphology of ZnO nanoparticles produced from the microwave-assisted technique in the presence of spinach extract for 30, 60, 90, and 120 minutes, respectively. Sample SA2 images showed low crystallinity due to the fast preparation process. As the preparation time increases, the crystallinity of the produced ZnO nanoparticles increases. The images of samples SA3, SA4, and SA5 showed agglomerated particles from smaller flower-like shape particles with an average size ranging from 50 nm to 100 nm. Energy dispersive X-ray spectrophotometry (EDX) results (Figure 3) accompanied by SEM showed that the produced nanoparticles are composed of Zn and O with Zn : O ratio in agreement.

3.2. X-Ray Diffraction Spectroscopy (XRD) of ZnO Nanoparticles. Study of standard data JCPDS 76-0704 confirmed that the synthesized materials are hexagonal ZnO phase (wurtzite structure). The pattern was indexed with hexagonal unit cell structure with P63mc, and the lattice parameters are given in Table 1. No diffraction peaks arising from any impurity can be detected in the pattern which confirms that they are the grown products of pure ZnO. The deviation of the lattice parameters is caused due to the presence of various point defects such as zinc antisites, oxygen vacancies, and extended defects, such as threading dislocation. ZnO prefers (002) direction for growth; this is attributed to the fact that the (002) plan has the lowest surface free energy [5, 20]. Both temperature and time affect the migration of atoms toward the preferred orientation. The behavior observed in XRD pattern could be attributed to the rate of heating which is considered much slower in conventional method than that in microwave-assisted technique that gives the needed time for the atoms to get their preferred orientation. The crystallite size of the produced nanoparticles was calculated by using the Scherr formula [21].

$$D = \frac{0.9\,\lambda}{\beta \cos \theta} \tag{1}$$

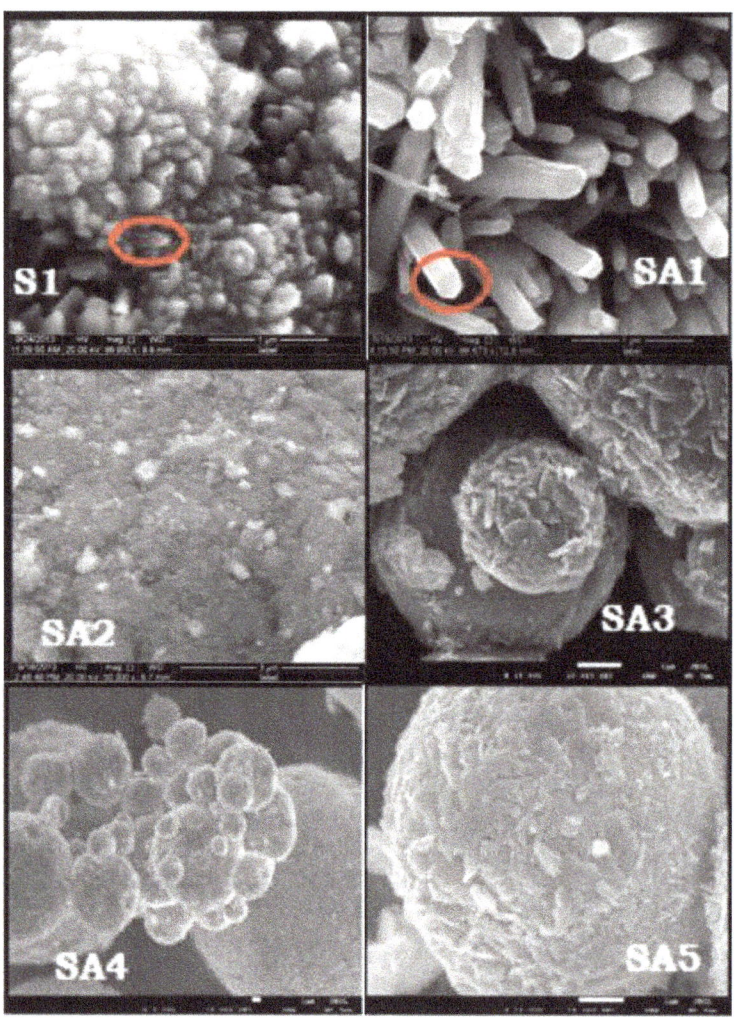

Figure 2: Scanning electron microscope spectroscopy images (S1: zero sample, SA1: conventional heating; SA2, SA3, SA4, and SA5 samples with assisted microwave technique).

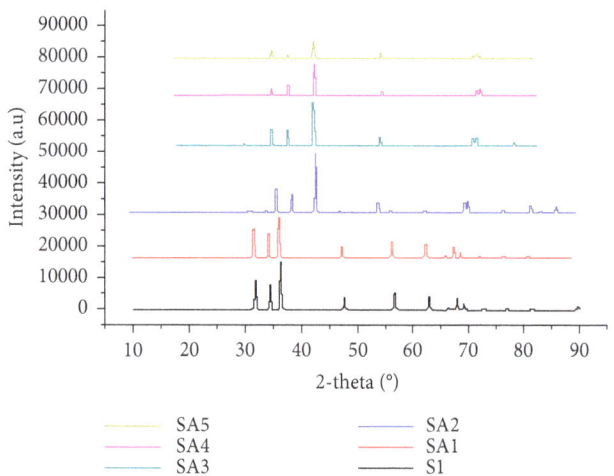

Figure 3: XRD patterns of ZnO nanoparticles.

where λ is the wavelength of the X-ray (0. 1541 A°), the FWHM (full width at half maximum) of the more intense peak, and the diffraction angle, and D is the particle diameter size.

3.3. Application of Catalytic Activity

3.3.1. The Photocatalytic Degradation of the Methyl Orange (MO)

(1) Effect of Catalyst Loading under Sunlight. A series of experiments was carried out to assess the optimum catalyst loading by varying the amount of catalyst from 0.1 g to 0.35 g of which the dye solution was prepared. The percentages of photodegradation under sunlight in 48°C with ZnO were illustrated in Figure 4. It was found that all microwave samples give almost the same percentage removal at a certain concentration, so we will discuss here only SA1. It was observed that with increasing the catalyst amount, the photodegradation increases. However, when the amount of catalyst exceeds the optimum amount (0.3 g), the photodegradation efficiency decreases. This could be attributed to the fact that

TABLE 1: The variation of lattice parameters with a variation of experimental conditions.

Sample	Volume (A°)³	a (A°)	b (A°)	c (A°)	c/a (A°)	Crystalline size (A°)
S1	47.65	3.2494 (14)	3.2494 (14)	5.211 (3)	1.4914	270 (8)
SA1	47.68	3.2500 (5)	3.2500 (5)	5.2056 (9)	1.6017	488 (12)
SA2	47.73	3.2517 (9)	3.2517 (9)	5.2125 (18)	1.6030	221.1 (7)
SA3	47.63	3.2494 (17)	3.2494 (17)	5.208 (3)	1.6028	224.97 (6)
SA4	47.51	3.2473 (14)	3.2473 (14)	5.202 (2)	1.6020	600 (7)
SA5	47.57	3.2481 (14)	3.2481 (14)	5.207 (2)	1.6030	234 (2)

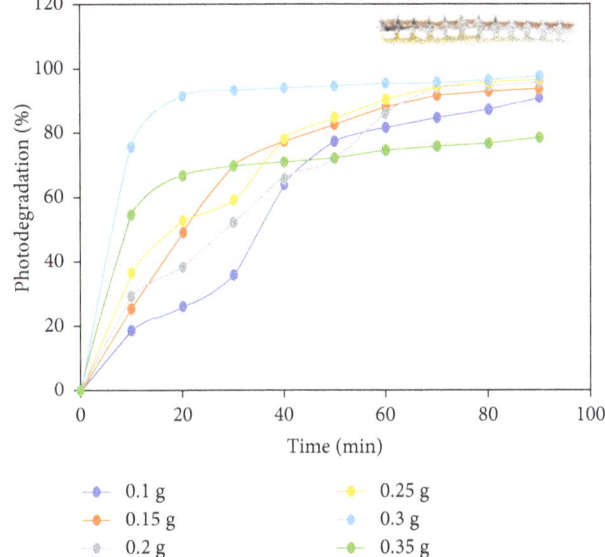

FIGURE 4: Variation of % of photodegradation versus time interval for the photodegradation of MO with a different dosage.

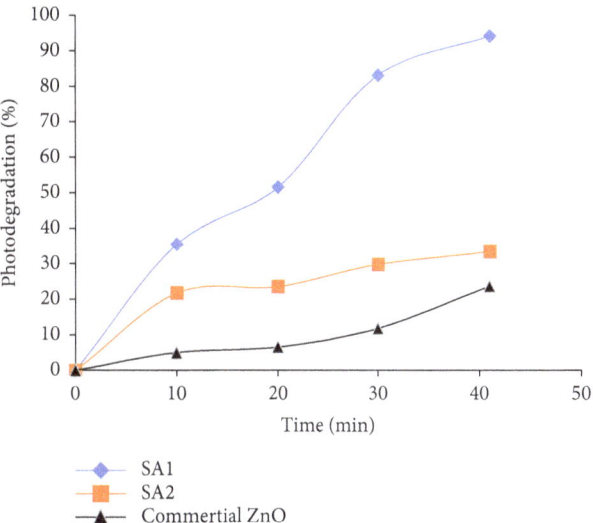

FIGURE 5: Variation of % of photodegradation versus time for various different ZnO catalyst (0.3 g) under sunlight.

the number of active sites increased as the amount of the catalyst increased, till to a certain point after which the higher concentrations of the catalyst increase the turbidity of the MO suspension and the penetration of sunlight, and UV light decreases thereby increasing the scattering effect and lowering the photodegradation rate.

(2) Effect of Different Prepared Method of Zinc Oxide Catalyst under Sunlight. The effect of zinc oxide catalyst prepared by different methods on the dye degradation was performed by varying the kind of ZnO nanoparticles of SA1, SA2, and zinc oxide commercial (0.18 ml≈0.3 g) with constant catalyst loading (0.3 g) and 10 ppm of MO dye. It was observed that the prepared ZnO nanocatalysts were excellent candidates with the commercial ZnO catalyst with very promising activity, and the sample of ZnO obtained by using conventional heating technique was the most active one in degradation of MO dye (Figure 5). The high photocatalytic activity of SA1 sample could be explained by the fact that the slow rate of heating (accompanied with the traditional heating technique) provides the time needed for the ZnO nanoparticles to attain the preferred orientation. As it was noticed from SEM images, in this case, the elongated rod structure with

small diameter increases the surface area and consequently the photocatalytic activity.

(3) Photocatalytic Activity of Zinc Oxide Catalyst Obtained under Different Parameters under a UV Lamp. The relationship between photodegradation efficiency of the dyes and time under UV lamp is presented in Figure 6. It is clearly observed that the sample of ZnO catalyst obtained with conventional heating technique S has 91% photodegradation efficiency to MO dye.

(4) The Photocatalytic Degradation of the Herbicide "Paraquat". As it was obvious from previous photocatalytic studies, step SA1 is the most active sample among the prepared ZnO photocatalyst even it gives higher activity than that of the commercial ZnO photocatalyst; thus, this study shows the photocatalytic degradation of aqueous paraquat solutions with zinc oxide (SA1) and TOC analysis illuminated under sunlight in (48°C). The ZnO sample (SA1) showed in Figure 7 features high photocatalytic activity in paraquat photodegradation. It is well known that as the concentration of the pollutant increases, the photocatalytic activity decreases due to the decrease of the tendency of the irradiating light beam to meet the catalyst particles. The paraquat

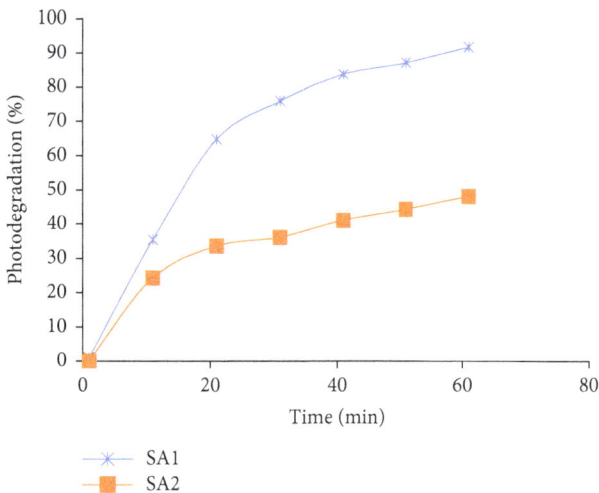

FIGURE 6: Variation of % of photodegradation versus time for various catalyst of ZnO (0.3 g) under a UV lamp.

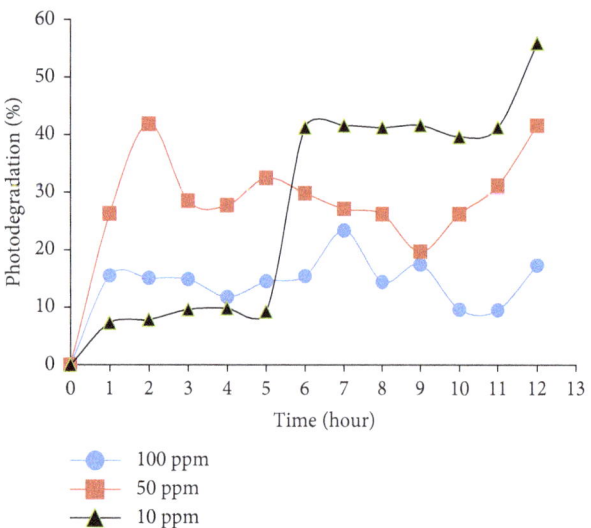

FIGURE 7: Variation of % of photodegradation versus for time of sample SA1 (0.3 g) under sunlight.

particle to catalyst particle ratio increases with increasing the concentration of paraquat. This behavior decreases the electron/hole pair formation probability, decreasing the photocatalytic activity.

(5) Total Organic Carbon (TOC). TOC analysis (total organic carbon) is one of the most important monitoring tests for tracking the removal of organic pollutants from water as the TOC content in water decreases the pollutant concentration decrease. For ZnO sample (SA1), Figure 8 showed high photodegradation activity at paraquat concentration 100 ppm which leads to the decrease of TOC from 62.6 ppm to 35 ppm in 12 hours. This photodegradation activity supported the decrease in TOC with sample SA1. This behavior could be devoted to their short diameter, high crystalline, and high surface area.

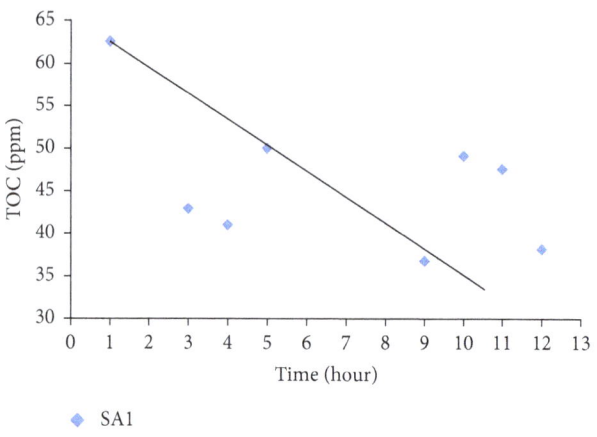

FIGURE 8: TOC values of SA1 versus time comparison photodegradation of paraquat (100 ppm) under sunlight.

4. Conclusions

The preparation of zinc oxide nanocatalyst using a green chemistry route as a catalyst (spinach extract) and also the catalytic activity evaluation of the prepared samples was studied. It is clearly concluded that the obtained ZnO nanocatalysts represent an excellent candidate with the commercial ZnO catalyst with high catalytic activity in the degradation of the methyl orange dye (MO) as well as paraquat herbicide. The produced ZnO catalyst samples were found to depend on the preparation parameters in their shape, size, and consequently catalytic activity. Among the obtained samples prepared with different heating techniques (traditional and microwave-assisted), the most promising sample is the one prepared under traditional heating technique (SA1). This behavior could be devoted to that the fact that the rate of heating in the traditional way is much slower than that in a microwave-assisted way. The slower rate of heating, in this case, provides the needed time for the produced ZnO catalyst to get the preferred crystalline orientation and in this case for the rod structure to be elongated with small diameter (150 nm); this increases the surface area and consequently increases the catalytic activity. All the used characterization and evaluation tools support the conclusion that chlorophyll is an excellent catalyst which acts under mild conditions and is favored economically.

Conflicts of Interest

The authors declare that they have no conflicts of interest.

Acknowledgments

This work was funded by the King Abdulaziz City for Science and Technology (KACST), under Grant no. 177-34. The authors, therefore, acknowledge with thanks the KASCT for technical and financial support.

References

[1] L. L. Welbes and A. S. Borovik, "Confinement of metal complexes within porous hosts: development of functional materials for gas binding and catalysis," *Accounts of Chemical Research*, vol. 38, no. 10, pp. 765–774, 2005.

[2] X. Wang, J. Zhou, J. Song, J. Liu, N. Xu, and Z. L. Wang, "Piezoelectric field effect transistor and nanoforce sensor based on a single ZnO nanowire," *Nano Letters*, vol. 6, no. 12, pp. 2768–2772, 2006.

[3] M. Law, L. E. Greene, J. C. Johnson, R. Saykally, and P. Yang, "Nanowire dye-sensitized solar cells," *Nature Materials*, vol. 4, no. 6, pp. 455–459, 2005.

[4] J. H. Lim, C. K. Kang, K. K. Kim, I. K. Park, D. K. Hwang, and S. J. Park, "UV electroluminescence emission from ZnO light-emitting diodes grown by high-temperature radiofrequency sputtering," *Advanced Materials*, vol. 18, no. 20, pp. 2720–2724, 2006.

[5] K. Ikegami, T. Yoshiyama, K. Maejima, H. Shibata, H. Tampo, and S. Niki, "Optical dielectric constant inhomogeneity along the growth axis in ZnO-based transparent electrodes deposited on glass substrates," *Journal of Applied Physics*, vol. 105, no. 9, article 093713, 2009.

[6] J. Schrier, D. O. Demchenko, Lin-Wang, and A. P. Alivisatos, "Optical properties of ZnO/ZnS and ZnO/ZnTe heterostructures for photovoltaic applications," *Nano Letters*, vol. 7, no. 8, pp. 2377–2382, 2007.

[7] A. B. Djurišić, X. Chen, Y. H. Leung, and A. Man Ching Ng, "ZnO nanostructures: growth, properties and applications," *Journal of Materials Chemistry*, vol. 22, no. 14, pp. 6526–6535, 2012.

[8] J. Bao, M. A. Zimmler, F. Capasso, X. Wang, and Z. F. Ren, "Broadband ZnO single-nanowire light-emitting diode," *Nano Letters*, vol. 6, no. 8, pp. 1719–1722, 2006.

[9] V. Sharma, "Sol-gel mediated facile synthesis of zinc-oxide nanoaggregates, their characterization and antibacterial activity," *IOSR Journal of Applied Chemistry*, vol. 2, no. 6, pp. 52–55, 2012.

[10] J.-S. Huang and C.-F. Lin, "Controlled growth of zinc oxide nanorod array in aqueous solution by zinc oxide sol-gel thin film in relation to growth rate and optical property," in *2008 8th IEEE Conference on Nanotechnology*, pp. 135–138, Arlington, TX, USA, August 2008.

[11] L. M. AL-Harbi, E. H. El-Mossalamy, H. M. Arafa, A. Al-Owais, and M. A. Shah, "Growth of zinc oxide (ZnO) nanorods and their optical properties," *Modern Applied Science*, vol. 5, no. 2, 2011.

[12] A. M. Ibrahim, M. M. Abd El-Latif, and M. S. Gohr, "Water / alcohol mediated preparation of ZnO hollow sphere," *Egyptian Journal of Chemistry*, vol. 58, no. 4, pp. 475–484, 2015.

[13] H. Zhang, X. Ma, J. Xu, J. Niu, and D. Yang, "Arrays of ZnO nanowires fabricated by a simple chemical solution route," *Nanotechnology*, vol. 14, no. 4, pp. 423–426, 2003.

[14] J. V. D. S. Araújo, R. V. Ferreira, M. I. Yoshida, and V. M. D. Pasa, "Zinc nanowires synthesized on a large scale by a simple carbothermal process," *Solid State Sciences*, vol. 11, no. 9, pp. 1673–1679, 2009.

[15] J. D. Rocha, A. R. Coutinho, and C. A. Luengo, "Biopitch produced from eucalyptus wood pyrolysis liquids as a renewable binder for carbon electrode manufacture," *Brazilian Journal of Chemical Engineering*, vol. 19, no. 2, pp. 127–132, 2002.

[16] T. Fernandes, S. Soares, T. Trindade, and A. Daniel-da-Silva, "Magnetic hybrid nanosorbents for the uptake of paraquat from water," *Nanomaterials*, vol. 7, no. 3, p. 68, 2017.

[17] M. G. Sorolla II, M. L. Dalida, P. Khemthong, and N. Grisdanurak, "Photocatalytic degradation of paraquat using nano-sized Cu-TiO$_2$/SBA-15 under UV and visible light," *Journal of Environmental Sciences*, vol. 24, no. 6, pp. 1125–1132, 2012.

[18] S. Shanmugam, J. Xu, and C. Boyer, "Utilizing the electron transfer mechanism of chlorophyll a under light for controlled radical polymerization," *Chemical Science*, vol. 6, no. 2, pp. 1341–1349, 2015.

[19] U. Maitra, S. R. Lingampalli, and C. N. R. Rao, "Artificial photosynthesis and the splitting of water to generate hydrogen," *Current Science*, vol. 106, no. 4, pp. 518–527, 2014.

[20] Y. Sun, D. J. Riley, and M. N. R. Ashfold, "Mechanism of ZnO nanotube growth by hydrothermal methods on ZnO film-coated Si substrates," *The Journal of Physical Chemistry B*, vol. 110, no. 31, pp. 15186–15192, 2006.

[21] B. D. Cullity, *Elements of X-Ray Diffraction*, Addison-Wesley, 2001.

14

Simulations of a Single-Phase Flow in a Compound Parabolic Concentrator Reactor

Tzayam Pérez[ID][1] and José L. Nava[2]

[1]*Departamento de Ing. Química, Noria Alta, Universidad de Guanajuato, 36050 Guanajuato, GTO, Mexico*
[2]*Departamento de Ingeniería Geomática e Hidráulica, Av. Juárez 77, Zona Centro, Universidad de Guanajuato, 36000 Guanajuato, GTO, Mexico*

Correspondence should be addressed to Tzayam Pérez; t.perezsegura@ugto.mx

Academic Editor: Leonardo Palmisano

This paper deals with the analysis and interpretation of flow visualization and residence time distribution (RTD) in a compound parabolic concentrator (CPC) reactor using computational fluid dynamics (CFD). CFD was calculated under turbulent flow conditions solving the Reynolds averaged Navier–Stokes (RANS) equation expressed in terms of turbulent viscosity and the standard k–ε turbulent model in 3D. A 3D diffusion-convection model was implemented in the CPC reactor to determine the RTD. The fluid flow visualization and RTD were validated with experimental results. The CFD showed that the magnitude of the velocity field remains almost uniform in most of the bulk reactor, although near and inside the 90° connectors and the union segments, the velocity presented low- and high-speed zones. Comparisons of theoretical and experimental RTD curves showed that the k–ε model is appropriate to simulate the nonideal flow inside the CPC reactor under turbulent flow conditions.

1. Introduction

Compound parabolic concentrator (CPC) reactors have been widely employed in UV light processes of disinfection and water treatment [1], such as photo-Fenton processes [2–5], TiO_2 photocatalytic processes [6–9], or direct irradiation processes [10]. One advantage of this technology is that the source of UV light could be taken from the sun, decreasing operational cost.

Several works show the efficiency of this technology in terms of pollutant concentration decay, during photo-Fenton or solar photoelectro-Fenton water treatment process [2–5, 11]. However, CPC studies that describe its performance from a chemical engineering standpoint are scarce. Among these, the nonideal flow analysis distinguishes, because it allows understanding the bulk reactor behavior and thereby to propose more efficient designs [12].

Most of the homogeneous (photo-Fenton) and heterogeneous (TiO_2 photocatalytic) reactions that take place inside the CPC are performed under turbulent flow conditions to

guarantee complete mixing conditions and an efficient mass transfer. The CPC is a coil-type tubular reactor, and one way to experimentally characterize the hydrodynamics inside it is through residence time distribution (RTD) analysis [13–16] or by flow visualization technique [17–20]. Benhabiles et al. [21] developed an experimental RTD study in a five-tube CPC reactor; these authors estimated Péclet numbers (Pe) and axial dispersion coefficients (D_a) with the analytical solution of the axial dispersion model at different volumetric flows. They found that for small values of Pe, the flow presents back mixing and dispersion in the radial direction.

The mathematical models can consider that the flow mixing occurs in the axial coordinate [22–25] and that the flow is completely mixed as the plug-flow model [24, 25]. In practice, flow behavior deviates from these owing to nonuniform velocity profile originated by turbulent diffusivity, by short circuiting, by passing and channeling of fluid, and by the presence of stagnant regions of fluid within the reactor. The above deviations can be determined by the diffusion-convection model, although these are scarcely

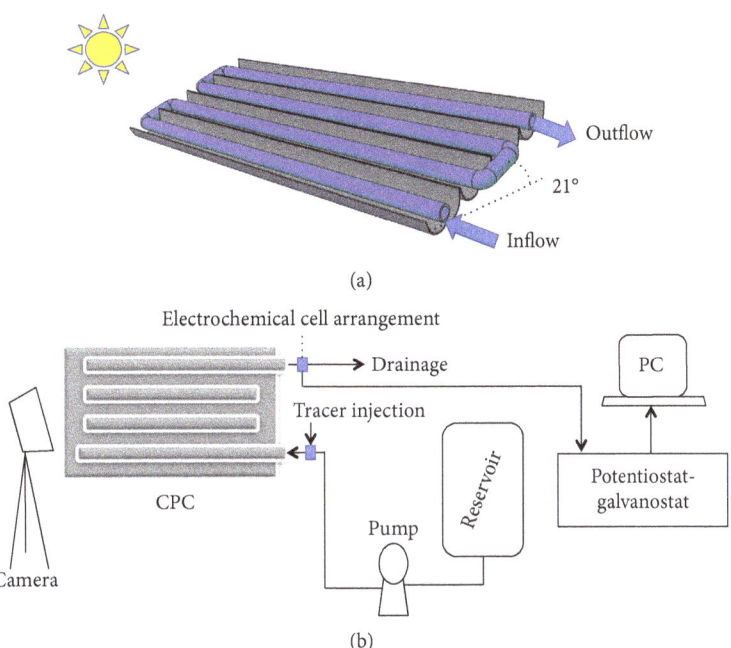

FIGURE 1: (a) Scheme of the CPC reactor. (b) Scheme of the system performed to carry out the experimental RTD studies and the flow visualizations.

modelled. A model for the CPC geometry must include the inclination of the reactor (see Figure 1(a)) owing to it may cause noncomplete mixing conditions.

This paper deals with the analysis and interpretation of flow visualization and RTD in a CPC reactor using CFD in 3D. CFD was calculated under turbulent flow conditions solving the Reynolds averaged Navier–Stokes (RANS) equations expressed in terms of turbulent viscosity and the standard k–ε turbulent model. Theoretical RTD curves were compared with experimental RTD data in order to validate the proposed model. Moreover, experimental and theoretical nonideal flow visualization was also performed.

2. Description of the CPC Reactor

Figure 1(a) shows a scheme of the CPC reactor which was used as a basis for the computational geometry in order to establish the simulation domain. It consists of 4 acrylic tubes of 90.5 cm length and 1.9 cm inner diameter connected with PVC tube segments and 90° PVC connections of 2.54 cm inner diameter. The parabolic reflectors are made of aluminum with a 180° acceptance angle and a concentration ratio of 1. The system has 1400 cm^3 of capacity and an inclination of 21° from the ground corresponding to the latitude of Guanajuato state, in central Mexico.

3. Formulation of the Numerical Simulation

Simulations in 3D of turbulent flow and a tracer inside the reactor were carried out. Figure 2(a) shows the simulation domain, and it was considered as "zigzag" pipes of constant diameter equal to 1.9 cm. The wall roughness was assumed to have a negligible effect. Table 1 shows the parameters for the simulations.

3.1. Turbulent Flow. Under turbulent flow conditions, the equations of the model for an incompressible fluid (water) can be stated as follows. The Reynolds averaged Navier–Stokes and the continuity equations are

$$(\rho \mathbf{u} \cdot \nabla)\mathbf{u} = -\nabla P + \nabla \cdot \left(\mu + \mu_T\right)\left(\nabla \mathbf{u} + (\nabla \mathbf{u})^T\right), \quad (1)$$

$$\nabla \cdot (\rho \mathbf{u}) = 0, \quad (2)$$

where $\mathbf{u}$ is the velocity vector, P is the pressure, μ is the dynamic viscosity, and ρ is the fluid density, and the so-called Reynolds stresses can be expressed in terms of a turbulent viscosity μ_T, according to the standard k–ε turbulence model

$$\mu_T = \rho C_\mu \frac{k^2}{\varepsilon},$$

$$\rho(\mathbf{u} \cdot \nabla)k = \nabla \cdot \left[\left(\mu + \frac{\mu_T}{\sigma_k}\right)\nabla k\right] + P_k - \rho\varepsilon, \quad (3)$$

$$\rho(\mathbf{u} \cdot \nabla)\varepsilon = \nabla \cdot \left[\left(\mu + \frac{\mu_T}{\sigma_\varepsilon}\right)\nabla\varepsilon\right] + C_{e1}\frac{\varepsilon}{k}P_k - C_{e2}\rho\frac{\varepsilon^2}{k},$$

where κ is the turbulent kinetic energy, ε is the turbulent energy dissipation rate, P_k is the energy production term $(P_k = \mu_T[\nabla\mathbf{u} : (\nabla\mathbf{u} + (\nabla\mathbf{u})^T)])$, and C_μ (0.09), C_{e1} (1.44), C_{e2} (1.92), σ_k (1), and σ_ε (1.3) are dimensionless constant values that are obtained by data fitting for a wide range of turbulent flows [26–28].

This model is applicable at high Reynolds numbers; for this reason, the near-wall regions, where the velocity is relative to the wall, decrease quickly and these regions are inaccessible by this model. To solve this problem, wall functions

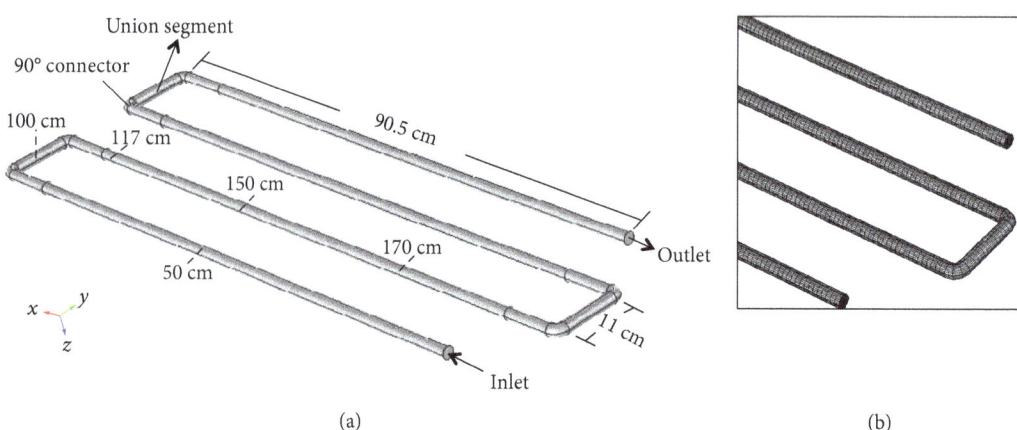

(a)   (b)

Figure 2: (a) Simulation domain established to implement the tracer simulation. The marked distances ($L = 50$ cm, $L = 100$ cm, $L = 117$ cm, $L = 150$ cm, and $L = 170$ cm) represent the regions where velocity profiles were taken at $z = 0$ cm in the center of the tube. (b) Inset of the mesh employed in the simulations.

Table 1: Parameters used in the numerical simulation at 293 K.

Initial concentration of Cu^{2+}, c_0 (mol·cm^{-3})	0.05
Dynamic viscosity of water, μ (gr·cm^{-1}·s^{-1})	0.01
Diffusion coefficient of Cu^{2+}, D (cm^2·s^{-1})	5×10^{-4}

based on a universal velocity distribution are usually required. In a turbulent layer, these functions are described by the following equation [26–28]:

$$u^+ = 2.5 \ln y^+ + 5.5, \qquad (4)$$

where u^+ is the normalized velocity component inside the logarithmic boundary layer and y^+ is the dimensionless distance from the wall, $y^+ = \rho u_\tau y / \mu$, where u_τ is the friction velocity, $u_\tau = C_\mu^{1/4}\sqrt{k}$, and y is the thickness from the wall [26–28].

To solve (1), (2), and (3), the corresponding boundary conditions are as follows [26]:

(1) A normal inflow velocity at the inlet, $\mathbf{u} = -\mathbf{n} U_0$; where $\mathbf{n}$ is the unit normal vector; in this work, the approximation for the inlet values of k_0 and ε_0 was obtained from the turbulent intensity I_T and the turbulent length scale L_T, by means of the following simple assumed forms: $k_0 = 3/2(U_0 I_T)^2$ and $\varepsilon_0 = C_\mu^{3/4} k^{3/2}/L_T$, where I_T and L_T were fixed at 0.05 and 0.0665 cm, respectively [27]. The turbulent intensity for fully turbulent flows has dimensionless values between 0.05 and 0.1. The turbulent length scale can be determined in pipes as a function of the radius by means of $L_T = 0.07r$, where r in this work is the inlet radius of 0.95 cm.

(2) A normal stress equal to a pressure at the outlet, $[-P + ((\mu + \mu_T)(\nabla\mathbf{u} + (\nabla\mathbf{u})^T))]\mathbf{n} = -\mathbf{n}P_0$, with $\nabla\varepsilon \cdot \mathbf{n} = 0$ and $\nabla k \cdot \mathbf{n} = 0$. This last equation expresses that the turbulent characteristic of whatever is outside the computational domain is guided by the flow

inside the computational domain. Such an assumption is physically reasonable as long as relatively small amounts of fluid enter the system [29].

(3) A velocity u^+ given by (4) at a distance y^+ from a solid surface, for all other boundaries.

After verifying the solution at different values of y^+ and step sizes, the value of y^+ was fixed at 11.1. This value is in the fully turbulent region ($5 < y^+ < 30$), where the turbulent stresses and fluxes are more important. The corresponding values of y and u_τ are 60 μm and 0.31 cm·s^{-1}, respectively. Equations (1), (2), and (3) were solved numerically in stationary regime in 3D through finite elements by using the commercial software COMSOL Multiphysics® (4.4) at different inflow velocities, U_0: 17.6 cm·s^{-1} (Re = 3344), 22.2 cm·s^{-1} (Re = 4218), and 24.4 cm·s^{-1} (Re = 4636). A simulation domain of 41,500 mesh elements was considered (Figure 2(b)). The numerical accuracy of the magnitude of the local velocity vectors was tested by enlarging and diminishing the size of the tetrahedral elements by a factor of 2 and by verifying that the solution of the magnitude of the local velocity vectors did not show significant differences as the convergence criterion was changed to below 10^{-5}. The simulation run times were typically from 30 to 45 minutes depending on the flow rate. A computer with 2 intel® Xeon 2.3 GHz processors, 96 GB of RAM, and 64 bits of operative system was employed.

3.2. Tracer Simulation. The time-dependent behavior of a tracer inside the reactor could be described by the general form of the diffusion-convection equation [30]

$$\frac{\partial c}{\partial t} = D\nabla^2 c - \mathbf{u} \cdot \nabla c, \qquad (5)$$

where c is the concentration of the tracer, t is the time, D is the diffusion coefficient, and $\mathbf{u}$ is the velocity vector obtained by the solution of (1).

It is important to mention that in some cases, mass transport models are involved with turbulent flow conditions, the term of turbulent diffusivity that appears in (5). This term is associated to jet flows, eddies, and local flow deviations caused by the inherent turbulent flow inside the system [31, 32]. Worthy of mentioning, we performed several simulations including the turbulent diffusivity coefficient (not shown herein) in order to develop a more rigorous model. However, when this term was included, the theoretical results did not fit with the experimental RTD studies. The explanation of the above behavior is discussed in the analysis of results and discussion in Section 5.

In order to simulate the tracer injection in an instant of time, a Gaussian pulse function was employed [33]

$$y(t) = \frac{1}{\sigma\sqrt{2\pi}} e^{-(t-t=0)^2/2\sigma^2}, \tag{6}$$

where σ is the standard deviation and t is the time which was fixed in an interval of time from 0 to 6 seconds. After several trials and verifying the solution, the standard deviation was fixed at 2 s. In this paper, the diffusion coefficient was fixed at $D = 5 \times 10^{-4}$ $cm^2 \cdot s^{-1}$.

In order to solve (5) and considering complete mixing conditions before the inlet and after the outlet of the reactor (Figure 2(a)), the boundary and the initial conditions established are as follows:

(i) Before tracer injection inside the reactor at $t = 0$, $c(x, y, z, t) = 0$

(ii) An initial concentration at the inlet, $c = c_0 y(t)/y(t = 0)$

(iii) Cero flux at the outlet and at the walls, $D\nabla c = 0$

Equation (5) was solved numerically in 3D through finite elements by using the commercial software COMSOL Multiphysics (4.4) at different inflow velocities: 17.6, 22.2, and 24.4 cm·s⁻¹. Equation (5) is in transient state; therefore, it needs an interval of time to be solved, which was established from $t = 0$ s to $t = 140$ s at 0.1 s of step size. A simulation domain of 41,500 mesh elements was considered. The simulation run times were typically from 20 to 30 minutes depending on the flow rate.

After solving (5), theoretical RTD curves and computational animations were performed in order to compare with experimental RTD curves and flow visualizations at different inflow velocities.

4. Experimental Details

The experimental flow visualization and RTD were performed in order to compare the theoretically obtained CFD and RTD results. Experimental tracer tests were filmed by a high-resolution camera and those were compared with computational tracer animations.

4.1. Flow Visualization. In order to visualize experimental flow patterns inside the CPC reactor, 5 mL of colored tracer (Blue 1 at a concentration of 5 g·L⁻¹) was injected few centimeters before the CPC inlet and filmed (Figure 1(b)). It is important to mention that the density of the tracer solution

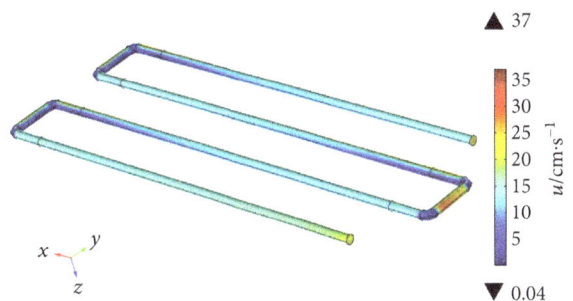

FIGURE 3: Velocity magnitude field plot inside the CPC reactor for an inflow velocity of 22.2 cm·s⁻¹.

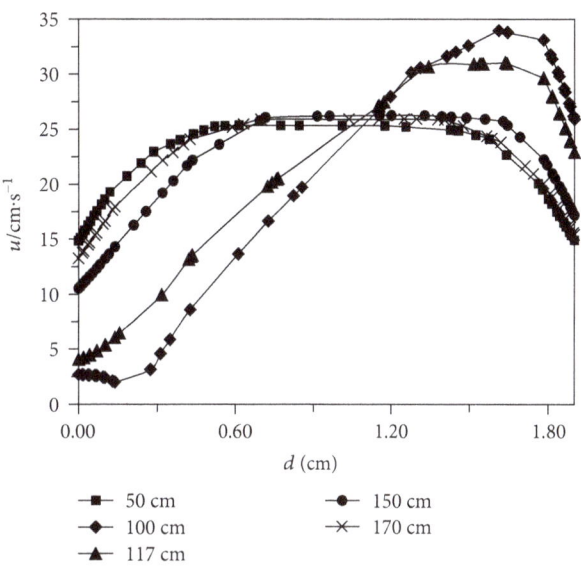

FIGURE 4: Velocity magnitude profiles inside the CPC reactor as a function of the diameter in the center of the tube at $z = 0$ cm. These profiles where taken at different distances, $L = 50$ cm (black square), $L = 100$ cm (black diamond), $L = 117$ cm (black triangle), $L = 150$ cm (black circle), and $L = 170$ cm (×) from the inlet of the reactor for an inflow velocity of 22.2 cm·s⁻¹.

does not affect the flow pattern inside the reactor. Moreover, the tracer volume is significantly smaller than the volume of the reactor (1400 mL). A 29-frames-per-second and 1920 × 1080-pixel-resolution camera was used. The tracer tests were performed at different inflow velocities of 17.6, 22.2, and 24.4 cm·s⁻¹.

4.2. RTD Studies. To determine the mixing flow pattern in the liquid phase, the stimulus response technique was employed. 5 mL of 0.05 M copper sulfate was injected as tracer at the inlet of the CPC and the cupric ions were detected at the exit (Figure 1(b)). The measurement of Cu(II) was carried out employing a typical two-electrode cell arrangement using two copper wires as electrodes; the cupric ions were quantified by typical transient current at a holding cell potential of −1.9 V. The current response was recorded every 0.5 - seconds by a potentiostat-galvanostat model SP-150 from BioLogic™ with EC-Lab® software. It is important to mention

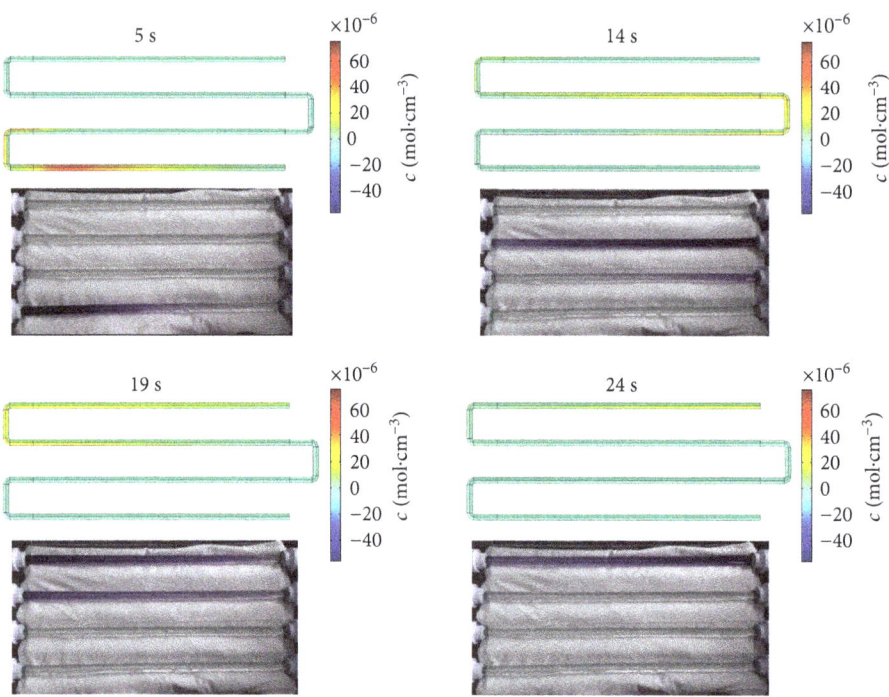

Figure 5: Comparisons of different pictures taken from the flow visualizations and the computational animations at different times of 5, 14, 19, and 24 seconds for an inflow velocity of 22.2 cm·s^{-1}.

that at such cell potential of -1.9 V, a limiting current of copper deposition governs the cathodic process ensuring that the response only depends on the cupric ion concentration.

For the nonideal flow, elements of fluid can take different routes throughout the CPC reactor, spending different periods of time inside the reactor. The RTD curve, $E(t)$, describes the distribution of these periods of time for the stream of fluid leaving the CPC, (7). The $E(t)$ function is normalized and the area under the curve reaches a value of 1, according to (8) [12]

$$\int_0^\infty E(t)dt = 1, \tag{7}$$

$$E(t) = \frac{I(t)}{\int_0^\infty I(t)dt}, \tag{8}$$

where $I(t)$ is the time-dependent current response. RTD studies were carried out at different inflow velocities of 17.6, 22.2 and 24.4 cm·s^{-1}.

5. Analysis of Results and Discussion

5.1. Turbulent Flow. Figure 3 shows a domain velocity field plot for a characteristic inflow velocity of 22.2 cm·s^{-1} inside the CPC reactor. Here, it could be observed that the velocity field remains almost uniform in most of the bulk reactor. However, near and inside the 90° connectors and the union segments, the velocity field presents low- and high-velocity zones due to the inherent CPC geometry, showing that the CPC cannot be considered as an ideal plug flow reactor.

Identical patterns (not shown herein) were obtained for the other inflow velocities, 17.6 and 24.4 cm·s^{-1}.

Figure 4 shows velocity profiles as a function of the tube diameter for a characteristic inflow velocity of 22.2 cm·s^{-1}. These curves were constructed in order to describe the hydrodynamic behavior of Figure 3. The velocity profiles inside the inner diameter were constructed at different distances from the CPC inlet as shown in Figure 2(a). The profile constructed at distance of 50 cm develops an ideal turbulent profile with a maximum of around 25 cm·s^{-1} and a minimum of around 15 cm·s^{-1} and a plug flow region of around $0.4 \le d \le 1.6$ cm. The velocity profile for the union segment at a distance of 100 cm shows the flow deviations generated by the 90° connector. In this profile, it could be observed that the union segment presents low velocity values of around $0 \le d \le 0.13$ cm with a minimum value of 2 cm·s^{-1}; then, the velocity increases until it reaches a maximum of 34 cm·s^{-1} at $d = 1.6$ cm. After, the velocity decreases until 26 cm·s^{-1} at $d = 1.9$ cm. The above behavior is very similar for the profile at a distance of 117 cm; here, it could be seen that this behavior is less pronounced that the profile performed at 100 cm, which means that the flow deviations attenuate away from the 90° connection as can be observed at distances of 150 and 170 cm. The behavior described in the profiles repeats in every lap of the CPC confirming the need to analyze the RTD. Similar plots not shown herein were obtained for the other inflow velocities at 17.6 and 24.4 cm·s^{-1}, obtaining the same pattern.

From the analysis of Figures 3 and 4, it could be confirming the convenience of using a mass transport model without the turbulent diffusion coefficient, described in Section 3.2. The above is because the system does not generate significant turbulent flow deviations such as eddies or jet flows, as it can be observed in Figure 3; this behavior confirms the omission of the turbulent diffusion coefficient of (5). Moreover, the velocity profiles develop a pseudoplug flow behavior in almost all regions inside the reactor (Figure 4); this last showed that the system is appropriate to develop a well homogenous environment to carry out the photoreactions inside the system. The above was proved in the analysis of results and discussion of the flow visualization and RTD described below. Nevertheless, a deeper analysis with different photocatalytic reactions is needed to support this statement. This later was beyond the scope of this paper.

5.2. Flow Visualization. Figure 5 shows comparisons of the experimental flow visualization and the tracer simulation at different times for the characteristic inflow velocity of $22.2\,\mathrm{cm\cdot s^{-1}}$. Here, it could be observed qualitatively that the tracer simulation agrees with the experimental visualization technique, because the tracer follows the same flow patterns at the same instants of time. For the time of 5 seconds, the tracer is completely mixed before entering the 90° connection. After the first union segment, at the time of 14 seconds, the tracer presents an elongation generated by the 90° connection. This elongation is more marked passing the third and fourth union segments at the time of 19 and 24 seconds, respectively; therefore, it agrees with those discussed in Figures 3 and 4; the 90° connections generate slow-velocity zones and flow deviations. Similar experimental flow visualization and the tracer simulation were obtained for 17.6 and $24.4\,\mathrm{cm\cdot s^{-1}}$ (not shown), obtaining the same pattern to that obtained at $22.2\,\mathrm{cm\cdot s^{-1}}$.

5.3. RTD. Figure 6 presents comparisons of theoretical and experimental RTD curves for the different inflow velocities as a function of the dimensionless residence time (t/τ), where τ is the spatial residence time given by the ratio between the length of the CPC and the inflow velocity $(= L/U_0)$. This figure shows close agreement between experiments and simulations (error $< 5\%$). Worthy of mentioning, the most fluid elements leave the system around the expected residence time $(t/\tau = 1)$. For such cases, the residence time is about 24 s and the dispersion of the tracer in the pipe is dominated by the convective transport, where the turbulent flow predominates (Re > 3000).

From the analysis carried out here, a CPC reactor develops nonideal flow distribution in the 90° connections, which could influence the efficiency of this technology to be employed in UVA light processes of water treatment, such as photo-Fenton process. This later should serve as a starting point for future research.

6. Conclusions

This work presented a way to analyze the hydrodynamic behavior inside a CPC reactor. A 3D diffusion-convection model was implemented considering that the flow is not completely mixed inside the CPC reactor. A single-phase flow analysis was performed solving the RANS equations with the standard k–ε turbulent model, for Re > 3000.

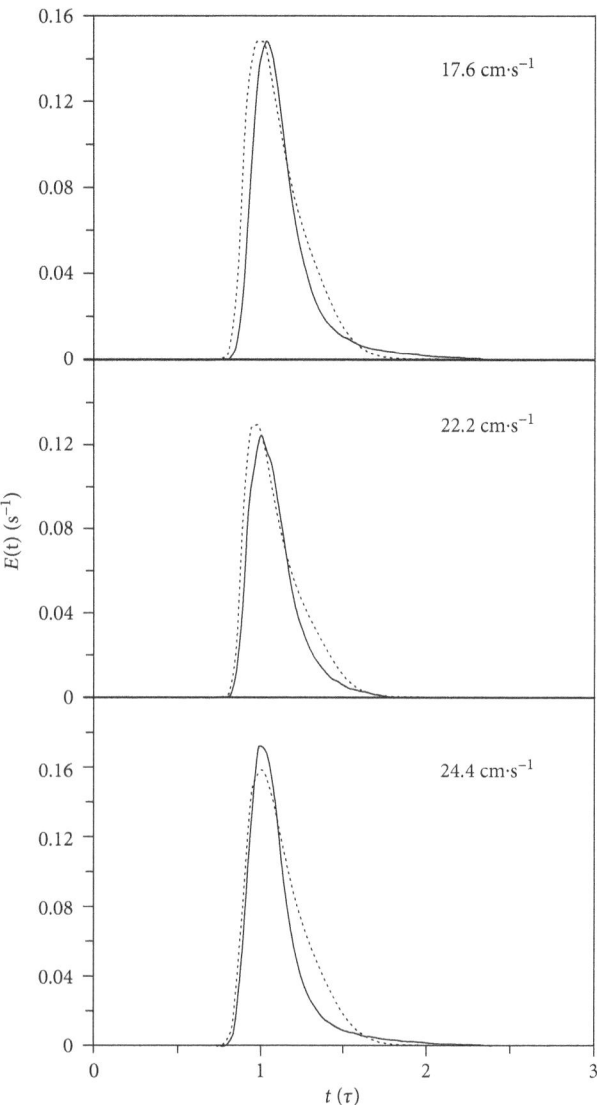

FIGURE 6: Comparisons of experimental (solid line) and theoretical (dot line) RTD curves as a function of the dimensionless residence time at different inflow velocities showed inside the figure.

The analysis of the turbulent flow simulations showed that near and inside the 90° connectors and the union segments, the reactor presents low- and high-velocity zones owing to the inherent CPC geometry, showing that the CPC cannot be considered as an ideal plug flow reactor. An appropriate CPC reactor design must consider the flow behavior deviations caused by the perpendicular union segments of the CPC. Then, other connection angles should be tested to improve the hydrodynamics within the "zigzag tubes" and, consequently, to improve the mixing conditions in the photoreactor.

The comparison of flow visualization studies and the computational animations showed good agreement for the inflow velocities of 17.6, 22.2, and $24.4\,\mathrm{cm\cdot s^{-1}}$. Excellent agreement between theoretical and experimental RTD curves was obtained. This later shows the robustness of the proposed model, envisaging a future application in photo-Fenton-like processes for water treatment. The mathematical model proposed here can be helpful for scaling up the CPC photoreactor, where photoreactions should be adapted, although a deeper analysis with different photocatalytic reactions is needed to support this assertion.

Conflicts of Interest

The authors declare that there is no conflict of interest regarding the publication of this paper.

Acknowledgments

The authors thank the University of Guanajuato for funding through Project no. 100/2018 and Project PRODEP no. UGTO-PTC-615.

References

[1] D. Spasiano, R. Marotta, S. Malato, P. Fernandez-Ibañez, and I. Di Somma, "Solar photocatalysis: materials, reactors, some commercial, and pre-industrialized applications. A comprehensive approach," *Applied Catalysis B: Environmental*, vol. 170-171, pp. 90–123, 2015.

[2] T. Pérez, S. Garcia-Segura, A. El-Ghenymy, J. L. Nava, and E. Brillas, "Solar photoelectro-Fenton degradation of the antibiotic metronidazole using a flow plant with a Pt/air-diffusion cell and a CPC photoreactor," *Electrochimica Acta*, vol. 165, pp. 173–181, 2015.

[3] R. Salazar, E. Brillas, and I. Sirés, "Finding the best Fe^{2+}/Cu^{2+} combination for the solar photoelectro-Fenton treatment of simulated wastewater containing the industrial textile dye Disperse Blue 3," *Applied Catalysis B: Environmental*, vol. 115-116, pp. 107–116, 2012.

[4] T. Perez, I. Sires, E. Brillas, and J. L. Nava, "Solar photoelectro-Fenton flow plant modeling for the degradation of the antibiotic erythromycin in sulfate medium," *Electrochimica Acta*, vol. 228, pp. 45–56, 2017.

[5] G. Coria, T. Pérez, I. Sirés, E. Brillas, and J. L. Nava, "Abatement of the antibiotic levofloxacin in a solar photoelectro-Fenton flow plant: modeling the dissolved organic carbon concentration-time relationship," *Chemosphere*, vol. 198, pp. 174–181, 2018.

[6] I. García-Fernández, I. Fernández-Calderero, M. I. Polo-López, and P. Fernández-Ibáñez, "Disinfection of urban effluents using solar TiO_2 photocatalysis: a study of significance of dissolved oxygen, temperature, type of microorganism and water matrix," *Catalysis Today*, vol. 240, no. A, pp. 30–38, 2015.

[7] S. García-Segura and E. Brillas, "Applied photoelectrocatalysis on the degradation of organic pollutants in wastewaters," *Journal of Photochemistry and Photobiology C: Photochemistry Reviews*, vol. 31, pp. 1–35, 2017.

[8] D. M. El-Mekkawi, N. Nady, N. A. Abdelwahab, W. A. A. Mohamed, and M. S. A. Abdel-Mottaleb, "Flexible bench-scale recirculating flow CPC photoreactor for solar photocatalytic degradation of methylene blue using removable TiO_2 immobilized on PET sheets," *International Journal of Photoenergy*, vol. 2016, Article ID 9270492, 9 pages, 2016.

[9] E. M. Saggioro, A. S. Oliveira, T. Pavesi et al., "Solar CPC pilot plant photocatalytic degradation of indigo carmine dye in waters and wastewaters using supported-TiO_2: influence of Photodegradation parameters," *International Journal of Photoenergy*, vol. 2015, Article ID 656153, 12 pages, 2015.

[10] J. Ndounla et al., "Relevant impact of irradiance (vs. dose) and evolution of pH and mineral nitrogen compounds during natural water disinfection by photo-Fenton in a solar CPC reactor," *Applied Catalysis B: Environmental*, vol. 148-149, pp. 144–153, 2014.

[11] S. Garcia-Segura, E. B. Cavalcanti, and E. Brillas, "Mineralization of the antibiotic chloramphenicol by solar photoelectro-Fenton: from stirred tank reactor to solar pre-pilot plant," *Applied Catalysis B: Environmental*, vol. 144, pp. 588–598, 2014.

[12] H. S. Fogler, *Elements of Chemical Reaction Engineering*, Prentice Hall, New Jersey, USA, 2005.

[13] P. Ghosh, K. Vahedipour, M. Lin, J. H. Vogel, C. Haynes, and E. von Lieres, "Computational fluid dynamic simulation of axial and radial flow membrane chromatography: mechanisms of non-ideality and validation of the zonal rate model," *Journal of Chromatography A*, vol. 1305, pp. 114–122, 2013.

[14] F. Jovic, V. Kosar, V. Tomasic, and Z. Gomzi, "Non-ideal flow in an annular photocatalytic reactor," *Chemical Engineering Research and Design*, vol. 90, no. 9, pp. 1297–1306, 2012.

[15] Q. Li, M. C. Henstridge, C. Batchelor-McAuley, N. S. Lawrence, R. S. Hartshorne, and R. G. Compton, "Glassy carbon tubular electrodes for the reduction of oxygen to hydrogen peroxide," *Physical Chemistry Chemical Physics*, vol. 15, no. 20, pp. 7854–7865, 2013.

[16] P. Rojahn, V. Hessel, K. D. P. Nigam, and F. Schael, "Applicability of the axial dispersion model to coiled flow inverters containing single liquid phase and segmented liquid-liquid flows," *Chemical Engineering Science*, vol. 182, pp. 77–92, 2018.

[17] M. Ansari, D. E. Turney, R. Yakobov et al., "Chemical hydrodynamics of a downward microbubble flow for intensification of gas-fed bioreactors," *AIChE Journal*, vol. 64, no. 4, pp. 1399–1411, 2018.

[18] D. Patel, F. Ein-Mozaffari, and M. Mehrvar, "Using tomography to visualize the continuous-flow mixing of biopolymer solutions inside a stirred tank reactor," *Chemical Engineering Journal*, vol. 239, pp. 257–273, 2014.

[19] W.-F. Li, K.-J. du, G.-S. Yu, H.-F. Liu, and F.-C. Wang, "Experimental study of flow regimes in three-dimensional confined impinging jets reactor," *AICHE Journal*, vol. 60, no. 8, pp. 3033–3045, 2014.

[20] G. Wang, Z. Li, M. Yousaf, X. Yang, and M. Ishii, "Experimental study on vertical downward air-water two-phase flow in a large diameter pipe," *International Journal of Heat and Mass Transfer*, vol. 118, pp. 919–930, 2018.

[21] O. Benhabiles, N. Chekir, and W. Taane, "Determining of the residence time distribution in CPC reactor type," *Energy Procedia*, vol. 18, pp. 368–376, 2012.

[22] S. A. Martínez-Delgadillo, H. R. Mollinedo P., M. A. Gutiérrez, I. D. Barceló, and J. M. Méndez, "Performance of a tubular electrochemical reactor, operated with different inlets, to remove Cr(VI) from wastewater," *Computers & Chemical Engineering*, vol. 34, no. 4, pp. 491–499, 2010.

[23] F. F. Rivera, M. R. Cruz-Díaz, E. P. Rivero, and I. González, "Analysis and interpretation of residence time distribution experimental curves in FM01-LC reactor using axial disper-

sion and plug dispersion exchange models with closed–closed boundary conditions," *Electrochimica Acta*, vol. 56, no. 1, pp. 361–371, 2010.

[24] A. Safari, J. Safdari, H. Abolghasemi, M. Forughi, and M. Moghaddam, "Axial mixing and mass transfer investigation in a pulsed packed liquid-liquid extraction column using plug flow and axial dispersion models," *Chemical Engineering Research and Design*, vol. 90, no. 2, pp. 193–200, 2012.

[25] P. Trinidad, C. Ponce de León, and F. C. Walsh, "The application of flow dispersion models to the FM01-LC laboratory filter-press reactor," *Electrochimica Acta*, vol. 52, no. 2, pp. 604–613, 2006.

[26] T. Pérez, C. Ponce de León, F. C. Walsh, and J. L. Nava, "Simulation of current distribution along a planar electrode under turbulent flow conditions in a laboratory filter-press flow cell," *Electrochimica Acta*, vol. 154, pp. 352–360, 2015.

[27] H. K. Versteeg and W. Malalasekera, *An Introduction to Computational Fluid Dynamics: The Finite Volume Method*, Prentice Hall, London, 1995.

[28] P. S. Bernard and J. M. Wallace, *Turbulent Flow: Analysis, Measurement, and Prediction*, John Wiley & Sons, New Jersey, 2002.

[29] D. C. Wilcox, *Turbulence Modeling for CFD*, DCW Industries Inc., La Cañada Flintridge, CA, USA, 1998.

[30] R. B. Bird, W. E. Stewart, and E. N. Lightfood, *Transport Phenomena*, John Wiley & Sons, New York, 2002.

[31] M. R. Cruz-Díaz, E. P. Rivero, F. J. Almazán-Ruiz, Á. Torres-Mendoza, and I. González, "Design of a new FM01-LC reactor in parallel plate configuration using numerical simulation and experimental validation with residence time distribution (RTD)," *Chemical Engineering and Processing: Process Intensification*, vol. 85, pp. 145–154, 2014.

[32] E. P. Rivero, M. R. Cruz-Díaz, F. J. Almazán-Ruiz, and I. González, "Modeling the effect of non-ideal flow pattern on tertiary current distribution in a filter-press-type electrochemical reactor for copper recovery," *Chemical Engineering Research and Design*, vol. 100, pp. 422–433, 2015.

[33] M. J. Rivera, M. Trujillo, V. Romero-García, J. A. López-Molina, and E. Berjano, "Numerical resolution of the hyperbolic heat equation using smoothed mathematical functions instead of Heaviside and Dirac delta distributions," *International Communications in Heat and Mass Transfer*, vol. 46, pp. 7–12, 2013.

15

Synthesis of MgFe$_2$O$_4$/Reduced Graphene Oxide Composite and Its Visible-Light Photocatalytic Performance for Organic Pollution

Fengmin Wu, Wenlu Duan, Mei Li, and Hang Xu (ID)

School of Chemical Engineering and Pharmaceutics, Henan University of Science and Technology, Luoyang 471023, China

Correspondence should be addressed to Hang Xu; xhinbj@126.com

Academic Editor: P. Davide Cozzoli

Recently, binary metal oxides have been proven to be the most investigated semiconductors due to their high activity for the removal of organic pollutants. In this paper, to improve the photocatalytic efficiency of MgFe$_2$O$_4$, a MgFe$_2$O$_4$/reduced graphene oxide (MFO/rGO) photocatalyst was synthesized by a facile generalized solvothermal method. The morphology, structure, and photocatalytic activities in the degradation of methyl orange (MO) reaction were systematically characterized by scanning electron microscopy, transmission electron microscopy, X-ray diffraction, and UV-vis absorption spectroscopy, respectively. The results showed that the MFO/rGO composite exhibited enhanced photocatalytic performance in the photodegradation of MO under visible-light irradiation and reached a maximum degradation rate of 99% within 60 min of irradiation. This excellent photocatalytic performance is attributed to the introduction of rGO in the composite, which can effectively reduce the photoproduction of the electron-hole pair recombination rate. The excellent photocatalytic activity reveals that the MFO/rGO composite photocatalyst is a promising photocatalyst with good visible-light response and has potential applications in the field of water treatment.

1. Introduction

Recently, environmental pollution has become one of the outstanding social problems. Photocatalytic technology has attracted great interest as a promising pathway for solving energy supply and environmental pollution problems [1, 2]. However, the traditional photocatalysts, typically TiO$_2$ (3.2 eV) or ZnO (3.3 eV), with a wide band gap can only exhibit excellent photocatalytic activity under ultraviolet light irradiation, which significantly limits their practical applications [3–5]. Faster kinetics recombination of photogenerated electrons and holes is another chief reason for lower photocatalytic performance [6]. It takes patience to develop efficient visible-light irradiation photocatalysts in order to utilize solar light, of which about 45% is composed of visible light. Thus, the development of visible-light photocatalysts has become one of the most critical topics in photocatalytical research today.

Magnesium ferrite (MgFe$_2$O$_4$, MFO) is a semiconductor with a spinel structure, which can absorb visible light due to its narrow band gap (approximately 2.0 eV) and is not sensitive to photoanodic corrosion. MFO has been used as a photocatalyst for the degradation of 2-propanol [7]. Besides, it was found that MFO/TiO$_2$ has a high photocatalytic activity for the degradation of the RhB contaminant [8]. Therefore, MFO is considered as one of the efficient photocatalyst candidates for air or water treatment. According to previous reports, to reduce the recombination of photogenerated electrons and holes, the surface of photocatalysts coated with carbon nanotubes (CNTs) or carbon nanofibers (CNFs) has been considered as a promising strategy [9]. In the process of photocatalysis, CNTs or CNFs can act as an excellent electron-acceptor/transport donor to effectively facilitate the migration of photoinduced electrons and hinder the quick recombination in electron transfer, which enhances photocatalytic performance.

As a rising star in the carbon family, reduced graphene oxide (rGO) has attracted a great deal of attention in recent years due to its excellent electronic properties (zero gap semiconductor where the conduction band and the valence band touch each other), outstanding ability as an electron acceptor and transport, chemical stability, and high surface area, which have been used to obtain hybrid materials with superior photocatalytic performance [10]. For instance, tremendous improvement has been reported for metal oxides [11], metal sulfides [12], and nonmetal g-C$_3$N$_4$ [13] by using rGO as a synergistic catalyst material. However, to the best of the authors' knowledge, there has been no report of the modification of MFO with rGO for preparing new photocatalysts.

In this paper, a novel MFO/rGO photocatalyst has been synthesized by a facile solvothermal method for the first time. The as-prepared samples were characterized, and the photocatalytic activity under visible-light irradiation ($\lambda \geq 420$ nm) was estimated by the photodegradation of methyl orange (MO). The results indicated that MFO/rGO nanocomposites exhibited much higher photocatalytic performance than the pure MFO. At last, the mechanism of enhanced photocatalytic activities of MFO/rGO is proposed.

2. Experimental

2.1. Catalyst Preparation. All reagents were of analytical grade and used without further purification. Commercial graphite power was used as the raw materials for synthesis of graphene oxide (GO) by a modified Hummer's method according to previous reports [14].

1.75 mmol of Mg(NO$_3$)$_2$·6H$_2$O and 3.5 mmol of Fe(NO$_3$)$_3$·9H$_2$O were mixed and dissolved in 80 ml of distilled water after being sonicated for 30 min. Next, a certain amount (10 ml) of 10 mg ml^{-1} GO suspension was added into the above solution under magnetic stirring for 4 h. The pH value of the solution was adjusted by ammonia hydroxide to about 10. The mixture was stirred vigorously for an hour and then sealed in a Teflon lined stainless steel autoclave (100 ml capacity). The autoclave was heated to and maintained at 180°C for 12 h, and allowed to cool to room temperature. The grayish-black products were washed several times with ethanol and distilled water, and then dried in a vacuum oven at 80°C overnight. For comparison, bare MFO was also synthesized by a similar route without adding the GO suspension.

2.2. Characterization. The morphology and structure of the as-prepared samples were characterized by a field-emission scanning electron microscope (FESEM, Hitachi S-4800) equipped with an energy-dispersive spectrometer (EDS, Bruker QUANTAX-400), high-resolution transmission electron microscope (HRTEM, Tecnai G2 F20), and X-ray diffractometer (XRD, Holland PaNalytical PRO PW3040/60) with Cu Kα radiation ($V = 30$ kV, $I = 25$ mA, $\lambda = 1.5418$ Å, scanning rate: 10° min^{-1}). UV-vis absorption spectra of samples were recorded on a UV-vis spectrophotometer (Hitachi U-3310) with a wavelength range of 200–600 nm.

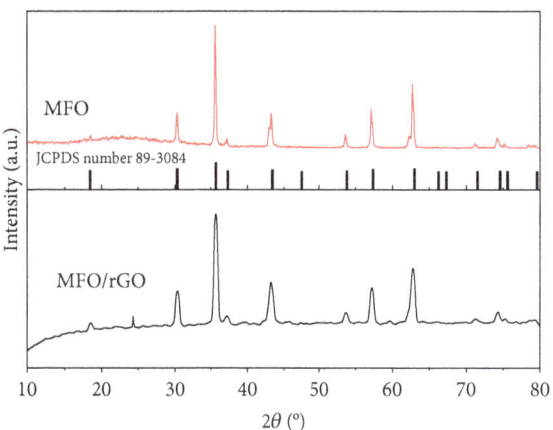

FIGURE 1: The XRD patterns of MFO and MFO/rGO samples.

2.3. Photocatalytic Tests. The photocatalytic performance was evaluated with the as-prepared sample powder (50 mg) suspended in MO (20 mg l^{-1}, 100 ml) with constant stirring. A 300 W Xe arc lamp (CEL-HXF 300, Beijing Jinyuan Co. Ltd.) was used as the light source and equipped with an ultraviolet cutoff filter to provide visible light ($\lambda \geq 420$ nm). Prior to irradiation, the suspensions were stirred in the dark for 30 min to ensure the adsorption/desorption equilibrium. At certain time intervals, 2 ml of the suspension was taken and centrifuged to remove the photocatalyst particles. The filtrate was then analyzed using a UV-vis spectrophotometer at 464 nm as a function of irradiation time by measuring the max absorbance of MO.

3. Results and Discussion

The XRD patterns of the products are shown in Figure 1. The dominant diffraction peaks in the pattern of MFO and MFO/rGO can be indexed to those from the spinel MgFe$_2$O$_4$ (JCDPS 89-3084). The diffraction peaks of the MFO are sharp and intense, revealing the highly crystalline character of the sample. After combination with rGO, the intensity of the diffraction peaks becomes weak. However, the crystalline spinel of the sample was not changed, indicating that the crystal structure of MFO remains stable. In addition, the characteristic diffraction peaks of GO (12.2°) are not found in the XRD of MFO/rGO composite samples, which indicates that GO is fully reduced to rGO during the process of reaction.

Figures 2(a) and 2(b) display the SEM images of the MFO and MFO/rGO samples. The MFO are composed of numerous hexahedron particles, each with a particle size of about 80 nm. The MFO/rGO shows ultrathin crumpled three dimensional (3D) nanosheets, and the MFO and GO are uniform and well-attached to each other. Figures 2(c) and 2(d) show the typical low-magnification TEM images of MFO/rGO, which indicate that the as-prepared material has a quadrangular morphology with the edge length in the range of 30–50 nm. Figures 2(e) and 2(f) show the high-resolution TEM images of MFO/rGO. The lattice spacing measured (Figure 2(f)) for the crystalline plane is 0.263 nm, corresponding to the (111) plane of the spinel MFO (JCPDS 89-

FIGURE 2: The SEM images of MFO (a) and MFO/rGO (b) and low-magnification (c, d) and high-resolution (e, f) TEM images of MFO/rGO.

3084). In addition, rGO serves as a 3D conductive support for MFO nanoparticles.

As can be seen from Figure 3(a), MO self-photodegradation (under light irradiation without the effect of catalysts) can be negligible; however, the MFO/rGO composite exhibits the best visible-light photocatalytic efficiency, which can decompose MO molecules in 60 min. In contrast, the MFO showed a poor degradation efficiency of ~40% after visible-light irradiation for 60 min. Interestingly, rGO also showed a degradation efficiency of about 20% owing to its absorbing ability to visible light [15]. Figure 3(b) shows the absorbance spectra of MO under rGO, MFO, and MFO/rGO photocatalysts at different times. The enhancement of the photocatalytic performance should be ascribed to the

prevention in electron-hole pair quick recombination due to the stepwise energy level structure in the composite and the increase in the light absorption with the presence of rGO [16]. Therefore, on the basis of the synergistic effect of the electron-acceptor and transport material rGO and the rGO can absorb visible light, and the MFO/rGO composite displays a highly efficient visible-light photocatalytic activity.

The mechanism of the photocatalytic degradation of MO is given as follows: Firstly, the semiconductor materials (MFO) absorb the photon under light irradiation, and the electron-hole pairs are generated at the photocatalyst surface (1). Then, photoinduced electrons are transferred from the MFO to the rGO, which could efficiently separate the photo-generated electrons and hinder the charge recombination in

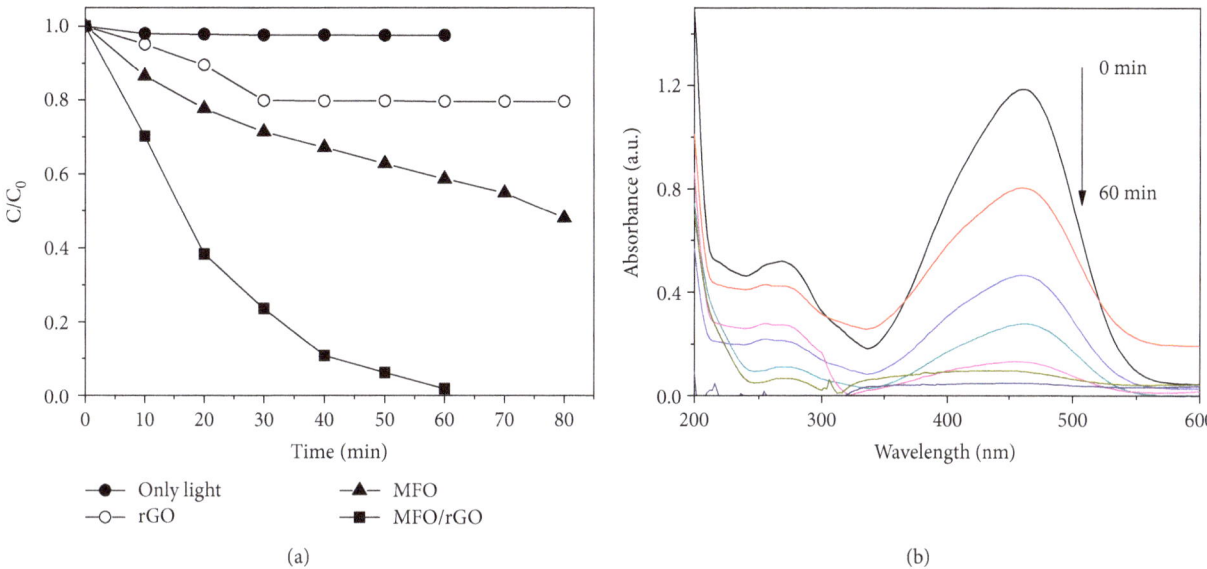

FIGURE 3: (a) Photocatalytic degradation profile of MO under visible-light irradiation with and without catalyst versus exposure time, (b) absorption spectra of the MO ($20\,\mathrm{mg\,l^{-1}}$) solution at different photocatalytic time under MFO/rGO photocatalytic.

the electron transfer processes. Afterwards, the high oxidation potential of the holes causes the direct oxidation of the MO and some reactive intermediates are generated (2). Finally, a hydroxyl radical can also be created by hydroxyl ions (OH^-) with holes (3). Furthermore, the superoxide anions are also formed by the molecular reduction of O_2, which may take place by the presence of electrons in the conduction band at the surface of the photocatalyst (4). The conduction band electrons are also responsible for the production of $^{\bullet}OH$ and $^{\bullet}O_2^-$, which have been identified as the main cause of MO degradation. Due to existence of rGO among MFO, which hinder the recombination of photogenerated electrons and holes, the photocatalytic performance is thus enhanced.

$$MFO + h\nu \longrightarrow MFO + e_{CB}^- + h_{VB}^+ \qquad (1)$$

$$h_{VB}^+ + MO \longrightarrow MO^* + \text{oxidation of MO} \qquad (2)$$

$$h_{VB}^+ + OH^- \longrightarrow {}^{\bullet}OH \qquad (3)$$

$$e_{CB}^+ + O_2 \longrightarrow {}^{\bullet}O_2^- \qquad (4)$$

4. Conclusion

In summary, a novel photocatalyst MFO/rGO composite was prepared via a facile generalized solvothermal method. The results demonstrate that the incorporation of rGO can enhance the photocatalytic activity for the degradation of MO. The reduced photogenerated electron-hole pair recombination and the enhanced visible-light absorbance with the introduction of rGO are mainly responsible for the enhanced photocatalytic activity of the MFO/rGO composite. The present synthetic strategy should be a promising fabrication technique for a simple, rapid, and effective method of the composite photocatalysts for the removal of organic pollutants.

Conflicts of Interest

The authors declare that they have no conflicts of interest.

Acknowledgments

The work was supported by the National Natural Science Foundation of China (no. 21006075) and Henan Provincial Science and Technology Foundation (nos. 152102210274 and 15A530008).

References

[1] J. Mu, C. Shao, Z. Guo et al., "High photocatalytic activity of ZnO-carbon nanofiber heteroarchitectures," *ACS Applied Materials & Interfaces*, vol. 3, no. 2, pp. 590–596, 2011.

[2] M. R. Hoffmann, S. T. Martin, W. Choi, and D. W. Bahnemann, "Environmental applications of semiconductor photocatalysis," *Chemical Reviews*, vol. 95, no. 1, pp. 69–96, 1995.

[3] Z. Zheng, J. Zhao, Y. Yuan et al., "Tuning the surface structure of nitrogen-doped TiO2 nanofibres—an effective method to enhance photocatalytic activities of visible-light-driven green synthesis and degradation," *Chemistry - A European Journal*, vol. 19, no. 18, pp. 5731–5741, 2013.

[4] X.-J. Lv, W.-F. Fu, H.-X. Chang et al., "Hydrogen evolution from water using semiconductor nanoparticle/graphene composite photocatalysts without noble metals," *Journal of Materials Chemistry*, vol. 22, no. 4, pp. 1539–1546, 2012.

[5] R. Huang, H. Ge, X. Lin et al., "Facile one-pot preparation of α-SnWO4/reduced graphene oxide (RGO) nanocomposite with improved visible light photocatalytic activity and anode performance for Li-ion batteries," *RSC Advances*, vol. 3, no. 4, pp. 1235–1242, 2013.

[6] R. M. Asmussen, M. Tian, and A. Chen, "A new approach to wastewater remediation based on bifunctional electrodes," *Environmental Science & Technology*, vol. 43, no. 13, pp. 5100–5105, 2009.

[7] H. G. Kim, P. H. Borse, J. S. Jang et al., "Fabrication of $CaFe_2O_4/MgFe_2O_4$ bulk heterojunction for enhanced visible light photocatalysis," *Chemical Communications*, no. 39, pp. 5889–5891, 2009.

[8] D.-H. Yoo, T. V. Cuong, V. H. Pham et al., "Enhanced photocatalytic activity of graphene oxide decorated on TiO_2 films under UV and visible irradiation," *Current Applied Physics*, vol. 11, no. 3, pp. 805–808, 2011.

[9] L.-P. Zhu, G.-H. Liao, W.-Y. Huang et al., "Preparation, characterization and photocatalytic properties of ZnO-coated multi-walled carbon nanotubes," *Materials Science and Engineering: B*, vol. 163, no. 3, pp. 194–198, 2009.

[10] J. Mu, B. Chen, M. Zhang et al., "Enhancement of the visible-light photocatalytic activity of In_2O_3-TiO_2 nanofiber hetero-architectures," *ACS Applied Materials & Interfaces*, vol. 4, no. 1, pp. 424–430, 2012.

[11] K. Zhou, Y. Zhu, X. Yang, X. Jiang, and C. Li, "Preparation of graphene—TiO_2 composites with enhanced photocatalytic activity," *New Journal of Chemistry*, vol. 35, no. 2, pp. 353–359, 2011.

[12] S. Min and G. Lu, "Sites for high efficient photocatalytic hydrogen evolution on a limited-layered MoS_2 cocatalyst confined on graphene sheets—the role of graphene," *Journal of Physical Chemistry C*, vol. 116, no. 48, pp. 25415–25424, 2012.

[13] Q. Xiang, J. Yu, and M. Jaroniec, "Preparation and enhanced visible-light photocatalytic H_2-production activity of graphene/C_3N_4 composites," *Journal of Physical Chemistry C*, vol. 115, no. 15, pp. 7355–7363, 2011.

[14] D. C. Marcano, D. V. Kosynkin, J. M. Berlin et al., "Improved synthesis of graphene oxide," *ACS Nano*, vol. 4, no. 8, pp. 4806–4814, 2010.

[15] X. He, T. Tang, F. Liu, N. Tang, X. Li, and Y. Du, "Photochemical doping of graphene oxide thin film with nitrogen for photoconductivity enhancement," *Carbon*, vol. 94, pp. 1037–1043, 2015.

[16] X. An, J. C. Yu, Y. Wang, Y. Hu, X. Yu, and G. Zhang, "WO_3 nanorods/graphene nanocomposites for high-efficiency visible-light-driven photocatalysis and NO_2 gas sensing," *Journal of Materials Chemistry*, vol. 22, no. 17, pp. 8525–8531, 2012.

16

Mathematical Modeling and Mechanism of VUV Photodegradation of H_2S in the Absence of O_2

Jian-hui Xu,[1] **Bin-bin Ding,**[2] **Xiao-mei Lv**(iD)**,**[1] **Shan-hong Lan,**[1] **Chao-lin Li,**[2] **and Liu Peng**[1]

[1]*School of Environmental and Civil Engineering, Dongguan University of Technology, Dongguan, Guangdong 523808, China*
[2]*Environmental Science and Engineering Center, Harbin Institute of Technology Shenzhen Graduate School, Shenzhen, Guangdong 518055, China*

Correspondence should be addressed to Xiao-mei Lv; lvxiaomei.daisy@gmail.com

Academic Editor: Juan M. Coronado

The existence of H_2S has limited the biogas energy promotion. The traditional photodegradation of H_2S is usually conducted in the presence of O_2, yet this is unsuitable for biogas desulfurization which should be avoided. Therefore, the ultraviolet degradation of H_2S in the absence of O_2 was investigated for the first time in the present study from a mathematical point of view. Light wavelength and intensity applied were $185\,nm$ and 2.16×10^{-12} Einstein/$cm^2 \cdot s$, respectively. Firstly, the mathematical model of H_2S photodegradation was established with MATLAB software, including the gas flow distribution model and radiation model of photoreactor, kinetics model, mass balance model, and calculation model of the degradation rate. Then, the influence of the initial H_2S concentration and gas retention time on the photodegradation rate were studied, for verification of the mathematical model. Results indicated that the photodegradation rate decreased with the increase in initial H_2S concentration, and the maximum photodegradation rate reached 62.8% under initial concentration of $3\,mg/m^3$. In addition, the photodegradation rate of H_2S increased with the increase in retention time. The experimental results were in good accordance with the modeling results, indicating the feasibility of the mathematical model to simulate the photodegradation of H_2S. Finally, the intermediate products were simulated and results showed that the main photodegradation products were found to be H_2 and elemental S, and concentrations of the two main products were close and agreed well with the reaction stoichiometric coefficients. Moreover, the concentration of free radicals of H• and SH• was rather low.

1. Introduction

During anaerobic digestion which is considered as one of the most important biomass-based renewable energy techniques to reclaim clean fuel of biogas [1], hydrogen sulfide (H_2S) is also produced in addition to CH_4 and CO_2, with content of 0.3%–0.4% [2]. H_2S is a foul acid gas, and it can result in the serious corrosion of pipeline, instruments, and equipment. In addition, H_2S endangers human health and causes environmental pollution. When the biogas containing H_2S is utilized as energy (such as burning, power generation, etc.), H_2S will be converted into SO_2 and cause serious air pollution [3]. Therefore, the existence of H_2S has limited the biogas energy promotion, and effective approaches are required for H_2S removal from biogas.

H_2S treatment methods include physical, chemical, biological, and combinatorial technologies. The direct degradation of H_2S for production of hydrogen and sulfur has been the research focus of domestic and foreign researchers, since it can effectively control the H_2S pollution produced during oil, gas, coal, and mineral processing and also achieve the recycling of hydrogen energy. The main methods of H_2S degradation for hydrogen and sulfur production include thermal degradation [4, 5], electrochemical degradation [6], photocatalytic degradation [7–9], and plasma degradation [10, 11]. Herein, photocatalytic degradation of H_2S is the most promising technology due to the high treatment efficiency and reaction rate.

Currently, researches on photocatalytic degradation of H_2S are mostly conducted under the condition of O_2 [12–14].

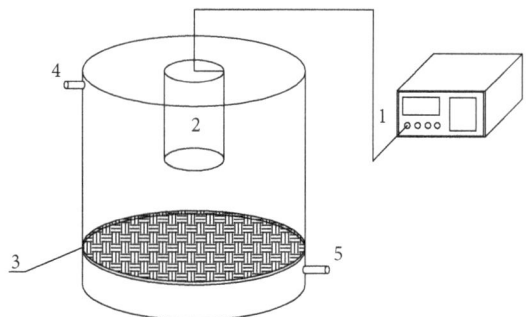

FIGURE 1: H₂S photodegradation reactor. (1) High-frequency generator, (2) electrodeless VUV lamp, (3) gas distribution device, (4) gas outlet, and (5) gas inlet.

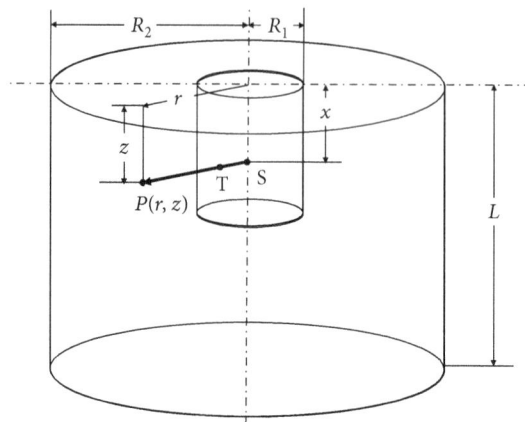

FIGURE 2: Cylindrical photoreactor with a cylindrical light source.

Nevertheless, there is potential of explosion with O_2 presented during photocatalytic degradation of H_2S from biogas. Moreover, a by-product of ozone (O_3) is produced and further treatment was required. Photocatalytic degradation of H_2S with the absence of O_2 can avoid the abovementioned problems, which is quite suitable for H_2S removal from biogas and other gas treatments requiring anaerobic conditions. In addition, conventional UV light sources for degradation, such as mercury vapor lamps and high-pressure xenon lamps, are incapable of producing intense light near UV or deep UV lights. The self-made high-frequency discharge vacuum ultraviolet (VUV) lamp has lots of advantages compared with the conventional UV lamps and microwave discharge electrodeless lamps, such as high efficiency, high radiation intensity, high ratio of 185 nm light, and long service life [12, 15].

Therefore, in the present study, the photodegradation of H_2S with VUV lamp in the absence of O_2 was studied. Firstly, the photodegradation model was built using the MATLAB software, and then the effect of initial H_2S concentration and retention time on H_2S degradation performance was investigated and compared with the modeling result. Moreover, the reaction kinetics and mechanism of H_2S degradation in the absence of O_2 were discussed. We hope the research result can help the effective H_2S removal under anaerobic conditions.

2. Materials and Methods

2.1. Reactor and Experiments. Figure 1 schematically depicts the H_2S photodegradation reactor used in this study. The reactor was designed as cylindrical in shape to avoid the dead space for photodegradation and the potential uneven gas distribution within the reactor. A cylindrical VUV lamp was placed on top of the reactor, and its connection with the reactor was sealed with corrosion-resistant and high temperature-resistant silica gel. A high-frequency generator was connected with the VUV lamp via an external circuit. A porous plate-like gas distribution device was installed at the gas inlet, and thus, the incoming gas can evenly flow upwards. The gas outlet was installed at the top of the reactor. To avoid short flow, the gas inlet and outlet were placed on different sides of the reactor. For the sake of shielding the high-frequency electromagnetic radiation and preventing

the corrosion of the reactor caused by H_2S, the body of the reactor was made of SUS304 stainless steel.

The diameter and height of the reactor were 15 and 14 cm, respectively. The total volume of the reactor was 2.5 L, with the effective volume of 2.0 L when the cylindrical UV lamp with a height of 5.8 cm and diameter of 4 cm was placed.

Figure 2 is diagrammatic sketch of the cylindrical photoreactor with a cylindrical light source for radiation field modeling, where $I(r, z)$ is the light intensity at point $P(r, z)$ within the reactor, $S_{L,\lambda}$ is the UV intensity at wavelength λ, and r and z are the horizontal distance and vertical distance of a random particle from the center of the UV lamp, respectively. L is the length of the UV lamp. μ_λ is the absorption coefficient of the medium in the reactor at the wavelength of λ. R_1 and R_2 are the radii of the cylindrical UV lamp and the cylindrical reactor, respectively.

2.2. Experimental Methods. In order to investigate the degradation efficiency of H_2S with only the high-frequency electrodeless VUV lamp without addition of O_3, OH, and photocatalyst, the reactor was firstly dried, and Ar was flushed into the reactor with a 10 L/min flow rate for an hour to expel the residual O_2 and H_2O in the pipeline. During the whole experiment, Ar gas was continuously flushed to exclude the effects of O_3 and OH on H_2S degradation.

2.3. Analytical Methods. H_2S concentration was determined with methylene blue spectrophotometry under a wavelength of 660 nm [16].

3. Results and Discussion

3.1. Mathematical Modeling of H_2S Photodegradation

3.1.1. Gas Flow Distribution Model. The rate of gas flow was determined by both the annular space of the reactor and the retention time. As the diameter of the reactor and the UV lamp were 15 cm and 4 cm, respectively, the rate of the gas flow was calculated as 20–40 L/min when the retention time was set as 3–6 s. Considering that the concentration of H_2S in the gas influent was low, the Re constant could be calculated based on the physical properties of the incoming

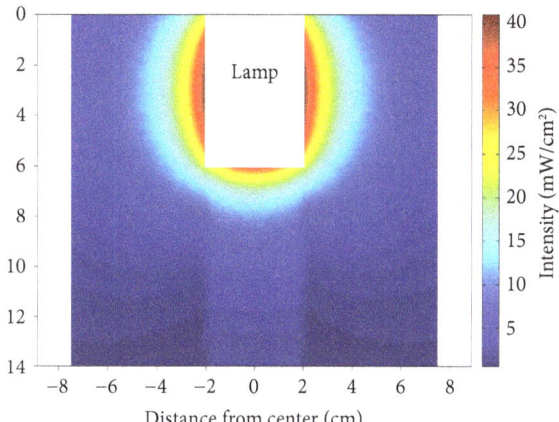

FIGURE 3: The light intensity distribution in the photoreactor.

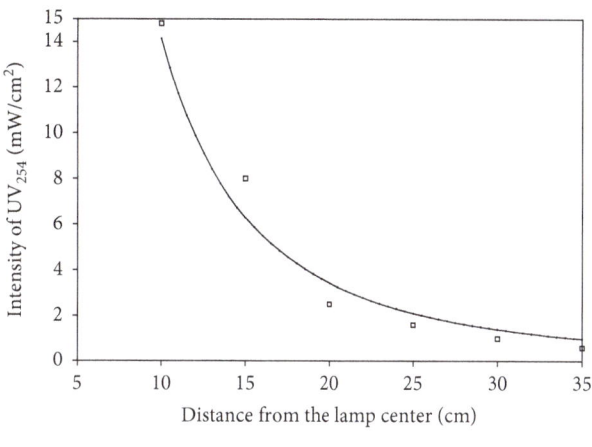

—— Simulation
▫ Measurement

FIGURE 4: The experimental data and simulation data of the UV intensity with the change of space.

gas. The Re calculated at room temperature with Ar as carrier gas was 203–406.

The velocity (v) of a particle at the radius of R away from the center axis of the annular reactor can be determined according to the following equation [13]:

$$v = \frac{\left(2U\left(R_2^2 - R_1^2\right)/\ln R_2/R_1\right)\ln\left(R/R_1\right) - 2U\left(R^2 - R_1^2\right)}{R_1^2 + R_2^2 - \left(\left(R_2^2 - R_1^2\right)/\ln R_2/R_1\right)}, \quad (1)$$

where U is the known average velocity and R_1, R_2, and R are the radius of the UV lamp, the radius of the reactor, and the horizontal distance between the particle and the center axis of the annular reactor, respectively.

Based on (1), the simulated maximum velocity was no more than 5.8 m/s (as shown in Figure S1), and then the calculated Re was also no more than 609. The above Re values were all far below the critical value of laminar flow (2300) and indicated a typical laminar flow pattern in the reactor (the Re calculation was provided in the Supplementary Materials).

3.1.2. Radiation Field Model. The light intensity ($I(r, z)$) at point $P(r, z)$ within the reactor can be expressed with (2) [17].

$$I(r, z)_\lambda = \frac{S_{L,\lambda}}{\pi^2} \int_0^L \frac{r}{\left[r^2 + (z - x)^2\right]^{3/2}} \times \exp$$
$$\cdot \left[-\mu_\lambda \left(r^2 + (z - x)^2\right)^{1/2}\left(1 - \frac{R_1}{r}\right)\right] dx. \quad (2)$$

The simulated three-dimensional model of the light intensity by the cylindrical VUV lamp at the center of the photoreactor with diameter and height of 15 cm and 14 cm, respectively, is shown in Figure 3. The white area was occupied by the VUV lamp with diameter and height of 4 cm and 5.8 cm, respectively, as scaled in the figure coordinate axis. According to Figure 3, light intensity decreased rapidly with the increase in the distance from the center of the light source.

In order to verify the result of the light intensity distribution model in the reactor, a UV radiation meter was applied to measure the UV_{254} intensity at 5–35 cm away from

the lamp wick of the cylindrical lamp. In the meantime, the light intensity at the same position was calculated based on the UV light radiation field model and then was compared with the measured value to verify the results of the model. As shown in Figure 4, the results indicated that the calculated value and the measured values were close, indicating that (2) was feasible to predict the radiation intensity distribution.

3.1.3. Photochemical Reaction Kinetics Model. The photochemical reaction kinetics equation involving component i can generally be presented as the following equation.

$$\frac{dC_i}{dt} = \Phi_i \cdot I \cdot \left(1 - 10^{\varepsilon b C_i}\right), \quad (3)$$

where C_i is the molar concentration of component i (mol/L), Φ_i is the quantum efficiency, ε is the molar absorption coefficient (L/mol·cm), b is the optical length (cm), and I is the light radiation intensity (Einstein/cm²·s). It should be noted that the VUV light absorption by other components of the initial biogas, including the typical CH_4 and CO_2, could be neglected [18].

Equation (3) has an exponential term, making the calculation complicated. When $\varepsilon b C_i < 0.02$, (3) can be reasonably approximated to the following expression:

$$\frac{dC_i}{dt} = 2.303 \cdot \Phi_i I \varepsilon b C_i. \quad (4)$$

Asili and De Visscher [13] proposed another equation for calculating the photochemical reaction rate r_i (molecule/cm³·s):

$$-r_i = \Phi_i \cdot C_i \cdot \sigma_i \cdot E_p, \quad (5)$$

where C_i is the molar concentration of component i (molecule/cm³), Φ_i is the quantum efficiency, σ_i is the absorption cross section of component i, and E_p is the photon flux (photon/cm²·s).

The comparison of (4) and (5) revealed that both equations reflect the same reaction behavior, though (4) describes the concentration change macroscopically while (5) describes it microscopically. Both equations describe the chemical changes that occur when the reacting components absorb a certain amount of energy.

3.1.4. Mass Balance Model.

Firstly, the Peclet number Pe in the photoreactor was calculated, revealing the numerical zone of 132–263. Pe represented the relative proportion of convection to diffusion. When Pe is larger than 40, the dominant mass transfer type was convection at the airflow direction. As the main gas flow pattern was a laminar flow, the main mass transfer pattern at the vertical direction was diffusion. According to the empirical equation, for H_2S-air mixed gas, the diffusion coefficient at the vertical direction can be expressed as [19]

$$D_{H_2S\text{-air}} = 0.7914 T^{1.75}/P, \qquad (6)$$

where $D_{H_2S\text{-air}}$ is the diffusion coefficient (cm^2/s), T is the gas temperature (K), and P is the gas pressure.

For a random component i of the photochemical reaction system in the annular photoreactor, its partial differential equation of mass conservation can be represented as

$$v \cdot \frac{\partial C_i}{\partial L} = D_i \cdot \frac{\partial^2 C_i}{\partial R^2} + \frac{D_i}{R} \cdot \frac{\partial C_i}{\partial R} + r_i, \qquad (7)$$

where v is the velocity of the gas molecule at current position (cm/s), C_i is the molar concentration of the component i (mole/cm^3), L is the vertical distance between a random position and the gas inlet (cm), R is the horizontal distance between a random position and the linear light source, D_i is the diffusion coefficient of the component i at the radical direction (cm^2/s), and r_i is the photochemical reaction rate (mole/$cm^3 \cdot s$).

The boundary condition of the partial differential equation (7) was assumed as the no-flow boundary, and the inlet gas concentration was considered as constant.

3.1.5. Calculation Model of Degradation Rate.

For a random component i in the photochemical reaction system, the degradation degree η_i can be expressed as

$$\eta_i = 1 - \frac{\bar{C}_i}{C_{i,\text{initial}}}, \qquad (8)$$

where $\bar{C}_i$ is the random component i in the photochemical reaction system at the gas outlet and $C_{i,\text{initial}}$ is the initial concentration of i at the inlet. For a tubular reactor, the average concentration of the random component i at the gas outlet can be expressed as

$$\bar{C}_i = \frac{\int_{R_1}^{R_2} v \cdot 2\pi R \cdot dR}{\int_{R_1}^{R_2} v \cdot 2\pi R \cdot dR}, \qquad (9)$$

where v is the velocity of the gas molecule at the present position (cm/s), R_1 and R_2 are the radius of the VUV lamp

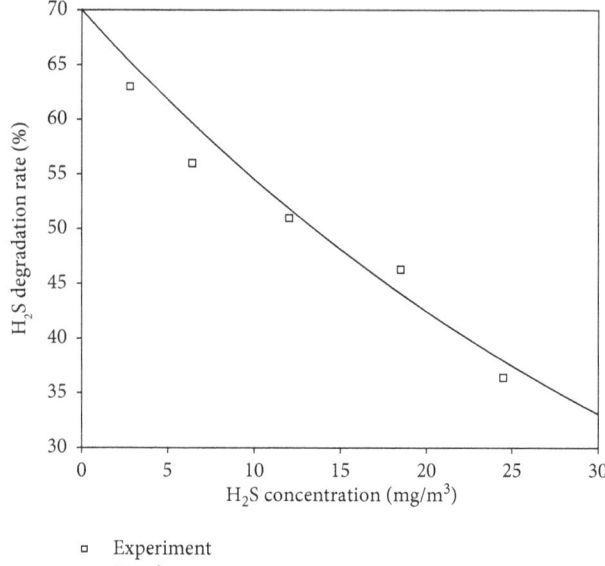

□ Experiment
—— Simulation

FIGURE 5: Influence of initial concentration of H_2S on the degradation rate.

and the reactor (cm), and R is the horizontal distance between a random position and the linear light source.

The velocity of the component i in the reactor can be calculated using (1) and (2) and can be used to acquire the light intensity of a random position in the reactor. Then, the acquired data can be used in (4) and (7).

Equation (4) is an ordinary differential equation (ODE). Although (7) is a partial differential equation (PDE), it can be converted into a series of ODEs using MATLAB ODE15s for equation solution. Then, (8) and (9) are used for solving the degradation rate of the target component and correlate with experimental data.

3.2. Major Influence Factors on H_2S Photodegradation

3.2.1. Initial H_2S Concentration.

The influence of the initial H_2S concentration on the degradation rate under retention time of 6 s is profiled in Figure 5. Results showed that the degradation rate continuously decreased with the increase in the initial H_2S concentration, which was similar to the situation with O_2 present [12]. As shown, when the initial H_2S concentration was 3 mg/m^3, the maximum degradation rate was about 62.8%. When the initial H_2S concentration reached 25 mg/m^3, the degradation rate decreased to 36.2%. UV light can only pass a relatively short distance in vacuum, and also there were no oxidative free radicals to assist degradation [20]; therefore, the degradation rate was comparatively lower than that with the presence of O_2 [12, 21]. Moreover, no SO_4^{2-} in the photodegradation products of H_2S was detected although it was confirmed as the main product in a previous study [22], and H_2 and elemental sulfur were the possible products. Yet, solid S was also not observable throughout the whole experiment. The first reason was the rather low particle settling velocity calculated by Stokes formula, and another reason was

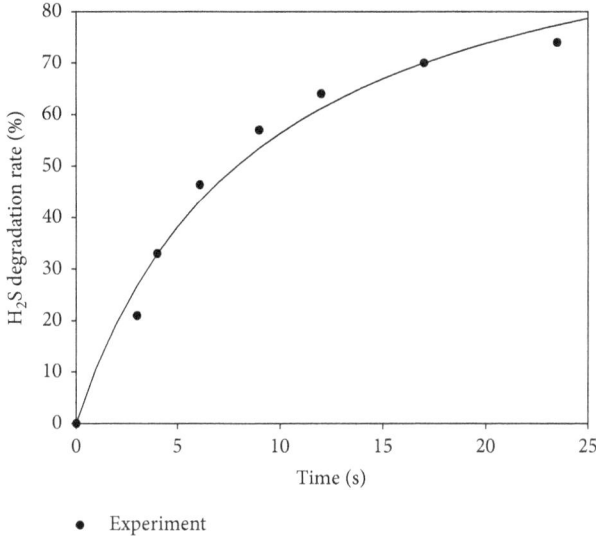

FIGURE 6: The relationship between H_2S degradation rate and retention time of direct photolysis by UV with the absence of oxygen.

the very low H_2S concentration and high airflow velocity. Therefore, it was difficult for the sulfur particle to settle on the internal reactor walls and no deposition of elemental sulfur was observed.

3.2.2. Gas Retention Time.

Under the initial H_2S concentration of 12 mg/m^3, the simulation of the H_2S degradation profile based on modeling and the experimental data are compared in Figure 6. The experimental results indicated that the degradation rate was significantly affected when gas retention time in the reactor was less than 12 s. Specifically, when gas retention time was 3 s and 6 s, the degradation rate was 22.0% and 48.1%, respectively, indicating that the degradation efficiency proportionally increased with gas retention time. With higher gas retention time, the increasing rate of the degradation efficiency became slower. For instance, when gas retention time was 24 s, the degradation efficiency was 74% and only increased by 30% compared with that at gas retention time of 6 s. However, it still could not reach 100% degradation with the further increase in gas retention time.

The dead zone was placed in a position far away from the lamp, and so H_2S was hard to be degraded thoroughly. Longer retention time was beneficial for increasing the degradation rate; however, the energy consumption was also increased. Therefore, the appropriate retention time should be selected considering the initial H_2S concentration, equipment cost, and operating cost.

The root-mean-square error (RMSE) for this model simulation was as small as 0.0622. A t-test was conducted for comparing the simulation data and experimental data, and the test result accepted the null hypothesis with a possibility of 96.72%, indicating that the simulation data and the experimental data were consistent. This proves that the model to simulate the degradation efficiency in the photodegradation reactor was feasible.

3.3. Mechanism of H_2S Photodegradation Using Only VUV Light.

Wilson et al. [23] studied the direct degradation of H_2S by photon of a near ultraviolet band. Results showed that a photon with a wavelength less than 270 nm could directly photodegrade H_2S into H· and SH·, and SH· could be degraded into H· and S· with a photon wavelength less than 230 nm, as shown in (10) and (11). The possible reactions during the direct photodegradation of H_2S are listed as follows:

$$H_2S + h\nu(\lambda \leq 270\,nm) \rightarrow H\cdot + SH\cdot \tag{10}$$

$$HS\cdot + h\nu(\lambda \leq 230\,nm) \rightarrow H\cdot + S \tag{11}$$

$$H\cdot + H_2S \rightarrow H_2 + HS\cdot \tag{12}$$

$$H\cdot + H\cdot \rightarrow H_2 \tag{13}$$

$$H\cdot + HS\cdot \rightarrow H_2 + S \tag{14}$$

$$HS\cdot + HS\cdot \rightarrow S + H_2S \tag{15}$$

By referring to the chemical dynamics database of the National Institute of Standards and Technology (NIST) and the related literatures, the rate constants can be acquired for the above reactions and are summarized in Table 1. A photon participates in reactions (10) and (11), and the molar absorption coefficients for wavelengths of 185 nm and 254 nm can be obtained by converting the absorptivity data summarized in Table 1.

Based on (10), (11), (12), (13), (14), and (15), the degradation rate equation of the various intermediates during hydrogen sulfide photodegradation in the absence of O_2 could be established. As photon is presented in (10) and (11), let $k_{obs} = 2.030\Phi_i\varepsilon b$ ((4)), denoting k_{obs1}, k_{obs2}, k_3, k_4, k_5, and k_6 as the rate constants for (10), (11), (12), (13), (14), and (15). The reaction rate equations for each component can be expressed as follows:

$$\frac{d[H_2S]}{dt} = -k_{obs1}\cdot[H_2S] - k_3\cdot[H_2S][H\cdot] + k_6\cdot[SH\cdot]^2, \tag{16}$$

$$\frac{d[H\cdot]}{dt} = k_{obs1}\cdot[H_2S] + k_{obs2}\cdot[SH\cdot] - k_3\cdot[H_2S][H\cdot] - k_4\cdot[H\cdot]^2 - k_5\cdot[SH\cdot][H\cdot], \tag{17}$$

$$\frac{d[SH\cdot]}{dt} = k_{obs1}\cdot[H_2S] - k_{obs2}\cdot[SH\cdot] - k_5\cdot[SH\cdot][H\cdot] - k_6\cdot[SH\cdot]^2, \tag{18}$$

$$\frac{d[H_2]}{dt} = k_3\cdot[H_2S][H\cdot] + k_4\cdot[H\cdot]^2 + k_5\cdot[H\cdot][SH\cdot], \tag{19}$$

$$\frac{d[S]}{dt} = k_{obs2}\cdot[SH\cdot] + k_5\cdot[SH\cdot][H\cdot] + k_6\cdot[SH\cdot]^2. \tag{20}$$

Then, the reaction rate equations for H_2S, H·, SH·, H_2, and S (from (16), (17), (18), (19), and (20)) were integrated into the UV-photodegrading reactor model, and the H_2S photodegradation effect without O_2 was simulated under the following conditions: the initial H_2S concentration was set as 12 mg/m^3 with Ar as the carrier gas. The concentration

TABLE 1: The chemical reaction of direct photolysis of H_2S and the corresponding photolysis rate constants.

Reaction	Quantum efficiency	Light absorptivity at 185 nm (L/mol·cm)	Light absorptivity at 254 nm (L/mol·cm)	Rate constant (L/mol·s)	Reaction order	References
$H_2S + h\nu \rightarrow H\cdot + SH$	1	1004.66	1.96	/	/	[24]
$HS\cdot + h\nu \rightarrow H\cdot + S$	1	20.5	0	/	/	[24]
$H\cdot + H\cdot \rightarrow H_2$	/	/	/	$2.498E + 10$	2	[25]
$H_2S + H\cdot \rightarrow H_2 + SH$	/	/	/	$4.786E + 08$	2	[26]
$H\cdot + SH\cdot \rightarrow H_2 + S$	/	/	/	$8.0E + 09$	2	[27]
$SH\cdot + SH\cdot \rightarrow S + H_2S$	/	/	/	$8.368E + 9$	2	[28]

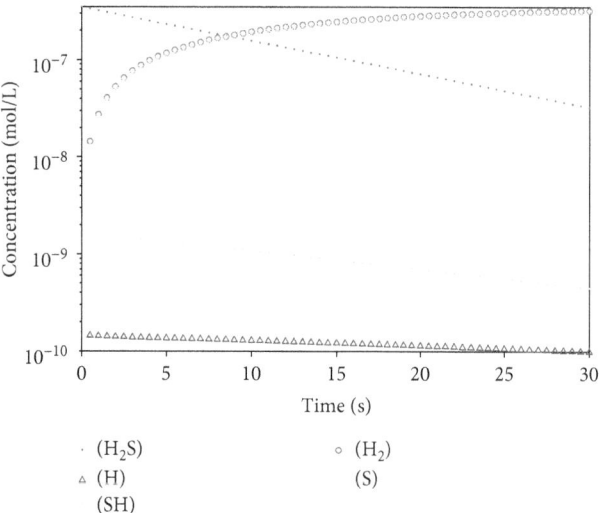

FIGURE 7: The simulated concentration changes of components of H_2S photodegradation by UV without the presence of oxygen.

evolution of H_2S, $H\cdot$, $SH\cdot$, H_2, and S were calculated 2 cm away from the UV light with MATLAB software and demonstrated by the logarithmic scale in Figure 7.

It was revealed that the final products of H_2S photodegradation were mainly H_2 and S, and their concentrations were close to each other, which was in accordance with the stoichiometric coefficients of the photodegradation reaction. Moreover, the simulated concentrations of $H\cdot$ and $SH\cdot$ radicals were rather low, which were about two orders of magnitude lower than H_2 and S.

Finally, we tried to build the analytical equation connecting the rate of H_2S consumption and its concentration and light intensity using steady-state approximation for radicals. Firstly, let (16) be equal to zero, as shown in (21).

$$\frac{d[H_2S]}{dt} = 0. \tag{21}$$

Then, (21) and (22) could be obtained combining (16) and (21), as follows.

$$[H_2S]_{s-t}\left(k_{obs1} + k_3[H\cdot]\right) = k_6[SH\cdot]^2, \tag{22}$$

$$[H_2S]_{s-t} = \frac{k_6[SH\cdot]^2}{k_{obs1} + k_3[H\cdot]}. \tag{23}$$

Then, k_{obs1} was substituted with (4), and the analytical equation connecting the rate of H_2S consumption and its concentration and light intensity using steady-state approximation for the radical could be obtained as (24).

$$[H_2S]_{s-t} = \frac{k_6[SH\cdot]^2}{2.303 \cdot \Phi_i I \varepsilon b C_i + k_3[H\cdot]}. \tag{24}$$

4. Conclusions

In the present study, the mathematical model of ultraviolet degradation of H_2S with the absence of O_2 was established, and the influence of the initial H_2S concentration and gas retention time on the photodegradation rate were studied and for verification of the model. The main findings were as follows.

(1) The photodegradation rate decreased with the increase in initial H_2S concentration, and the maximum photodegradation rate was about 62.8% under initial concentration of 3 mg/m³.

(2) The photodegradation rate increased with the increase in H_2S retention time.

(3) Experimental results were in good accordance with the modeling results.

(4) The main photodegradation products were H_2 and elemental S based on mathematical modeling. Concentrations of both products were close and agreed well with the reaction stoichiometric coefficients.

Conflicts of Interest

The authors declare that they have no conflicts of interest.

Authors' Contributions

Jian-hui Xu and Bin-bin Ding contributed equally to this work and should be considered co-first authors.

Acknowledgments

The authors gratefully acknowledge the financial support of the Development of Social Science and Technology in Dongguan (Key) (2017507101426).

Supplementary Materials

Calculation of Reynolds number (Re): the calculation formula for Re was $Re = \rho v d / \mu$, where ρ, v, and μ were the gas density ($1.169 \, g/cm^3$), velocity, and viscosity coefficient ($18.448 \, \mu Pa \cdot s$) of Ar gas, respectively, and d was the equivalent diameter of the photoreactor. The diameter (d) was $0.15 \, cm$. The velocity (v) was 0.019–$0.038 \, m/s$ based on the gas flow rate ($v = Q/S$, where Q was 20–$40 \, L/min$ and S was obtained based on the diameter of $15 \, cm$). Then the calculated Re was about 203–406. In addition, the velocity based on the simulation of (1) was no more than $0.058 \, m/s$ (as shown in Figure S1), and Re was about 609. Both the above Re values indicated a typical laminar flow pattern in the reactor. Figure S1: the simulated gas flow velocity by (1). *(Supplementary Materials)*

References

[1] G. Silvestre, B. Fernandez, and A. Bonmati, "Significance of anaerobic digestion as a source of clean energy in wastewater treatment plants," *Energy Conversion and Management*, vol. 101, pp. 255–262, 2015.

[2] A. M. Montebello, M. Mora, L. R. López et al., "Aerobic desulfurization of biogas by acidic biotrickling filtration in a randomly packed reactor," *Journal of Hazardous Materials*, vol. 280, pp. 200–208, 2014.

[3] X. Song, W. Yao, B. Zhang, and Y. Wu, "Application of Pt/CdS for the photocatalytic flue gas desulfurization," *International Journal of Photoenergy*, vol. 2012, Article ID 684735, 5 pages, 2012.

[4] N. O. Guldal, H. E. Figen, and S. Z. Baykara, "New catalysts for hydrogen production from H_2S: preliminary results," *International Journal of Hydrogen Energy*, vol. 40, no. 24, pp. 7452–7458, 2015.

[5] K. Akamatsu, M. Nakane, T. Sugawara, T. Hattori, and S. Nakao, "Development of a membrane reactor for decomposing hydrogen sulfide into hydrogen using a high-performance amorphous silica membrane," *Journal of Membrane Science*, vol. 325, no. 1, pp. 16–19, 2008.

[6] A. A. Anani, Z. Mao, R. E. White, S. Srinivasan, and A. J. Appleby, "Electrochemical production of hydrogen and sulfur by low-temperature decomposition of hydrogen sulfide in an aqueous alkaline solution," *Journal of the Electrochemical Society*, vol. 137, no. 9, pp. 2703–2709, 1990.

[7] X. Zong, J. Han, B. Seger et al., "An integrated photoelectrochemical–chemical loop for solar-driven overall splitting of hydrogen sulfide," *Angewandte Chemie*, vol. 53, no. 17, pp. 4399–4403, 2014.

[8] X. Bai, Y. Cao, and W. Wu, "Photocatalytic decomposition of H_2S to produce H_2 over CdS nanoparticles formed in HY-zeolite pore," *Renewable Energy*, vol. 36, no. 10, pp. 2589–2592, 2011.

[9] H. Yan, J. Yang, G. Ma et al., "Visible-light-driven hydrogen production with extremely high quantum efficiency on Pt–PdS/CdS photocatalyst," *Journal of Catalysis*, vol. 266, no. 2, pp. 165–168, 2009.

[10] E. Linga Reddy, J. Karuppiah, V. M. Biju, and C. Subrahmanyam, "Catalytic packed bed non-thermal plasma reactor for the extraction of hydrogen from hydrogen sulfide," *International Journal of Energy Research*, vol. 37, no. 11, pp. 1280–1286, 2013.

[11] S. John, J. C. Hamann, S. S. Muknahallipatna, S. Legowski, J. F. Ackerman, and M. D. Argyle, "Energy efficiency of hydrogen sulfide decomposition in a pulsed corona discharge reactor," *Chemical Engineering Science*, vol. 64, no. 23, pp. 4826–4834, 2009.

[12] J. Xu, C. Li, P. Liu, D. He, J. Wang, and Q. Zhang, "Photolysis of low concentration H_2S under UV/VUV irradiation emitted from high frequency discharge electrodeless lamps," *Chemosphere*, vol. 109, pp. 202–207, 2014.

[13] V. Asili and A. De Visscher, "Mechanistic model for ultraviolet degradation of H_2S and NO_x in waste gas," *Chemical Engineering Journal*, vol. 244, pp. 597–603, 2014.

[14] R. Portela, M. C. Canela, B. Sanchez et al., "H_2S photodegradation by TiO_2/M-MCM-41 (M = Cr or Ce): deactivation and by-product generation under UV-A and visible light," *Applied Catalysis B: Environmental*, vol. 84, no. 3-4, pp. 643–650, 2008.

[15] J. Xu, C. Li, Q. Zhang, D. He, P. Liu, and Y. Ren, "Destruction of toluene by the combination of high frequency discharge electrodeless lamp and manganese oxide-impregnated granular activated carbon catalyst," *International Journal of Photoenergy*, vol. 2014, Article ID 365862, 9 pages, 2014.

[16] J. C. C. Santos and M. Korn, "Exploiting sulphide generation and gas diffusion separation in a flow system for indirect sulphite determination in wines and fruit juices," *Microchimica Acta*, vol. 153, no. 1-2, pp. 87–94, 2006.

[17] T. Yokota and S. Suzuki, "Estimation of light absorption rate in a tank type photoreactor with multiple lamps inside," *Journal of Chemical Engineering of Japan*, vol. 28, no. 3, pp. 300–305, 1995.

[18] H. G. Baldovi, J. Albero, B. Ferrer, D. Mateo, M. Alvaro, and H. García, "Gas-phase photochemical overall H_2S splitting by UV light irradiation," *ChemSusChem*, vol. 10, no. 9, pp. 1996–2000, 2017.

[19] R. B. Bird, W. E. Stewart, and E. N. Lightfoot, *Transport Phenomena*, John Wiley & Sons, New York, NY, USA, 2002.

[20] Z. Li, H. Li'an, and Y. Linsong, "The experimental investigations of dielectric barrier discharge and pulse corona discharge in air cleaning," *Plasma Science and Technology*, vol. 5, no. 5, pp. 1961–1964, 2003.

[21] L. Huang, L. Xia, X. Ge, H. Jing, W. Dong, and H. Hou, "Removal of H_2S from gas stream using combined plasma photolysis technique at atmospheric pressure," *Chemosphere*, vol. 88, no. 2, pp. 229–234, 2012.

[22] M. C. Canela, R. M. Alberici, and W. F. Jardim, "Gas-phase destruction of H_2S using TiO_2/UV-VIS," *Journal of Photochemistry and Photobiology A: Chemistry*, vol. 112, no. 1, pp. 73–80, 1998.

[23] S. H. S. Wilson, J. D. Howe, and M. N. R. Ashfold, "On the near ultraviolet photodissociation of hydrogen sulphide," *Molecular Physics*, vol. 88, no. 3, pp. 841–858, 1996.

[24] S. P. Sander, A. JPD, J. R. Barker et al., *Chemical Kinetics and Photochemical Data for Use in Atmospheric Studies: Evaluation Number 17*, JPL Publication, 2011.

[25] J. N. Bradley, S. P. Trueman, D. A. Whytock, and T. A. Zaleski, "Electron spin resonance study of the reaction of hydrogen atoms with hydrogen sulphide," *Journal of the Chemical Society, Faraday Transactions 1: Physical Chemistry in Condensed Phases*, vol. 69, no. 0, pp. 416–425, 1973.

[26] J. Peng, X. Hu, and P. Marshall, "Experimental and ab initio investigations of the kinetics of the reaction of H atoms with H_2S," *The Journal of Physical Chemistry A*, vol. 103, no. 27, pp. 5307–5311, 1999.

[27] G. Loraine, W. Glaze, and W. Glaze, "Destruction of vapor phase halogenated methanes by means of ultraviolet photolysis," *Proceedings of the Industrial Waste Conference*, vol. 47, pp. 309–316, 1993.

[28] R. A. Stachnik and M. J. Molina, "Kinetics of the reactions of mercapto radicals with nitrogen dioxide and oxygen," *The Journal of Physical Chemistry*, vol. 91, no. 17, pp. 4603–4606, 1987.

17

E. coli Bacteriostatic Action Using TiO₂ Photocatalytic Reactions

Thammasak Rojviroon ⓘ[1] and Sanya Sirivithayapakorn[2]

[1]Department of Civil Engineering, Faculty of Engineering, Rajamangala University of Technology Thanyaburi,
 Pathum Thani 12110, Thailand
[2]Department of Environmental Engineering, Faculty of Engineering, Kasetsart University, Bangkok 10900, Thailand

Correspondence should be addressed to Thammasak Rojviroon; thammasak@rmutt.ac.th

Academic Editor: Mohammad Al-Amin

This experimental research comparatively investigates the *Escherichia coli (E. coli)* bacterial inactivation of the TiO₂ photocatalytic thin films fabricated by the sol–gel dip-coating (SG) and low-temperature spray-coating (SP) techniques, with low-intensity ($12\,\mu$W·cm^{-2}) UVA-light-emitting diodes (UVA-LED) as the light source. The bacteriostatic experiments were undertaken using the nutrient broth (NB) and 0.85% NaCl with the initial *E. coli* concentrations of 10^2, 10^4, 10^6, and 10^8 CFU·mL^{-1}. Moreover, the essential physical characteristics of the SG-TiO₂ and SP-TiO₂ photocatalytic thin films were determined prior to the experimental bacterial inactivation. The findings showed that both photocatalytic thin films possessed the ideal physical characteristics, especially the SP-TiO₂ thin film. In addition, the viable cell counts, the cell morphology, and the bioluminescence-based adenosine triphosphate (ATP) indicated that both SG-TiO₂ and SP-TiO₂ thin films under UVA could effectively inhibit the proliferation of the *E. coli* cells in both NB and 0.85% NaCl.

1. Introduction

The recent decades have witnessed a growing interest in the development of innovative antibacterial technologies against pathogens in the aquatic environment. The phenomenon is attributable to the drawbacks inherent in the existing technologies, including the implementation challenge, the high operation and maintenance costs, and the carcinogenic effects [1–3]. Moreover, the long-term use of antibiotics could cause the bacteria to become antibiotic-resistant and render the drugs less effective in controlling the spread of diseases [4, 5].

One such innovative antibacterial technology is the photocatalytic reactions in the presence of a semiconducting solid catalyst, which generate the free radicals, that is, hydroxyl radicals (•OH) and superoxide radicals (•O₂⁻), which are naturally strong oxidizing agents [6, 7]. In fact, the photocatalytic technology has been utilized in numerous applications, including deodorization, bacterial and viral disinfection, and air and water decontamination [8–11]. The technology is also easy to implement as it

essentially requires an ultraviolet light source and a photocatalyst. The most commonly used photocatalyst is titanium dioxide (TiO₂) due to its high levels of photocatalytic activity, prolonged chemical stability, and low toxicity and production cost [12, 13].

Specifically, this experimental research comparatively investigates the TiO₂ photocatalytic thin films fabricated by the sol–gel dip-coating (SG) and low-temperature spray-coating (SP) techniques. In the experiment, the low-intensity UVA-light-emitting diodes (UVA-LED) were used as the light source because they are safer than UVB and UVC, easy to install, inexpensive, lightweight, and energy efficient [14, 15]. Prior to the experimental bacterial inactivation, the essential physical characteristics of the SG-TiO₂ and SP-TiO₂ photocatalytic thin films were determined, including the crystalline phase, bandgap energy, contact angle, morphology, adhesion, and acid-base corrosion resistance.

The experimental bacterial inactivation was carried out using the nutrient broth (NB) and 0.85% NaCl with the initial *E. coli* concentrations of 10^2, 10^4, 10^6, and 10^8 CFU·mL^{-1} treated with the SG- and SP-TiO₂ photocatalytic thin films

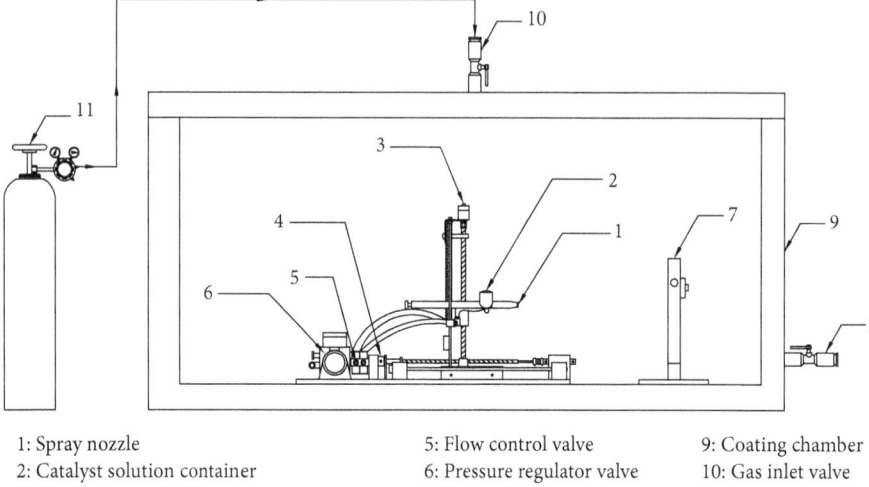

1: Spray nozzle
2: Catalyst solution container
3: Up-down nozzle driving motor
4: Forward-backward nozzle driving motor

5: Flow control valve
6: Pressure regulator valve
7: Spray platform
8: Gas outlet valve

9: Coating chamber
10: Gas inlet valve
11: n_2 gas container

FIGURE 1: The schematic of the spray-coating chamber.

TABLE 1: Physical characteristics and the corresponding analysis techniques.

Physical characteristics	Analytical technique/equipment
Crystalline structure	X-ray diffraction/Bruker model D8 Advance
Bandgap energy	Absorption spectrum fitting technique/Thermo Electron's Helios Alpha UV-Vis spectrometer
Contact angle	Sessile drop technique/DataPhysics OCA series TBU90E
Surface morphology	Atomic force microscopy/Asylum Research MFP-3D-BIO
Adhesion test	ASTM method D3359B-17
Corrosion test	Strong acid-base test

and UVA. For comparison, the bacteriostatic activity testing was also conducted in the absence of UVA or the photocatalytic thin films (the control). The bacterial inactivation performance was determined by the viable cell count using the standard plate count (SPC) method and the bioluminescence-based adenosine triphosphate (ATP) test.

2. Experimental

2.1. Preparation of TiO$_2$ Thin Films. The photocatalytic thin films were fabricated using the sol–gel dip-coating (SG) and low-temperature spray-coating (SP) techniques, whereby the TiO$_2$ solution was coated onto borosilicate glass slides ($40.0 \times 85.0 \times 0.3$ mm) and petri dishes (100 mm in diameter $\times$ 15 mm in height). The SG- and SP-TiO$_2$ thin films on the borosilicate glass slides and the petri dishes were for determination of the physical properties and the *E. coli* bacteriostatic action, respectively.

The SG-TiO$_2$ thin film was prepared using the TiO$_2$ acid-catalyzed sol–gel dip-coating process [16]. The sol–gel solution was prepared using titanium isopropoxide ($C_{12}H_{28}O_4Ti$, TTIP) and isopropanol (C_3H_7OH) in a volume ratio of 1:15 under a pH of 2-3 in a container flushed with N$_2$ gas at room temperature. The TiO$_2$ sol–gel solution was then transferred to a container flushed with N$_2$ gas for dip-coating. In this research, the SG-TiO$_2$ thin film consisted of five layers of film independently thermally treated: 100°C for the bottom layer and 200, 250, 350, and 500°C for the second, third, fourth, and top layers, respectively.

The low-temperature SP-TiO$_2$ thin film was produced using the modified TiO$_2$ composite colloid technique [17]. In the preparation of the TiO$_2$ composite colloids, 0.01 g of anatase TiO$_2$ nanoparticles was dispersed in 100 mL of ethanol and polyethylene glycol (PEG) and continuously stirred with a magnetic stirrer at room temperature for 1 hour. The pH of the solution was then adjusted to a pH of 2-3. The colloid solution was then stirred for another 1 hour at room temperature in a container flushed with N$_2$ and left at room temperature for 24 hours.

The TiO$_2$ colloid-based photocatalyst was then sprayed onto the substrates (i.e., the glass slide and petri dish) in a proprietary spray-coating chamber (Figure 1). The spray-coating process was carried out at room temperature flushed with N$_2$. In this research, the SP-TiO$_2$ thin film was made up of five film layers individually treated at 80°C for 2 hours.

2.2. Physical Characteristics of the SG- and SP-TiO$_2$ Thin Films. In this research, the crystalline phase of both thin films was analyzed using a D8 Advance X-ray diffractometer

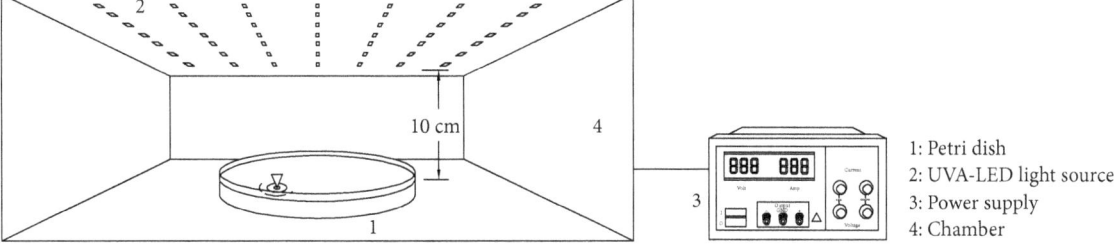

FIGURE 2: The schematic of the photocatalytic bacterial inactivation system.

(Bruker) under a Cu Kα radiation scan range of 10–80°. The bandgap energy was determined with a Helios Alpha UV-Vis spectrometer (Thermo Electron) with a wavelength range of 290–800 nm. The surface morphology of both SG- and SP-TiO$_2$ thin films was determined using an MFP-3D-BIO™ Atomic Force Microscope (AFM Asylum Research), and the contact angles were obtained by the sessile drop technique using an OCA series TBU90E instrument (DataPhysics). Specifically, the contact angles were the averages of five replications after reposing a 1 mL water droplet on the experimental thin films for 30 seconds.

The AFM images were analyzed by Gwyddion v.2.22 (http://gwyddion.net) for the grain size and the apparent surface area of the thin films. The adhesion between the thin film and the substrate (i.e., the glass slide) was evaluated in accordance with the ASTM D3359B-17 standard. The acid-base corrosion resistance of the experimental thin films was determined by dipping the coated substrates (i.e., the glass slides with the TiO$_2$ thin film) for 5 minutes independently in 1 M nitric acid and 1 M sodium hydroxide solution [12, 18]. Table 1 tabulates the physical characteristics of interest and the corresponding analysis techniques.

2.3. Photocatalytic Bacterial Inactivation. Gram-negative bacteria E. coli (strain TISTR 073) were used to evaluate the photocatalytic bacterial inactivation of both the SG- and SP-TiO$_2$ thin films. The stock E. coli in −80°C was inoculated into 10 mL brain heart infusion (BHI) broth and incubated at 37°C for 24 h. The product was then inoculated into 30 mL nutrient broth (NB) that contained 3 g beef extract, 5 g peptone, and 1000 mL distilled water and incubated at 37°C (i.e., the substrate enrichment condition) for the initial E. coli concentrations (N_0) of 10^2, 10^4, 10^6, and 10^8 CFU·mL^{-1} (E. coli in NB).

The procedure was repeated for another batch of the products with the E. coli concentrations of 10^2, 10^4, 10^6, and 10^8 CFU·mL^{-1} prior to centrifugation at 3000 rpm for 10 min at room temperature. The supernatant was discarded, and the subnatant (i.e., the E. coli cells) was harvested and washed three times with 0.85% NaCl. The bacterial cells were then resuspended in 30 mL 0.85% NaCl (i.e., the nonsubstrate enrichment condition) for the initial E. coli concentrations in 0.85% NaCl of 10^2, 10^4, 10^6, and 10^8 CFU·mL^{-1} (E. coli in 0.85% NaCl).

Afterward, 30 mL aliquots of the NB and 0.85% NaCl with the E. coli concentrations of 10^2, 10^4, 10^6, and 10^8 CFU·mL^{-1} were transferred to the petri dishes coated with the SG-TiO$_2$ or SP-TiO$_2$ thin film and placed under

the UVA light source (warm white UVA-LED) in a photoreactor for 180 minutes (Figure 2). The electric power was controlled by a 12 V 3 A power supply AC-DC adapter. The average UVA intensity in the photoreactor was 12 μW·cm^{-2} such that the energy falling onto the coating surface in an hour was about 1.2 W·h, measured by the UV Light Meter Model UV-340. In this experimental condition, the cool down was not necessary because the average temperature of the photoreactor increased slightly by 1–3°C. The samples were collected after 30, 60, 120, and 180 minutes to analyze the viable cells using the standard plate count (SPC) method. Prior to the viable cell count, a serial dilution was carried out by introducing 1 mL of each sample into 30 mL of 0.85% NaCl. Then 0.1 mL of the serial dilutions was transferred to the plate count agar (PCA) and incubated 24–48 hours at 37°C. The viable cells of E. coli were subsequently counted.

Moreover, the viable E. coli cells in NB and in 0.85% NaCl (with the initial bacterial concentrations of 10^2, 10^4, 10^6, and 10^8 CFU·mL^{-1}) treated with the SG- and SP-TiO$_2$ thin films and UVA were verified against the viable cell counts of the control with the identical initial E. coli concentrations (i.e., those treated with the thin films in the absence of UVA (dark) and those treated with UVA without the thin films). Figure 3 illustrates the overall scheme of this experimental research.

To further verify the photocatalytic bacterial inactivation performance of both thin films, the ATP bioluminescence assay was carried out for the adenosine triphosphate (ATP). The viable E. coli cells were determined by the cellular ATP content using the Lumitester PD-20 (Kikkoman Biochemifa, Japan), based on the detection of light generated by the ATP-dependent enzymatic conversion. Specifically, D-luciferin in the ATP was transformed into oxyluciferin by luciferase whereby the light was generated. The quantity of light emission was measured by the Lumitester and the result expressed as the relative light unit (RLU). The ATP of the viable E. coli cells (in RLU) was the averages of five replications of the E. coli cells in NB and in 0.85% NaCl, given the initial bacterial concentrations of 10^2, 10^4, 10^6, and 10^8 CFU·mL^{-1}, after 30, 60, 120, and 180 minutes.

3. Results and Discussion

3.1. Physical Characteristics of the Photocatalytic Thin Films. Figure 4 illustrates the X-ray diffraction (XRD) patterns of both experimental thin films, with the peak at the diffraction angle (2θ) of 25.0°. In addition, the analysis results revealed

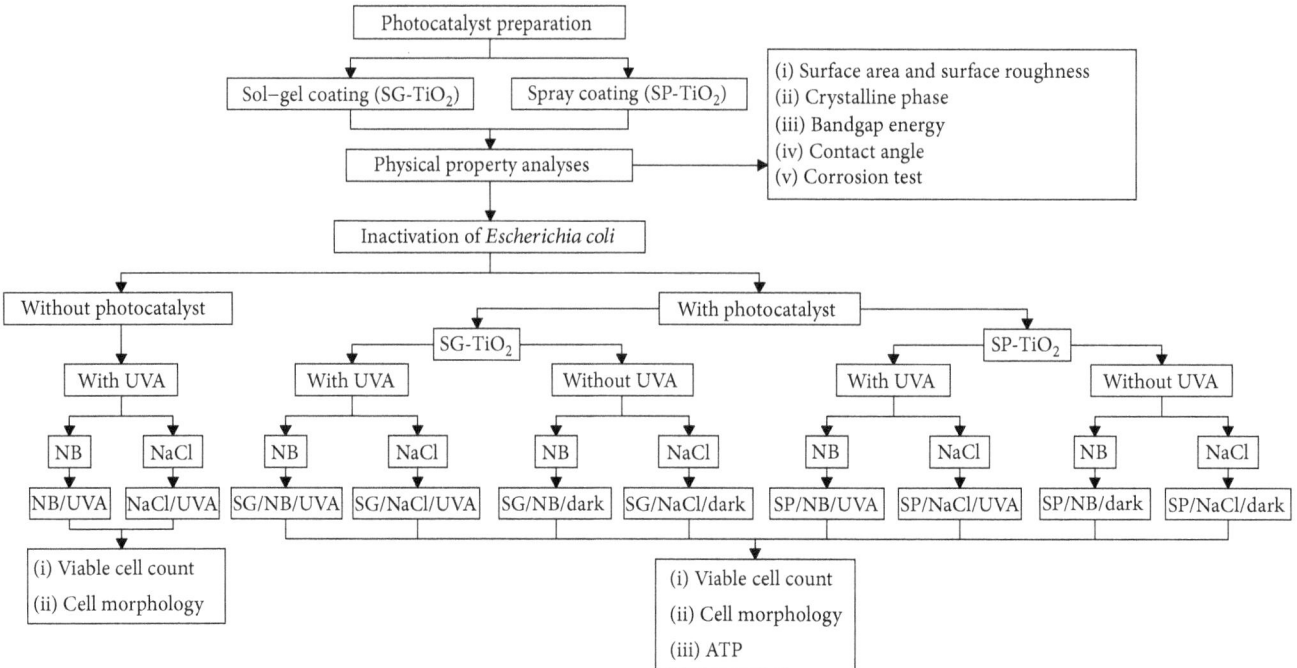

FIGURE 3: The overall scheme of this experimental research.

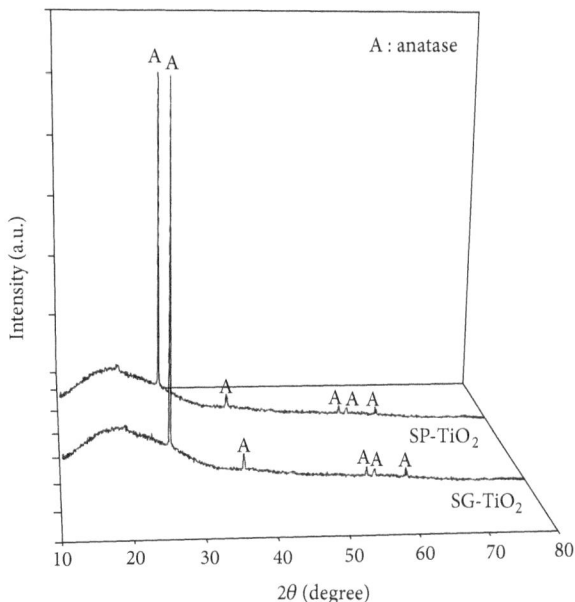

FIGURE 4: XRD patterns of the SG-TiO$_2$ and SP-TiO$_2$ thin films.

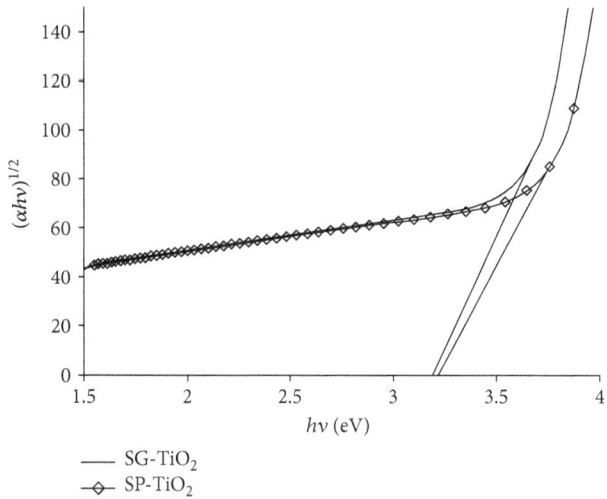

FIGURE 5: The energy bandgaps of the SG-TiO$_2$ and SP-TiO$_2$ thin films.

that only the anatase (101) phase (anatase TiO$_2$, ICSD code: 01-089-4921) was present (no rutile phase) in both thin films.

The bandgap energy (E_g) is calculated from the linear relationship of $(hv\alpha)^{1/2}$ against hv with extrapolation to zero, which is referred to as the Tauc plot and can be expressed as [13, 19], where h is Plank's constant (eV), v is the frequency of vibration, α is the absorption coefficient, A is a proportional constant, and E_g is the bandgap energy (eV).

$$(hv\alpha)^{1/2} = A(hv - E_g). \tag{1}$$

Figure 5 illustrates the Tauc plots of both experimental thin films, in which the dotted lines intercepted the x-axis (hv-intercept) at 3.19 and 3.22 eV for the SG-TiO$_2$ and SP-TiO$_2$ thin films, respectively. The findings indicated that the bandgap energies of both thin films were in the range of anatase TiO$_2$ [20, 21].

In Figure 6, the contact angles of the water droplet on the surface of the SG- and SP-TiO$_2$ thin films were 46.78° and 57.31°, indicating that both films were hydrophilic [22, 23]. Figures 7(a) and 7(b), respectively, depict the 2D and 3D AFM images of the SG- and SP-TiO$_2$ thin films, in which the TiO$_2$ particles were round and uniform and evenly distributed with the grain sizes of 35–90 nm and 25–80 nm

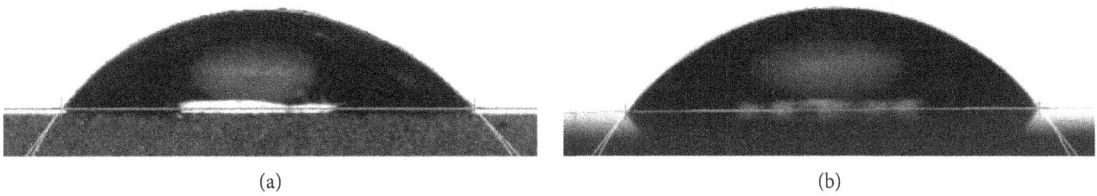

(a)   (b)

FIGURE 6: The contact angles of (a) SG-TiO$_2$ and (b) SP-TiO$_2$ thin films.

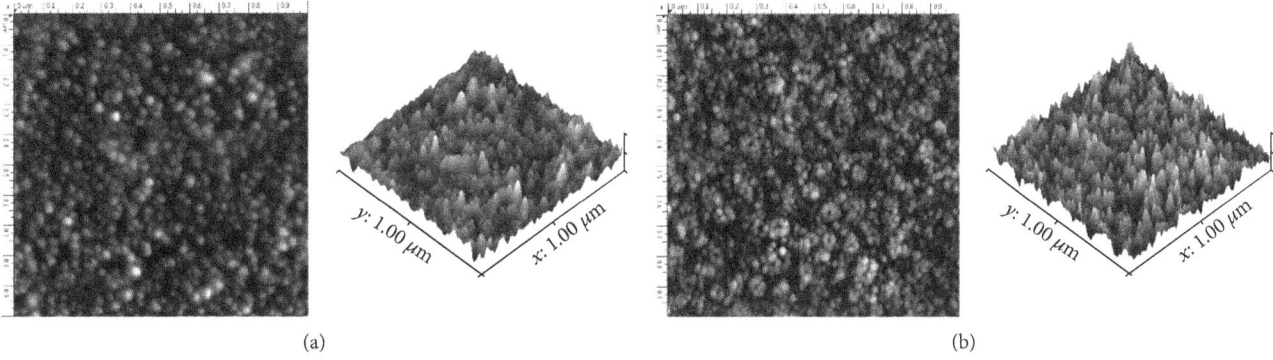

(a)   (b)

FIGURE 7: The 2D and 3D AFM images of (a) SG-TiO$_2$ and (b) SP-TiO$_2$ thin films.

TABLE 2: Physical properties of the SG-TiO$_2$ and SP-TiO$_2$ photocatalytic thin films.

Physical properties	Thin films	
	SG-TiO$_2$	SP-TiO$_2$
Crystalline phase	Anatase	Anatase
Bandgap energy (eV)	3.19	3.22
Contact angle (°)	46.78	57.31
Measured by Gwyddion analysis		
Grain size (nm)	35–90	25–80
RMS average roughness (nm)	1.75	0.88
Apparent surface area (m^2·m^{-2}) ❶	1.02	1.01
Total weight of TiO$_2$ on substrate (g·m^{-2}) ❷	0.45	0.38
Total apparent surface area per total weight of TiO$_2$ (m^2·g^{-1}) ❸ = ❶/❷	2.27	2.53
Adhesion		
Rank	Good	
Classification	4B	
Corrosion test	No visible damage	

for the SG- and SP-TiO$_2$ thin films. The root mean square (RMS) averages of the roughness of the SG- and SP-TiO$_2$ thin films were 1.75 and 0.88 nm, respectively, resulting in the smooth surface and elevated hydrophilicity, which in turn contributed to the reduced contact angle and increased polar interaction with the water droplet [24].

The hydrophilicity of both thin films transformed the oxidation state from Ti^{4+} to Ti^{3+}, while the photogenerated holes oxidized the O^{2-} anions to O$_2$. The expulsion of oxygen anions from Ti^{3+} generated $^{\bullet}$OH and $^{\bullet}$O$_2^{-}$ and produced

holes which play a crucial role in the photocatalytic activity and bacterial inactivation. In fact, the hydrophilicity, as expressed by the contact angle, could be used to approximate the photocatalytic performance of the TiO$_2$ thin films [25–27].

In Table 2, the total apparent surface areas per total weight of catalyst of the SG- and SP-TiO$_2$ thin films were 2.27 and 2.53 m^2·g^{-1}, respectively, indicating that the proposed low-temperature spray-coating (SP) technique increased the total apparent surface area. In addition, both thin films exhibited a good substrate adhesion, achieving the 4B classification. For the acid-base corrosion resistance, neither of the thin films showed visible damage, suggesting a high acid-base resistance.

Table 3 compares the physical characteristics of the SG- and SP-TiO$_2$ photocatalytic thin films with those of existing research studies using variable coating techniques. Unlike the other techniques whose curing temperatures were in the range of 250 to 500°C, the curing temperature of the SP-TiO$_2$ thin film of this research was only 80°C. Notably, the SP-TiO$_2$ thin film possessed the physical characteristics resembling those fabricated under the high temperature conditions. In addition, the low-temperature spray-coating technique requires smaller amounts of TiO$_2$ and is applicable to the substrates with low thermal resistance. The proposed spray-coating scheme could also be applied to materials with large surface areas at minimal costs and short fabrication time.

3.2. Photocatalytic Bacterial Inactivation.

The photocatalytic bacterial inactivation experiments were performed with four initial *E. coli* concentrations in NB and 0.85% NaCl of approximately 10^2, 10^4, 10^6, and 10^8 CFU·mL^{-1}. The SPC was used to quantify the viable cells under the substrate enrichment condition (in NB) and the nonsubstrate

TABLE 3: Comparison of the properties of the SG- and SP-TiO$_2$ photocatalytic thin films and of other studies.

Substrate	Catalyst	Amount of Ag (% wt)	Preparation Method	Temp (°C)	% crystalline Anatase	Rutile	Bandgap energy (eV)	Contact angle (°)	Number of coating layers	Grain size (nm)	Roughness RMS (nm)	Surface area/weight of catalyst (m$^2 \cdot$g^{-1})	Reference
Glass	SG-TiO$_2$	—	Sol–gel	500	100	0	3.19	46.78	5	35–90	1.75	2.27*	This study
Glass	SP-TiO$_2$	—	Spray	80	100	0	3.22	57.31	5	25–80	0.88	2.53*	This study
Glass	TiO$_2$	—	Spin	500	100	0	—	2	—	50–100	—	—	[28]
Silicon wafer	TiO$_2$	—	Spin	500	100	0	—	25	3	30	3.44	—	[29]
Glass	TiO$_2$	—	Sol–gel	500	100	0	3.27	—	5	15–100	2.62–5.74	2.58–10.27*	[30]
Glass	TiO$_2$	—	Sputtering	250	100	0	3.75	—	—	185	6.87	—	[31]

*The apparent surface area is measured by AFM.

TABLE 4: *E. coli* inactivation performance of the SG- and SP-TiO$_2$ thin films under UVA after 180 minutes given the various initial bacterial concentrations.

Initial bacterial concentration of *E. coli* (CFU·mL^{-1})	*E. coli* inactivation $\pm$ SD (%) at 180 min with $n = 5$ (number of replication)			
	SG/NB/UVA	SG/NaCl/UVA	SP/NB/UVA	SP/NaCl/UVA
10^2	86.18 ± 1.43	92.53 ± 1.58	84.46 ± 1.67	92.94 ± 1.38
10^4	74.80 ± 1.39	80.77 ± 1.33	72.31 ± 1.69	78.67 ± 1.66
10^6	65.14 ± 1.25	72.40 ± 0.60	65.20 ± 1.55	71.43 ± 1.71
10^8	44.59 ± 1.59	63.72 ± 2.01	47.25 ± 1.98	60.71 ± 2.05

enrichment condition (in 0.85% NaCl). For comparison, this research also determined the viable cell counts of the *control*, given the same initial *E. coli* concentrations (i.e., those treated with the thin films in the absence of UVA (dark) and those treated with UVA without the thin films).

Table 4 compares the *E. coli* inactivation performance of the SG- and SP-TiO$_2$ photocatalytic thin films under the UVA light, given the initial bacterial concentrations in NB and 0.85% NaCl of 10^2, 10^4, 10^6, and 10^8 CFU·mL^{-1}. The results revealed that the thin-film type (SG- or SP-TiO$_2$ thin film) had no significant impact on the bacterial inactivation performance. In addition, given the same *E. coli* concentration, the photocatalytic bacterial inactivation in 0.85% NaCl was higher than in NB because the 0.85% NaCl solution was unconducive to the bacterial proliferation. Nevertheless, both experimental thin films under the UVA light were effective in inhibiting the proliferation of the *E. coli* cells in NB (10^2, 10^4, 10^6, and 10^8 CFU·mL^{-1}).

Figures 8(a) and 8(b) illustrate the photocatalytic bacterial inactivation of the SG-TiO$_2$ thin film under UVA (SG/NB/UVA and SG/NaCl/UVA) relative to that of the control (SG/NB/dark, NB/UVA, SG/NaCl/dark, and NaCl/UVA), given the initial *E. coli* concentrations in NB and 0.85% NaCl of 10^2, 10^4, 10^6, and 10^8 CFU·mL^{-1}. Meanwhile, Figures 8(c) and 8(d) depict the bacteriostatic activity of the SP-TiO$_2$ thin film under UVA (SP/NB/UVA and SP/NaCl/UVA) vis-à-vis that of the control (SP/NB/dark, NB/UVA, SP/NaCl/dark, and NaCl/UVA), given the same initial *E. coli* concentrations.

The results revealed that both SG- and SP-TiO$_2$ thin films, with the UVA exposure, were able to inhibit the proliferation of *E. coli* under the substrate enrichment (NB) and nonsubstrate enrichment (0.85% NaCl) conditions. Specifically, in the absence of the UVA light or the photocatalytic thin films (SG/NB/dark, SG/NaCl/dark, SP/NB/dark, SP/NaCl/dark, NB/UVA, and NaCl/UVA), the *E. coli* growth was normal. On the other hand, with the UVA exposure and the thin films (SG/NB/UVA, SG/NaCl/UVA, SP/NB/UVA, and SP/NaCl/UVA), the bacterial abundance declined. The reduction in the *E. coli* cells indicated that both photocatalytic thin films, given the UVA exposure, could effectively inhibit the bacterial growth. In addition, the photocatalytic bacterial inactivation performance increased with the elongated UVA irradiation time.

The bacterial inactivation analysis revealed that neither TiO$_2$ nor UVA could independently inhibit the growth of *E. coli*. In fact, the concurrent deployment of the photocatalytic thin film (SG- or SP-TiO$_2$ thin film) and the UVA

light source is imperative to induce the bacteriostatic activity of the *E. coli* cells. The results showed that both thin films were good enough to inhibit the growth of *E. coli* cells (bacteriostatic) but not kill them (bactericidal). Since this prepared photocatalyst can be applied onto different surfaces, there are potential applications for surfaces that are needed to control the proliferation of bacteria, such as a kitchen counter, inside surface of a refrigerator, and door knobs, in order to gain benefit through a more eco- and environment-friendly process when compared to the use of harmful chemicals.

Figure 8 indicates the abundance of $^\bullet$OH species [32, 33]. $^\bullet$OH is bacteriostatic or even bactericidal and is an oxidizing agent stronger than chlorine, hydrogen peroxide, or even ozone [18, 34–36]. $^\bullet$OH is generated as the holes in the valence band oxidize H$_2$O molecules to generate $^\bullet$OH for the oxidation pathway and O$_2$ captures the electrons in the conduction band to produce $^\bullet$O$_2^-$ and subsequently generate $^\bullet$OH for the reduction pathway [32, 33]. The high photocatalytic bacteriostatic action was attributable to the oxidation of *E. coli* by $^\bullet$OH and $^\bullet$O$_2^-$. In Figure 8, the reaction kinetics of photocatalytic bacterial inactivation under a UVA-LED light can be described by pseudo first-order kinetics, and the highest kinetics constants for *E. coli* concentrations in NB and 0.85% NaCl of 10^2 CFU·mL^{-1} were 4.5×10^{-3} and 5.8×10^{-3} min^{-1} for the SG-TiO$_2$ thin film, respectively, and were 4.1×10^{-3} and 6.0×10^{-3} min^{-1} for the SP-TiO$_2$ thin film, respectively.

Moreover, the bacteriostatic activity was further verified with the FE-SEM images of the *E. coli* cells, in addition to the SPC method. Figures 9(a)–9(d) illustrate the morphology and structure of the *E. coli* cells in NB and 0.85% NaCl after 180 minutes (at termination) treated with the SG- and SP-TiO$_2$ thin films and UVA irradiation, given the initial bacterial concentration of 10^4 CFU·mL^{-1}, vis-à-vis the control (Figure 9(e)). The FE-SEM images showed that, with the photocatalytic thin films and UVA, the outer cell membranes exhibited the deformation or even destruction (Figures 9(a)–9(d)).

In Figures 9(a)–9(d), the *E. coli* cells were mostly either irreversibly deformed or collapsed, with fissures and pits visible on the cell membrane, indicating the loss of integrity and viability of the *E. coli* cells, consistent with [37–40]. In fact, the bacterial inhibition performance of the SG-TiO$_2$ thin film, given the UVA exposure, resembles that of the SP-TiO$_2$ thin film. The similarity could be attributed to the similar total apparent surface areas of the SG-TiO$_2$ (2.27 m^2·g^{-1}) and SP-TiO$_2$ (2.53 m^2·g^{-1}) thin films.

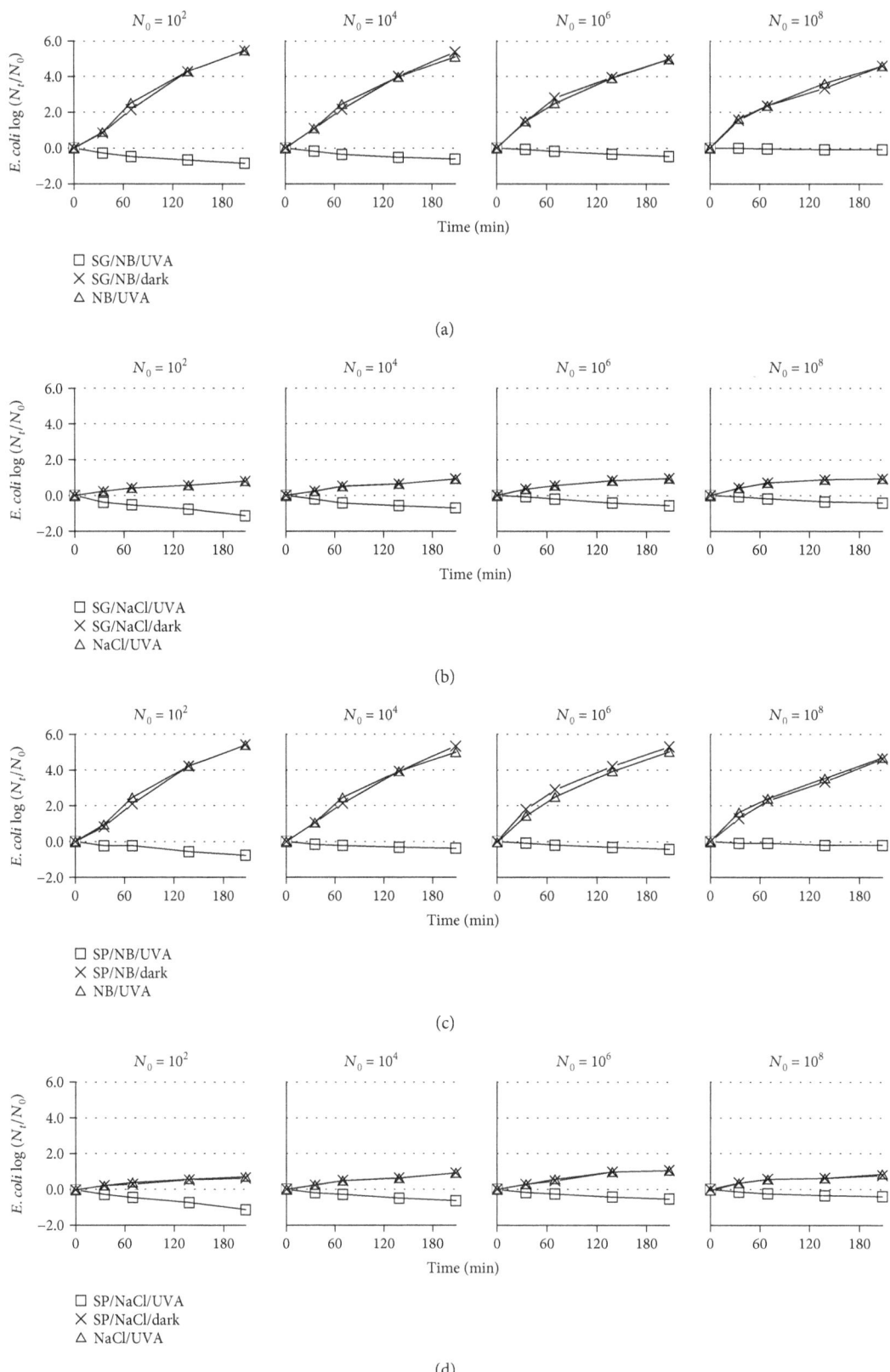

FIGURE 8: The photocatalytic bacterial inactivation, given the initial *E. coli* concentrations in NB and 0.85% NaCl of 10^2, 10^4, 10^6, and 10^8 CFU·mL^{-1}: (a) NB treated with SG-TiO$_2$ thin film and UVA relative to the control (SG/NB/dark and NB/UVA), (b) NaCl treated with SG-TiO$_2$ thin film and UVA relative to the control (SG/NaCl/dark and NaCl/UVA), (c) NB treated with SP-TiO$_2$ thin film and UVA relative to the control (SP/NB/dark and NB/UVA), and (d) NaCl treated with SP-TiO$_2$ thin film and UVA relative to the control (SP/NaCl/dark and NaCl/UVA).

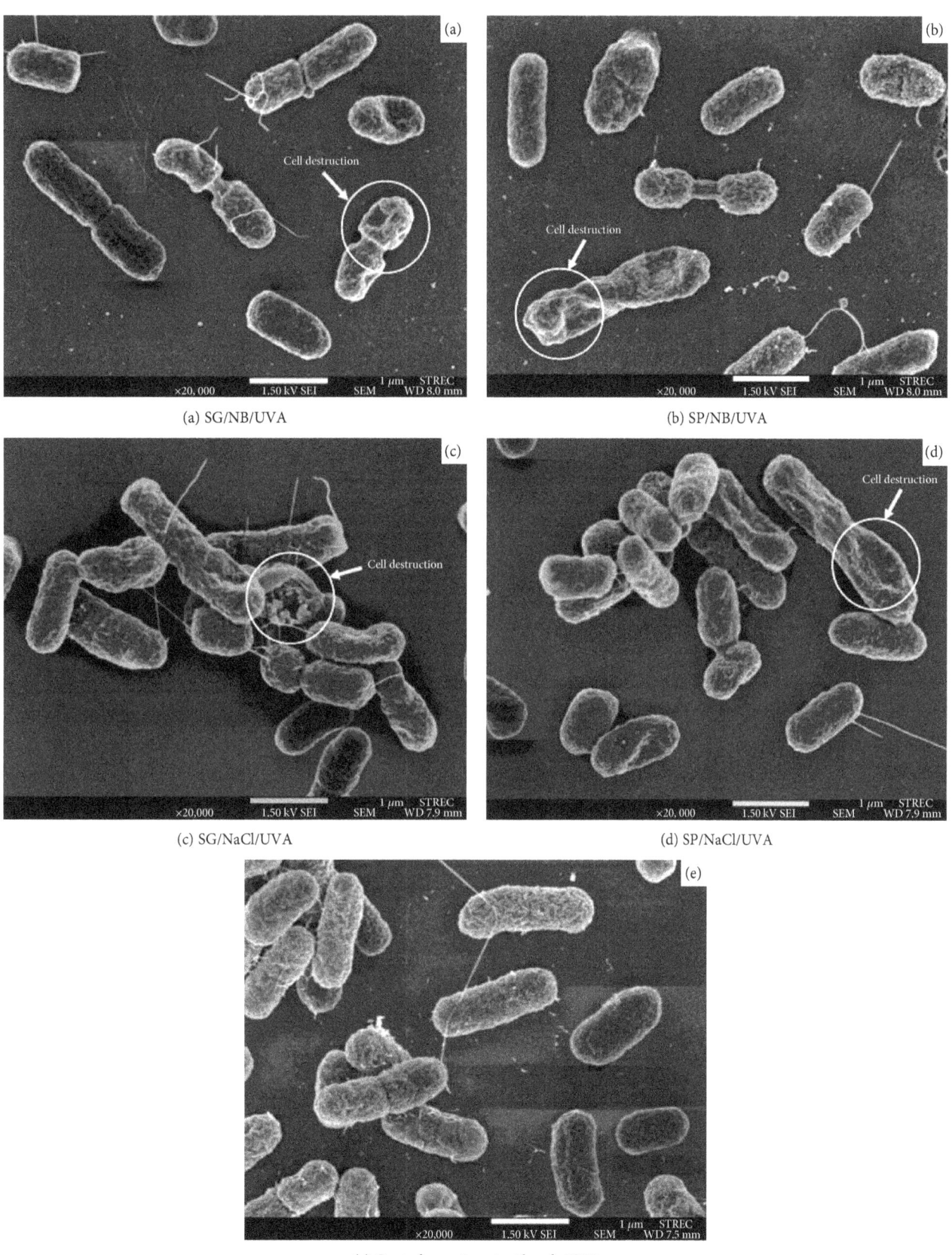

(a) SG/NB/UVA

(b) SP/NB/UVA

(c) SG/NaCl/UVA

(d) SP/NaCl/UVA

(e) Control experiment with only UVA

FIGURE 9: The FE-SEM images of *E. coli* in NB and 0.85% NaCl after 180 minutes treated with the SG- and SP-TiO$_2$ thin films and UVA irradiation vis-à-vis the control, given the initial bacterial concentration of 10^4 CFU·mL^{-1}.

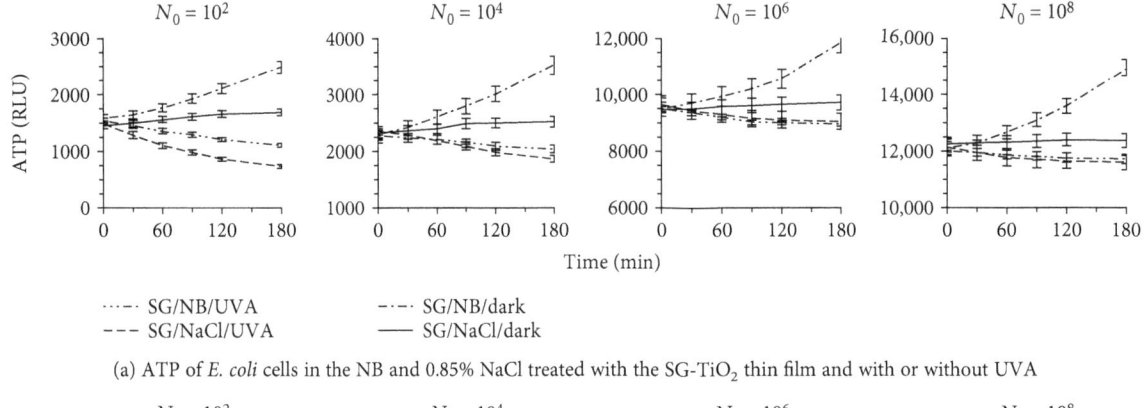

(a) ATP of *E. coli* cells in the NB and 0.85% NaCl treated with the SG-TiO$_2$ thin film and with or without UVA

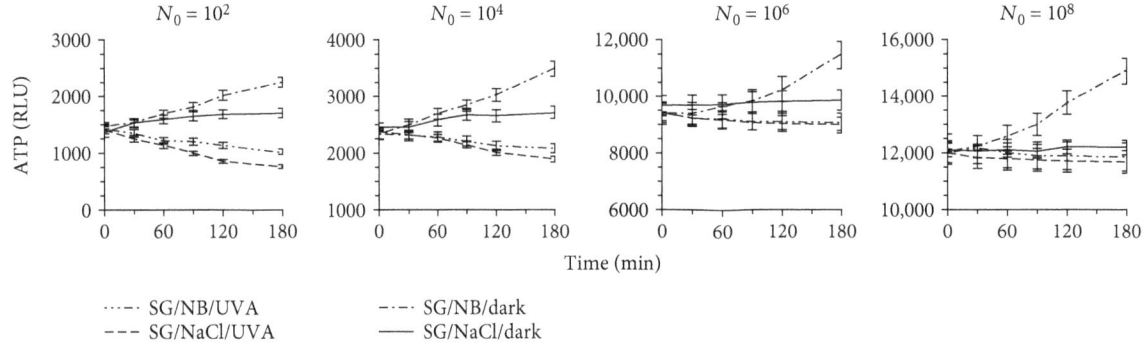

(b) ATP of *E. coli* cells in the NB and 0.85% NaCl treated with the SP-TiO$_2$ thin film and with or without UVA

FIGURE 10: The ATP of *E. coli* cells treated with the experimental thin films and with/without UVA.

In Figure 10, the ATP of the *E. coli* cells treated with the SG- and SP-TiO$_2$ thin films, with and without UVA, validated the SPC results and the bacteriostatic activity. Specifically, the ATP declined as the viable cells decreased, which subsequently led to the decline in the cellular metabolic rates and the eventual cell death.

In short, the concurrent use of the SG- or SP-TiO$_2$ thin film and UVA could effectively inhibit the proliferation of the *E. coli* cells in both NB and 0.85% NaCl. However, the elevated initial *E. coli* concentrations in NB and 0.85% NaCl lowered the photocatalytic bacterial inactivation performance, due to the restricted active photocatalytic surface site [41–43] and the subsequently lower $^\bullet OH$ and $^\bullet O_2^-$ [9, 44, 45].

4. Conclusion

The aim of this experimental research is to comparatively examine the *E. coli* bacterial inactivation of the SG-TiO$_2$ and SP-TiO$_2$ photocatalytic thin films under the low-intensity UVA light source. The bacteriostatic experiments were undertaken using the NB and 0.85% NaCl with the initial *E. coli* concentrations of 10^2, 10^4, 10^6, and 10^8 CFU·mL^{-1}. The bacteriostatic activity assessments were also carried out without UVA or the photocatalytic thin films (the control). The experimental results revealed that both SG-TiO$_2$ and SP-TiO$_2$ photocatalytic thin films possessed the ideal physical characteristics, especially the SP-TiO$_2$ thin film given its lower fabrication temperature (80°C), subsequent lower energy demand, minimal TiO$_2$ requirement, and applicability to large surface area objects. In addition, both

photocatalytic thin films could effectively inhibit the proliferation of *E. coli* under the low-intensity UVA irradiation, as evidenced by the lower viable cell counts. The bacterial inactivation performance was further verified by the FE-SEM images of deformed *E. coli* cells and the ATP measurement. Nevertheless, the *E. coli* inactivation efficiencies declined as the initial bacterial concentration increased due to the restricted active photocatalytic surface site and the subsequently lower $^\bullet OH$ and $^\bullet O_2^-$.

Conflicts of Interest

The authors declare that they have no conflicts of interest.

Acknowledgments

The authors would like to express their deep gratitude to the Thailand Research Fund for TRF Grant for New Researcher (2014): TRG 57802545, and to Rajamangala University of Technology Thanyaburi for the technical assistance.

References

[1] L. B. B. Ndong, M. P. Ibondou, Z. Miao et al., "Efficient dechlorination of chlorinated solvent pollutants under UV irradiation by using the synthesized TiO$_2$ nano-sheets in aqueous

phase," *Journal of Environmental Sciences*, vol. 26, no. 5, pp. 1188–1194, 2014.

[2] F.-X. Tian, B. Xu, Y.-L. Lin et al., "Chlor(am)ination of iopamidol: kinetics, pathways and disinfection by-products formation," *Chemosphere*, vol. 184, pp. 489–497, 2017.

[3] C. Kolb, R. A. Francis, and J. M. VanBriesen, "Disinfection byproduct regulatory compliance surrogates and bromide-associated risk," *Journal of Environmental Sciences*, vol. 58, pp. 191–207, 2017.

[4] P. S. M. Dunlop, M. Ciavola, L. Rizzo, D. A. McDowell, and J. A. Byrne, "Effect of photocatalysis on the transfer of antibiotic resistance genes in urban wastewater," *Catalysis Today*, vol. 240, pp. 55–60, 2015.

[5] N. D. Friedman, E. Temkin, and Y. Carmeli, "The negative impact of antibiotic resistance," *Clinical Microbiology and Infection*, vol. 22, no. 5, pp. 416–422, 2016.

[6] R. Nosrati, A. Olad, and S. Shakoori, "Preparation of an antibacterial, hydrophilic and photocatalytically active polyacrylic coating using TiO$_2$ nanoparticles sensitized by graphene oxide," *Materials Science and Engineering: C*, vol. 80, pp. 642–651, 2017.

[7] B. Liu, L. Mu, B. Han, J. Zhang, and H. Shi, "Fabrication of TiO$_2$/Ag$_2$O heterostructure with enhanced photocatalytic and antibacterial activities under visible light irradiation," *Applied Surface Science*, vol. 396, pp. 1596–1603, 2017.

[8] D. M. Tobaldi, C. Piccirillo, N. Rozman et al., "Effects of Cu, Zn and Cu-Zn addition on the microstructure and antibacterial and photocatalytic functional properties of Cu-Zn modified TiO$_2$ nano-heterostructures," *Journal of Photochemistry and Photobiology A: Chemistry*, vol. 330, pp. 44–54, 2016.

[9] R. Ahmad, Z. Ahmad, A. U. Khan, N. R. Mastoi, M. Aslam, and J. Kim, "Photocatalytic systems as an advanced environmental remediation: recent developments, limitations and new avenues for applications," *Journal of Environmental Chemical Engineering*, vol. 4, no. 4, pp. 4143–4164, 2016.

[10] C. Adán, J. Marugán, S. Mesones, C. Casado, and R. van Grieken, "Bacterial inactivation and degradation of organic molecules by titanium dioxide supported on porous stainless steel photocatalytic membranes," *Chemical Engineering Journal*, vol. 318, pp. 29–38, 2017.

[11] S. K. Misra, S. I. Andronenko, D. Tipikin, J. H. Freed, V. Somani, and O. Prakash, "Study of paramagnetic defect centers in as-grown and annealed TiO$_2$ anatase and rutile nanoparticles by a variable-temperature X-band and high-frequency (236 GHz) EPR," *Journal of Magnetism and Magnetic Materials*, vol. 401, pp. 495–505, 2016.

[12] J. Wang, W. Liu, H. Li et al., "Preparation of cellulose fiber-TiO$_2$ nanobelt–silver nanoparticle hierarchically structured hybrid paper and its photocatalytic and antibacterial properties," *Chemical Engineering Journal*, vol. 228, pp. 272–280, 2013.

[13] J. Singh, S. A. Khan, J. Shah, R. K. Kotnala, and S. Mohapatra, "Nanostructured TiO$_2$ thin films prepared by RF magnetron sputtering for photocatalytic applications," *Applied Surface Science*, vol. 422, pp. 953–961, 2017.

[14] M. R. Eskandarian, H. Choi, M. Fazli, and M. H. Rasoulifard, "Effect of UV-LED wavelengths on direct photolytic and TiO$_2$ photocatalytic degradation of emerging contaminants in water," *Chemical Engineering Journal*, vol. 300, pp. 414–422, 2016.

[15] L. C. Ferreira, M. S. Lucas, J. R. Fernandes, and P. B. Tavares, "Photocatalytic oxidation of Reactive Black 5 with UV-A LEDs," *Journal of Environmental Chemical Engineering*, vol. 4, no. 1, pp. 109–114, 2016.

[16] T. Rojviroon, O. Rojviroon, and S. Sirivithayapakorn, "Photocatalytic decolourisation of dyes using TiO$_2$ thin film photocatalysts," *Surface Engineering*, vol. 32, no. 8, pp. 562–569, 2016.

[17] W. Su, S. S. Wei, S. Q. Hu, and J. X. Tang, "Preparation of TiO$_2$/Ag colloids with ultraviolet resistance and antibacterial property using short chain polyethylene glycol," *Journal of Hazardous Materials*, vol. 172, no. 2-3, pp. 716–720, 2009.

[18] S. Guo, R. Huang, and H. Chen, "Application of water-assisted ultraviolet light in combination of chlorine and hydrogen peroxide to inactivate *Salmonella* on fresh produce," *International Journal of Food Microbiology*, vol. 257, pp. 101–109, 2017.

[19] J. F. Guayaquil-Sosa, B. Serrano-Rosales, P. J. Valadés-Pelayo, and H. de Lasa, "Photocatalytic hydrogen production using mesoporous TiO$_2$ doped with Pt," *Applied Catalysis B: Environmental*, vol. 211, pp. 337–348, 2017.

[20] C. Dette, M. A. Pérez-Osorio, C. S. Kley et al., "TiO$_2$ anatase with a bandgap in the visible region," *Nano Letters*, vol. 14, no. 11, pp. 6533–6538, 2014.

[21] G. Pathak, S. Pandey, R. Katiyar et al., "Analysis of photoluminescence, UV absorbance, optical band gap and threshold voltage of TiO$_2$ nanoparticles dispersed in high birefringence nematic liquid crystal towards its application in display and photovoltaic devices," *Journal of Luminescence*, vol. 192, pp. 33–39, 2017.

[22] D.-J. Huang and T.-S. Leu, "Fabrication of high wettability gradient on copper substrate," *Applied Surface Science*, vol. 280, pp. 25–32, 2013.

[23] J. C. Joud, M. Houmard, and G. Berthomé, "Surface charges of oxides and wettability: application to TiO$_2$–SiO$_2$ composite films," *Applied Surface Science*, vol. 287, pp. 37–45, 2013.

[24] M. Alzamani, A. Shokuhfar, E. Eghdam, and S. Mastali, "Influence of catalyst on structural and morphological properties of TiO$_2$ nanostructured films prepared by sol–gel on glass," *Progress in Natural Science: Materials International*, vol. 23, no. 1, pp. 77–84, 2013.

[25] J. Du, Q. Wu, S. Zhong et al., "Effect of hydroxyl groups on hydrophilic and photocatalytic activities of rare earth doped titanium dioxide thin films," *Journal of Rare Earths*, vol. 33, no. 2, pp. 148–153, 2015.

[26] D. Luca, D. Mardare, F. Iacomi, and C. M. Teodorescu, "Increasing surface hydrophilicity of titania thin films by doping," *Applied Surface Science*, vol. 252, no. 18, pp. 6122–6126, 2006.

[27] L. Huang, S. Jing, O. Zhuo, X. Meng, and X. Wang, "Surface hydrophilicity and antifungal properties of TiO$_2$ films coated on a Co-Cr substrate," *BioMed Research International*, vol. 2017, Article ID 2054723, 7 pages, 2017.

[28] F. Li, Q. Li, and H. Kim, "Spray deposition of electrospun TiO$_2$ nanoparticles with self-cleaning and transparent properties onto glass," *Applied Surface Science*, vol. 276, pp. 390–396, 2013.

[29] B. Yu, K. M. Leung, Q. Guo, W. M. Lau, and J. Yang, "Synthesis of Ag-TiO$_2$ composite nano thin film for antimicrobial application," *Nanotechnology*, vol. 22, no. 11, article 115603, 2011.

[30] T. Rojviroon and S. Sirivithayapakorn, "Properties of TiO$_2$ thin films prepared using sol-gel process," *Surface Engineering*, vol. 29, no. 1, pp. 77–80, 2013.

[31] S. Vyas, R. Tiwary, K. Shubham, and P. Chakrabarti, "Study the target effect on the structural, surface and optical properties of TiO_2 thin film fabricated by RF sputtering method," *Superlattices and Microstructures*, vol. 80, pp. 215–221, 2015.

[32] B. R. Cruz-Ortiz, J. W. J. Hamilton, C. Pablos et al., "Mechanism of photocatalytic disinfection using titania-graphene composites under UV and visible irradiation," *Chemical Engineering Journal*, vol. 316, pp. 179–186, 2017.

[33] J. Zhang and Y. Nosaka, "Mechanism of the OH radical generation in photocatalysis with TiO_2 of different crystalline types," *The Journal of Physical Chemistry C*, vol. 118, no. 20, pp. 10824–10832, 2014.

[34] J. Zheng, C. Su, J. Zhou, L. Xu, Y. Qian, and H. Chen, "Effects and mechanisms of ultraviolet, chlorination, and ozone disinfection on antibiotic resistance genes in secondary effluents of municipal wastewater treatment plants," *Chemical Engineering Journal*, vol. 317, pp. 309–316, 2017.

[35] X. Huang, Y. Qu, C. A. Cid et al., "Electrochemical disinfection of toilet wastewater using wastewater electrolysis cell," *Water Research*, vol. 92, pp. 164–172, 2016.

[36] B. Sun, M. Sato, and J. Sid Clements, "Optical study of active species produced by a pulsed streamer corona discharge in water," *Journal of Electrostatics*, vol. 39, no. 3, pp. 189–202, 1997.

[37] J. J. Murcia, E. G. Ávila-Martínez, H. Rojas, J. A. Navío, and M. C. Hidalgo, "Study of the *E. coli* elimination from urban wastewater over photocatalysts based on metallized TiO_2," *Applied Catalysis B: Environmental*, vol. 200, pp. 469–476, 2017.

[38] D. Scthi and R. Sakthivel, "ZnO/TiO_2 composites for photocatalytic inactivation of *Escherichia coli*," *Journal of Photochemistry and Photobiology B: Biology*, vol. 168, pp. 117–123, 2017.

[39] V. Binas, D. Venieri, D. Kotzias, and G. Kiriakidis, "Modified TiO_2 based photocatalysts for improved air and health quality," *Journal of Materiomics*, vol. 3, no. 1, pp. 3–16, 2017.

[40] J. Mac Mahon, S. C. Pillai, J. M. Kelly, and L. W. Gill, "Solar photocatalytic disinfection of *E. coli* and bacteriophages MS2, ΦX174 and PR772 using TiO_2, ZnO and ruthenium based complexes in a continuous flow system," *Journal of Photochemistry and Photobiology B: Biology*, vol. 170, pp. 79–90, 2017.

[41] J. Ma, C. Zhu, J. Lu et al., "Catalytic degradation of gaseous benzene by using TiO_2/goethite immobilized on palygorskite: preparation, characterization and mechanism," *Solid State Sciences*, vol. 49, pp. 1–9, 2015.

[42] N. R. Khalid, A. Majid, M. B. Tahir, N. A. Niaz, and S. Khalid, "Carbonaceous-TiO_2 nanomaterials for photocatalytic degradation of pollutants: a review," *Ceramics International*, vol. 43, no. 17, pp. 14552–14571, 2017.

[43] J. Ø. Hansen, R. Bebensee, U. Martinez et al., "Unravelling site-specific photo-reactions of ethanol on rutile TiO_2(110)," *Scientific Reports*, vol. 6, no. 1, article 21990, 2016.

[44] R. van Grieken, J. Marugán, C. Pablos, L. Furones, and A. López, "Comparison between the photocatalytic inactivation of Gram-positive *E. faecalis* and Gram-negative *E. coli* faecal contamination indicator microorganisms," *Applied Catalysis B: Environmental*, vol. 100, no. 1-2, pp. 212–220, 2010.

[45] T.-D. Pham and B.-K. Lee, "Effects of Ag doping on the photocatalytic disinfection of *E. coli* in bioaerosol by Ag-TiO_2/GF under visible light," *Journal of Colloid and Interface Science*, vol. 428, pp. 24–31, 2014.

18

Modeling and Optimization of BT and DBT Photooxidation over Multiwall Carbon Nanotube-Titania Composite by Response Surface Methodology

Molood Barmala ⑩[1] **and Mohammad Behnood**[2]

[1]*Department of Chemical Engineering, Dezful Branch, Islamic Azad University, Dezful, Iran*
[2]*Department of Petroleum and Chemical Engineering, Science and Research Branch, Islamic Azad University, Tehran, Iran*

Correspondence should be addressed to Molood Barmala; m.barmala@iaud.ac.ir

Academic Editor: Alberto Álvarez-Gallegos

This study investigates optimization of benzothiophene (BT) and dibenzothiophene (DBT) removal via a photocatalytic process by using central composite design (CCD) method. Temperature, pH, and p-25 to MWCNT ratio (g/g) in the composite structure are considered as design factors. According to the results, temperature has the greatest impact on removal rate. In optimal condition, after being exposed to UV lamps (9 W) for 20 min, 59.8% of the solutions' BT was removed, while DBT was completely removed. Although the generated structure band gap is 3.4, but due to the presence of MWCNTs in the structure, it is capable of absorbing visible light, and this leads to complete removal of DBT and 42% removal of BT under visible light radiation (in similar circumstances). Kinetics analysis of thiophene's reaction showed that, in the presence of visible light, first order removal rate constants for DBT and BT are 7.98 and 0.953 1/h, respectively.

1. Introduction

With dramatic growth of petrochemical and automotive industries, air pollution caused by sulfur dioxide has become one of the major global problems. Since 2006, America Environmental Protection Agency announced that permitted concentrations of sulfur in diesel fuels should be below 15 ppm [1–4].

Hydrogen desulfurization is considered as a common process for sulfur removal. The process requires large amounts of hydrogen and huge reactors [4–6]. Unfortunately, hydrogen process cannot remove thiophene and its derivatives which exist in heavy oil compounds [1, 2, 4, 6, 7].

To save energy and reduce costs, an alternative desulfurization process needs to be developed. In thiophenic compound removal, advanced oxidation processes, which begin with oxygen radicals, may take 3 forms: chemical, bio-, and photooxidation [8–10]. The great advantage of

photocatalytic removal compared to the previous two processes is that it occurs under ambient temperature and pressure, so after a reasonable time, an acceptable amount of pollutant is eliminated [6, 10].

When a semiconductor is irradiated by a light beam with energy equal or greater than its band gap, it can act as a photocatalyst [6, 8, 9]. In this case, an electron is transferred from the valence band to the conduction band. This transmission creates an electron in conduction band and a hole with positive charge in the valence band [8, 9, 11, 12]. These electrons and holes can be transferred to the catalyst surface and begin redox reactions [8, 9, 11].

Recombination of electrons and generated holes is considered as the most important factor limiting semiconductors' photocatalytic power [5, 9, 10]. In this case, the stimulated electron returns to the valence band without entering into reaction with the adsorbed component on the surface of the semiconductor and releases its energy as light or heat [9, 11]. Normally, only 1% of electrons and generated holes participate in photocatalytic reaction, and the rest may be recombined without participating in the reaction [13].

Due to efficiency and the feasibility of conducting photocatalytic reactions at ambient conditions, and no need for hydrogen, research studies on the use of photocatalysts in order to remove sulfur compounds are increasing. Given the significance of sulfur removal from fuels and the important role of photo catalysts in the process efficiency, it seems very necessary to find a way to increase removal process efficiency. There are various methods for increasing photocatalytic strength and reducing recombination rate. Some methods increase photocatalytic power by improving structural properties, such as increasing the surface area and porosity, while some others increase the potential to remove contaminants by chemical modification and addition of another component to semiconductor structure [5, 7, 9–11]. One way to increase efficiency is to perform a photocatalytic process using carbon nanotubes and combine it with semiconductor [5, 9, 14–19].

Vu et al. indicated that addition of nanotubes to titania reduces the possibility of electron and hole recombination, and combination of titania with nanotubes with a ratio of 20 : 1 has more photocatalytic power than raw titania. In this study, after 8 h of irradiation by ultraviolet lamp (250 W), 80% DBT removal was achieved [5].

Given that very few research studies are conducted on the use of nanotubes to increase the efficiency of photocatalytic removal of thiophenic compounds based on studies done so far, the aim of this study is to evaluate the effect of important factors on removal of BT and DBT, as sample compounds with low reactivity, by using MWCNT-titania composite.

Today, with the advancement of computer science, design and optimization have become common ways of research development. Thus, many studies have been conducted for optimization of photocatalytic construction using different methods of experimental design. Thus, in this research, CCD method is applied to optimize photocatalyst construction condition.

2. Materials and Methods

2.1. Photocatalyst Synthesis for Optimization of MWCNT : Titania Ratio in Sol. The composites are made using sol-gel method and ratios that were optimized in our previous research [20]. Tetraethyl orthotitanate (95%, Merck) was dissolved in isopropanol (obtained from Chemical Industry Co. of Dr. Mojalali IR) with molar ratio of 1 : 70. In order to stabilize pH level, the required amount of nitric acid (65%, Merck) solution with pH = 3 was added, and the mixture was stirred for 1 h using a magnetic stirrer during the sol preparation. Thereafter, MWCNT (RIPI, D = 10–20 nm) was added to the mixture and was stirred for another 3 h. The nanotube ratio was selected in such a way that the mass ratio of MWCNT to titania in the resultant composite will be equal to 0.65, 1, 1.3, and 2.6. The samples were then labeled as samples 1–4 (S1, S2, S3, and S4).

After this step, until when the gel is formed and dried, the obtained mixture was kept at room temperature. Then, the samples were placed in a furnace and heated at 5°C/min to reach 450°C. The samples were calcined for 1 h at this temperature. Then, they were cooled down at room temperature with the same rate.

2.2. Photocatalyst Synthesis for CCD Optimization. In this part of the research, MWCNT to titania ratio was constant and equal to the best ratio based on the previous section result, but we used different pH values (2, 2.4, 3, 3.6 and 4) for sol preparation. Also, proper amounts of p-25 (Degussa, BET = 50 m²/g, band gap = 3.4 ev) were added to the sol exactly after MWCNT addition and stirred for 3 h. This solution was kept at room temperature. After solvent evaporation, the samples were calcinated at different temperatures (300, 361, 450, 539, and 600°C). The heating and cooling rates were the same as in the previous section.

2.3. Test Method and Concentration Measurement. To assess photocatalytic power of the particles, 0.5 g of the catalyst was added to 100 ml of n-hexane (95%, Dr. Mojalali IR) solution with initial concentration of DBT (100%, Merck) or BT (100%, Merck) equal to 200 ppm. Then, the liquid was stirred in the dark until adsorption equilibrium is established. In this case, the solution's equilibrium concentration was considered as the initial concentration for the next step.

In the next step, 1 mM of hydrogen peroxide (30%, Merck), as an oxidizing agent, was added to the mixture. Then, it was exposed to UV lamp (UVC, Philips, 9 W) in a mirror box for 20 min. In order to measure concentration, samples were taken at each stage and photocatalyst was separated from the sample by centrifugation. Then, concentration was calculated by measuring absorption using a UV spectrophotometer.

2.4. Experimental Design and Statistical Analysis. In order to achieve desired results, there are various methods of experimental design which reduce the number of tests, determine the effects of the studied parameters on the target function, and ultimately specify the optimum conditions. In the current research, CCD, which is the most widely used RSM (response surface method), is applied to optimize BT and DBT removal.

There are many parameters that can affect the photocatalytic power of composites. But in this study, calcination temperature, sol pH, and p-25 to nanotubes ratio were considered as important parameters in determining the photocatalytic power of this structure. The rest of the conditions are assumed to be fixed in all experiments.

The photocatalytic removal rate of thiophenic compounds is defined as the system response and target parameter. Equation 1 is used for its calculation, where C_t is final concentration and C_0 is initial concentration (the obtained value after absorption equilibrium).

$$y = 100 \times \left(\frac{C_0 - C_t}{C_0} \right). \tag{1}$$

TABLE 1: Actual and coded values of independent variables.

Factor	Level				
	−1.68	−1	0	1	1.68
α: Temp (°C)	300	361	450	539	600
β: pH	2	2.4	3	3.6	4
γ: p-25 : MWCNT ratio (g/g)	0	1	2.5	4	5

TABLE 2: Different experiment conditions and their response values.

Experiment number	Factor			% removal	
	α	β	γ	BT	DBT
1	361	2.4	1	25	88.5
2	539	2.4	1	8.4	21.6
3	361	3.6	1	29.4	82.5
4	539	3.6	1	9.5	29.1
5	361	2.4	4	26.4	90.6
6	539	2.4	4	9.4	42
7	361	3.6	4	35.8	93.2
8	539	3.6	4	15.7	55.7
9	300	3	2.5	37.3	90.4
10	600	3	2.5	8.9	21.7
11	450	2	2.5	10.9	68.3
12	450	4	2.5	11.9	66.5
13	450	3	0	13.7	41.9
14	450	3	5	23.5	62
15	450	3	2.5	11.3	60
16	450	3	2.5	10.2	55.2
17	450	3	2.5	14.8	53.6
18	450	3	2.5	10.9	55

Ranges of change for every variable are determined based on previous studies and preliminary tests. Accordingly, temperature between 300°C (the temperature required for the formation of anatase phase) and 600°C (the nanotube burns at higher temperatures) [5], pH between 2 and 4 (due to the absence of gel formation in nonacidic pH [21]), p-25 to nanotube ratio between 0 and 5 (composite cannot be stimulated with visible light at higher ratios.) were considered. Thus, actual and coded values of every parameter are shown in Table 1.

Experiments are designed according to five levels of CCD. A design with these conditions and four replicas at center point includes 18 tests according to Table 2.

Data analysis was performed using design expert v.7 software, and the quadratic model was used to predict the parameters' behavior. The accuracy of a model can be assessed in different ways. One of the parameters used in this field was to compare the values obtained experimentally with values predicted by the model.

Also, in order to evaluate the parameters' effect and model validity, R^2, F value, p value, and main effects chart are employed. Of course, to assess the model accuracy, R^2 should not be considered as the only comparison criteria,

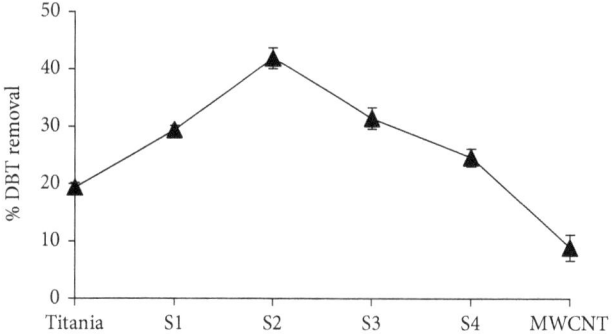

FIGURE 1: % DBT removal using composites with different MWCNT : titania ratio.

because it can be increased by adding a number of factors (even noninfluential factors); to do so, adjusted R^2 and proximity of these two values to each other should be employed, as well.

3. Results

3.1. Influence of MWCNT : Titania Ratio on Photooxidation of DBT. Figure 1 indicates percentage DBT removed versus photocatalyst with different MWCNT : titania ratio. For better understanding, results using MWCNT alone and using photocatalyst without MWCNT addition were also shown. MWCNT has photocatalytic potential, and its addition to composite increases photocatalytic power. This increment can be explained as a result of reduction in the possibility of electron and hole recombination. However, higher ratios may lead to light scattering in solution; though, they may inversely affect photocatalytic removal rate [5, 7, 16, 19]. Therefore, the optimum sample is S2 with mass ratio of MWCNT : titania equal to 1.

3.2. ANOVA. The 18 suggested experiments were performed under specified conditions, and BT and DBT removal results are shown in Table 2. The results were analyzed using ANOVA test. Equation 2 is the model that is proposed to predict BT and DBT removal. Coefficients of this equation are listed in Table 3. α, β, and γ are main factors (listed in Table 1) in actual value.

$$\%\text{removal} = A + B \times \alpha + C \times \beta + D \times \gamma + E \times \alpha \\ \times \beta + F \times \alpha \times \gamma + G \times \beta \times \gamma + H \quad (2) \\ \times \alpha^2 + I \times \beta^2 + J \times \gamma^2.$$

R^2, F value, and p value are given in Table 4. For both thiophenic compounds, high F value and p value < 0.05 indicate that the model is significant. In both cases, nonsignificant lack of fit means the model is properly fitted to the data [22, 23]. R^2 values > 0.95 means that the model properly predicts parameters' changes (in the studied range) and, in both cases, under 5% of total changes cannot be explained by the model [24] which can also be confirmed by proximity of R^2 and adjusted R^2 [22–24]. Figure 2 presents predicted values versus obtained values from the experiments, where

TABLE 3: Coefficients of (2).

Coefficient	Pollutant	
	BT	DBT
A	160.75	437.364
B	-0.55133	-0.6270
C	1.47975	-113.8877
D	-8.75249	-14.9496
E	-0.015085	0.05798
F	-0.0005606	0.031954
G	1.42952	2.07421
H	0.0005537	0.00012
I	0.75875	14.07153
J	0.12506	-0.21661

proximity of points to lines suggests little difference between predicted and obtained values [23, 24].

In addition, p values for the given parameters and their interaction are shown in Table 4. According to this table, α, β, and γ are 3 significant factors in BT removal, while only α and γ and their interaction are significant in DBT removal. Among these factors, temperature is most effective in both BT and DBT removal. According to the main effect graph, shown in Figure 3, the same result is obtained because temperature changes lead to 30% change in BT response and 80% change in DBT response.

Response surface plot, shown in Figure 4, indicates simultaneous effects of α and γ (when the third parameter is at the central point) on the DBTs' response.

According to the obtained results, the structure of photocatalytic strength decreases with increasing temperature. It has been demonstrated that by increasing the calcination temperature or calcination time, crystalline particles grow in size which results in decreased specific surface of the structure [3]. Decreased specific surface of the structure indicates reduction of active sites for the reaction. On the other hand, rising temperature converts anatase phase to rutile and therefore reduces photocatalytic power. For this reason, at high temperatures, positive impact of more amount of p-25 on the structure is obvious.

3.3. Optimization and Confirmation. In this process, optimization means obtaining a structure which results in the highest rate of BT and DBT removal. Based on the performed analysis, a structure which is made at pH = 4 with p-25 five times more than carbon nanotube that is calcined at 300°C can be an optimal structure. The model predicts that employing this structure may result in 100% DBT removal and 60.38% BT removal. This sample was synthesized, and within 20 minutes (average of four iterations), all DBT and 59.8% of BT were removed. Proximity of the obtained and predicted values by the model indicates suitability of the test design and the proposed model.

3.4. Characterization of the Optimized Sample. The result of XRD test (tube Co, $\lambda = 1.78897$ Å, step size $= 0.02°$/s) on the optimized sample is shown in Figure 5. Anatase

diffraction peaks was found at 25.1, 38.5, 47.8, and 70.2° corresponding to the reflections from 101, 112, 200, and 116 crystal planes [5, 7, 14–17, 25]. The peaks at 27.6, 40, 41, 44, and 85.8° are the diffractions of rutile 110, 200, 111, 210, and 400 [14, 25]. Nanotube (002) peak appears at 25.8° [18, 19] which is usually combined with anatase phase and cannot be specified separately.

Nanotube 100, 004, and 110 peaks appear at 42.2, 52, and 78° [16–19]. Brookite titania [26] or hydrate titania [27] peaks at 59.2 and 60.6 may be due to titania and acid presence in the sol. Accordingly, the produced titania is a combination of anatase and rutile. However, the observed rutile phase cannot be inferred as a result of anatase to rutile conversion at this temperature, yet it is because of a large amount of p-25 in the structure, so the resulting graph is actually similar to the p-25 graph to which nanotube peaks are added.

Using Scherrer equation (3), the size of the crystal is calculated [2, 17].

$$D = \frac{k\lambda}{\beta \cos \theta}, \tag{3}$$

where D is crystal diameter, k is a shape factor equal to 0.94, λ is X-ray wavelength equal to 1.79 Angstrom, θ is maximum peak angle, and β is bandwidth at half height of the graph's maximum peak. Considering Figure 4 and Scherrer equation, the particles' size is 36 nm. It is proved that adding nanotubes to the compound reduces the size of crystals when compared to raw titania and p-25 [14, 17].

Figure 6 shows light absorption of the optimized sample. To better compare the graphs, p-25 light absorption is also included in Figure 6. All attempts made to improve structural properties of the photocatalysts are to produce a catalyst which can be stimulated by visible light irradiation.

As shown in the figure, due to the high amount of p-25 in the combination, the band gap of the optimized sample is equal to p-25. However, compared with p-25, the sample absorbs light of the visible region, as well [7, 14–17, 19]. When combined with titania, nanotube creates an energy level under titania conducting band. Therefore, electron can also be stimulated when irradiated with visible light and moves between nanotubes and titania. It should be noted that extending the absorption edge of TiO_2 to the visible region is considered as one of the main objectives to have active photocatalyst in the visible light region [7, 10, 14, 15, 17].

3.5. Photolysis, Oxidation, and Photooxidation Comparison. The decrease in the concentration of BT and DBT under different conditions is presented in Table 5. According to the results, the photolysis process does not have the power to eliminate thiophene. The catalyst, by itself, induces a slight decrease in concentrations by adsorption. However, when catalyst and light are used, in addition to adsorption, oxidation also occurs, and it is evident that, when ultraviolet light is used due to more emitted energy, the electron and hole production rate increases, and therefore, the amount of removal is higher than the use of visible light. Hydrogen peroxide is not capable of oxidizing BT but oxidizes DBT to

TABLE 4: ANOVA results for quadratic model of BT and DBT photooxidation.

	Mean square	F value	p value (BT)	Mean square	F value	p value (DBT)
Model	168.16	28.11	<0.0001	982.27	37.24	<0.0001
α (Temp)	1078.50	180.27	<0.0001	7589.22	287.72	<0.0001
β (pH)	38.34	6.41	0.0352	15.98	0.61	0.4588
γ (p-25)	72.57	12.13	0.0083	641.56	24.32	0.0011
$\alpha\beta$	5.12	0.86	0.3820	75.65	2.87	0.1288
$\alpha\gamma$	0.045	0.0075	0.9330	146.21	5.54	0.0464
$\beta\gamma$	13	2.17	0.1786	27.38	1.04	0.3381
α^2	245.43	41.02	0.0002	11.71	0.44	0.5240
β^2	0.91	0.15	0.7067	313.08	11.87	0.0088
γ^2	100.15	16.74	0.0035	3.00	0.11	0.7444
Lack of fit	7.05	1.68	0.356	37.53	4.81	0.1131
R^2			0.9693			0.9767
Adj R^2			0.9349			0.9505

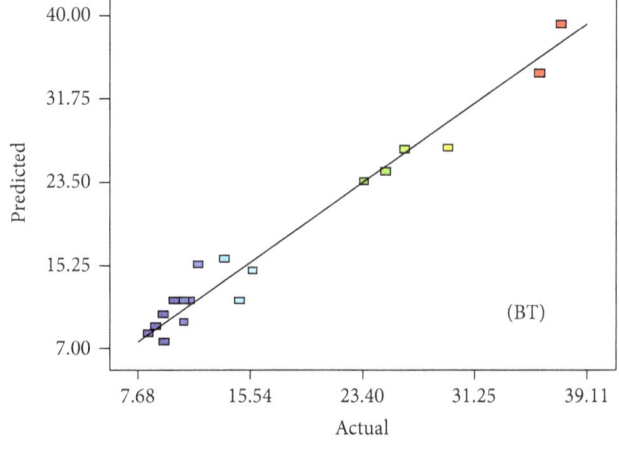

(BT)

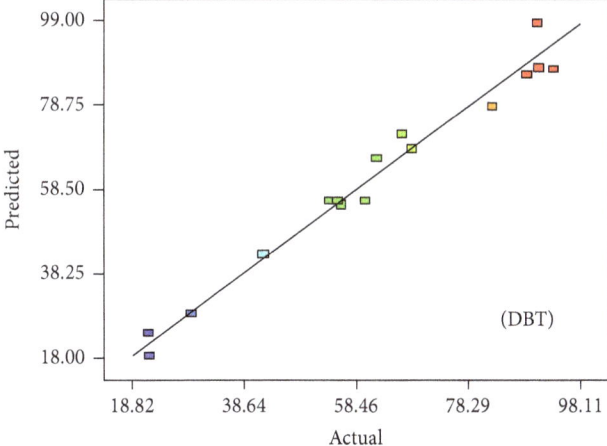

(DBT)

FIGURE 2: Predicted versus actual values for photooxidation of BT and DBT.

a small extent, and when solution is exposed to light, its oxidation power is slightly increased and, under ultraviolet light, it can also reduce the concentration of BT, but it is not significant at all. When the catalyst and hydrogen peroxide are added to the solution (chemical oxidation),

a significant change in concentrations occurs. As shown in the table, there is a reduction of 29.3% in the concentration of BT and 86% reduction in the concentrations of DBT after 20 min. This synergistic effect shows that in this case, the catalyst acts as an active site and increases the contact surface of hydrogen peroxide and thiophenes. If light is also added to this system, chemical oxidation becomes photooxidation, and the power of thiophene removal is increased. Based on the results obtained, the catalyst with visible light is also excited, and its removal ability is more than chemical oxidation (especially in the case of BT).

3.6. Kinetics of Photooxidation. When it became obvious that samples can absorb visible light, some tests were conducted using visible light for BT and DBT removal. The results showed that visible light can help in achieving total removal of DBT and 42% removal of BT. BT and DBT removal reactions can be regarded as a first-order equation [2–4, 6], in which reaction rate constants of BT and DBT removal rate are 0.953 and 7.98 1/h, respectively. For better comparison of results obtained, constant values obtained for BT and DBT removal rate in previous studies are listed in Table 6. Comparison of data in Table 6 shows dramatic increase of removal rate in the current research due to the addition of nanotube to titania. There is also an important difference between the current and previous research results; in this study, visible light is used, while the results presented in Table 6 were obtained with ultraviolet light.

4. Conclusion

In order to achieve maximum thiophenic compound removal rate, construction condition of titania-nanotube composite is optimized at 300° C, pH = 4, and titania-nanotube ratio equal to 5. The results showed that, added to increasing photocatalytic power, addition of nanotubes to the system stimulates the structure with visible light. Moreover, DBT removal is far easier than BT removal, because the optimized structure can completely remove

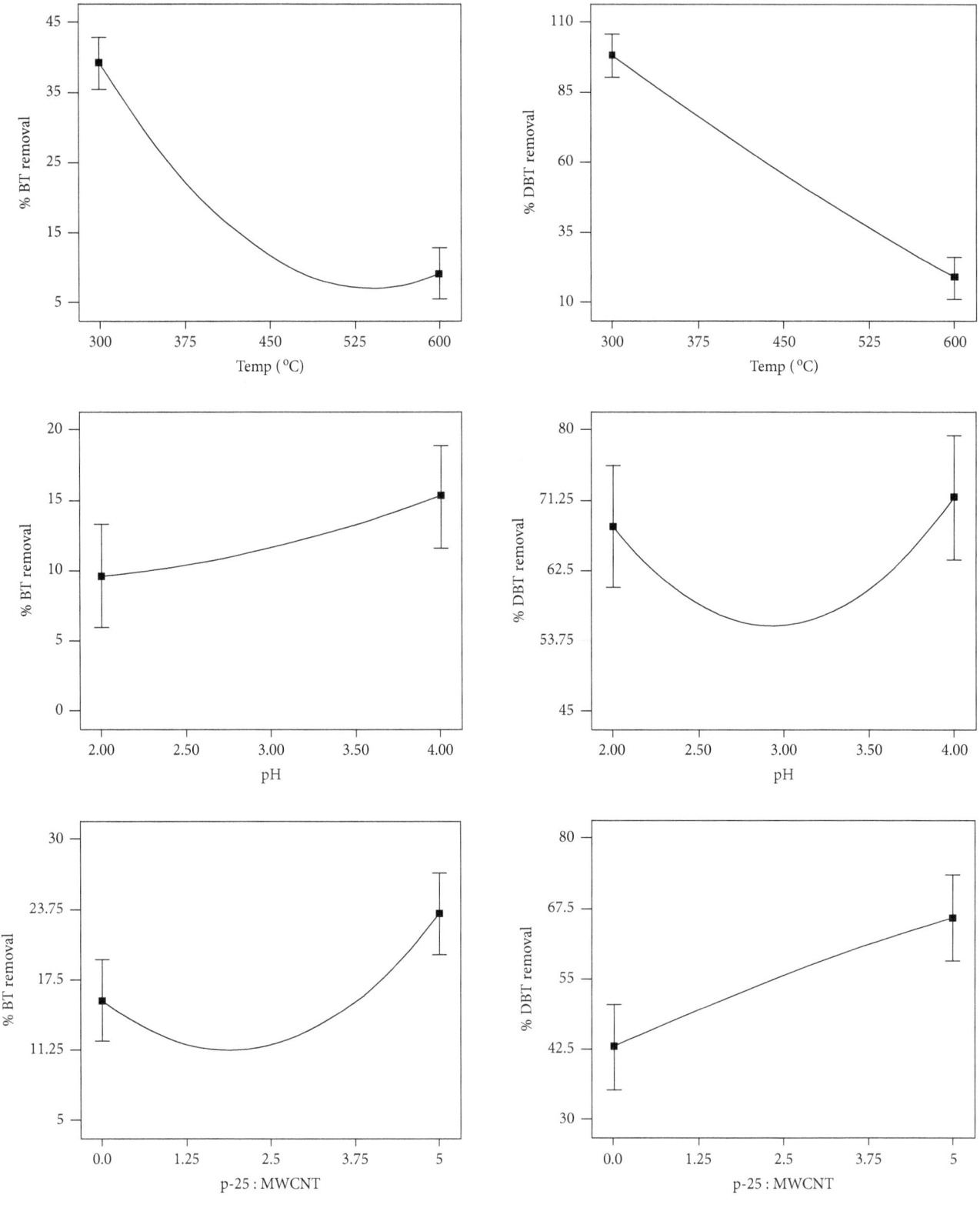

FIGURE 3: Main effect plots for the photooxidation of BT and DBT.

DBT within 20 min (both with ultraviolet light and with visible light); but changing ultraviolet to visible light in case of BT reduces its removal rate from 59.8 to 42%. Reaction rate constant of DBT and BT removal rate in a first-order equation are 7.98 and 0.953 1/h, respectively. These values are several times more than those obtained by previous researchers; the increase was caused principally by adding nanotube to the composite structure.

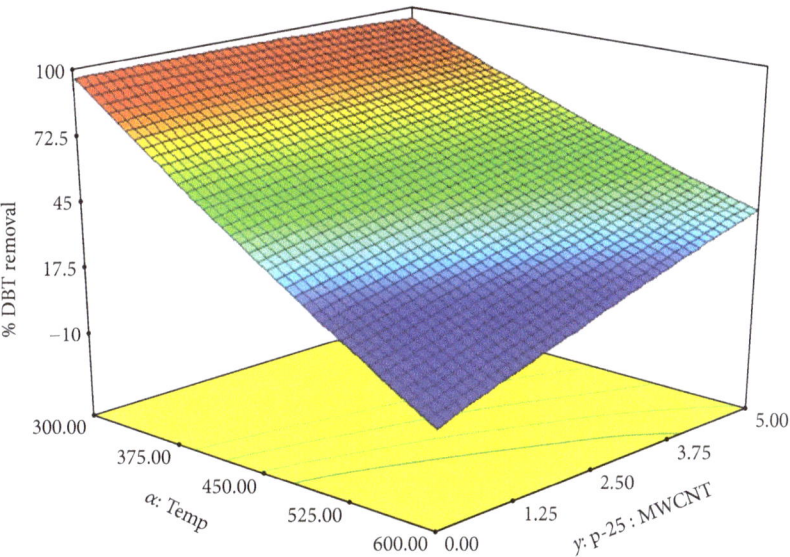

FIGURE 4: Contour plot for the degradation of DBT indicating the effect of interaction between temperature and p-25 : MWCNT.

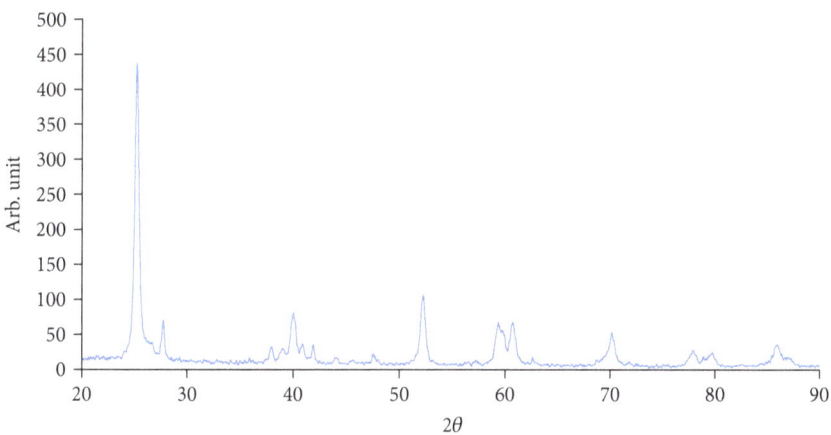

FIGURE 5: XRD pattern of optimum sample.

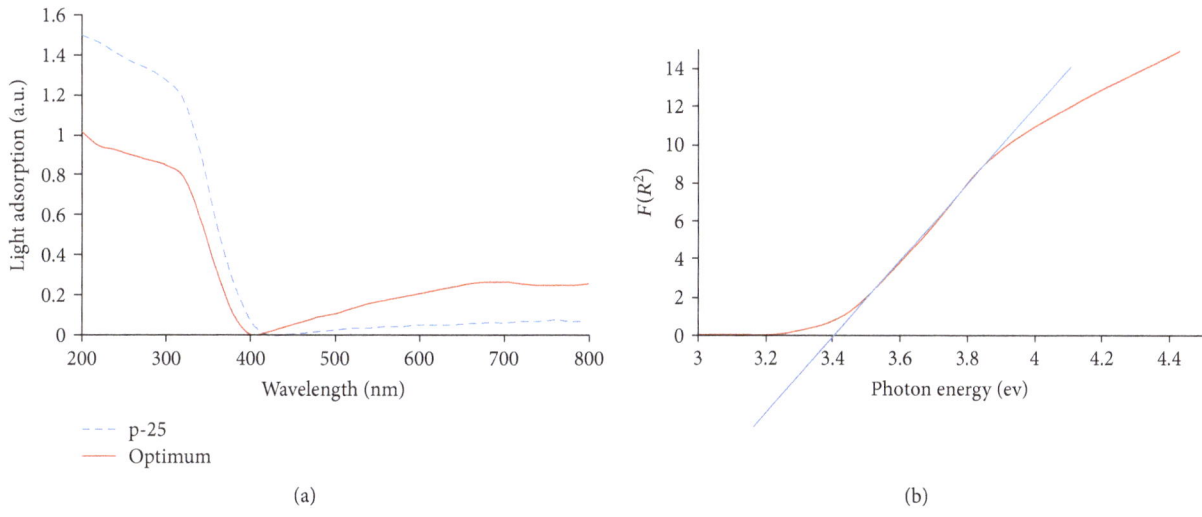

FIGURE 6: UV–Vis diffuse reflectance spectra of p-25 and optimum sample.

TABLE 5: % BT and DBT removal in different condition after 20 min.

Factor	% BT removal	% DBT removal
UV light	—	—
Visible light	—	—
Catalyst	2.6	9.8
Catalyst + UV light	9.5	14.2
Catalyst + visible light	6.4	11.9
H_2O_2	—	2.0
H_2O_2 + UV light	1.3	4.4
H_2O_2 + visible light	—	3.3
Catalyst + H_2O_2	29.3	86
Catalyst + H_2O_2 + UV light	59.8	100
Catalyst + H_2O_2 + visible	42	100

TABLE 6: Reaction rate constant in different researches.

Reference	Year	Catalyst	K_{BT} (1/h)	K_{DBT} (1/h)
[4]	2002	p-25		0.035
[2]	2006	TiO_2		0.9
[1]	2008	TiO_2		0.9
[28]	2008	p-25		0.6
[29]	2009	TS-1		0.48
[30]	2010	TS-1		0.48
[6]	2014	TiO_2	0.38	0.79
This research		TiO_2 + MWCNT	0.953	7.98

Conflicts of Interest

The authors declare that they have no conflicts of interest.

Acknowledgments

The researchers of the current study wish to thank Degussa Iran Co. for providing p-25 titania free of charge.

References

[1] F. T. Li, R. H. Liu, Z. M. Sun, and D. S. Zhao, "Photocatalytic oxidation kinetics of thiophene with nano-TiO_2 as photocatalyst," in *2008 2nd International Conference on Bioinformatics and Biomedical Engineering*, Shanghai, China, 2008.

[2] J. Robertson and T. J. Bandosz, "Photooxidation of dibenzothiophene on TiO_2/hectorite thin films layered catalyst," *Journal of Colloid and Interface Science*, vol. 299, no. 1, pp. 125–135, 2006.

[3] Z. Hasan, J. Jeon, and S. H. Jhung, "Oxidative desulfurization of benzothiophene and thiophene with WO_x/ZrO_2 catalysts: effect of calcination temperature of catalysts," *Journal of Hazardous Materials*, vol. 205-206, pp. 216–221, 2012.

[4] S. Matsuzawa, J. Tanaka, S. Sato, and T. Ibusuki, "Photocatalytic oxidation of dibenzothiophenes in acetonitrile using TiO_2: effect of hydrogen peroxide and ultrasound irradiation," *Journal of Photochemistry and Photobiology A: Chemistry*, vol. 149, no. 1-3, pp. 183–189, 2002.

[5] T. H. T. Vu, T. T. T. Nguyen, P. H. T. Nguyen et al., "Fabrication of photocatalytic composite of multi-walled carbon nanotubes/TiO_2 and its application for desulfurization of diesel," *Materials Research Bulletin*, vol. 47, no. 2, pp. 308–314, 2012.

[6] G. Dedual, M. J. MacDonald, A. Alshareef, Z. Wu, D. C. W. Tsang, and A. C. K. Yip, "Requirements for effective photocatalytic oxidative desulfurization of a thiophene-containing solution using TiO_2," *Journal of Environmental Chemical Engineering*, vol. 2, no. 4, pp. 1947–1955, 2014.

[7] E. S. Aazam, "Visible light photocatalytic degradation of thiophene using Ag–TiO_2/multi-walled carbon nanotubes nanocomposite," *Ceramics International*, vol. 40, no. 5, pp. 6705–6711, 2014.

[8] R. Munter, "Advanced oxidation process - current status and prospects," *Proceedings of the Estonian Academy of Sciences Chemistry*, vol. 50, no. 2, pp. 59–80, 2001.

[9] H. Park, Y. Park, W. Kim, and W. Choi, "Surface modification of TiO_2 photocatalyst for environmental applications," *Journal of Photochemistry and Photobiology C: Photochemistry Reviews*, vol. 15, pp. 1–20, 2013.

[10] T. P. Dhanya and S. Sugunan, "Preparation, characterization and photocatalytic activity of N doped TiO_2," *IOSR Journal of Applied Chemistry*, vol. 4, no. 3, pp. 27–33, 2013.

[11] M. Pelaez, N. T. Nolan, S. C. Pillai et al., "A review on the visible light active titanium dioxide photocatalysts for environmental applications," *Applied Catalysis B: Environmental*, vol. 125, pp. 331–349, 2012.

[12] V. Loryuenyong, K. Angamnuaysiri, J. Sukcharoenpong, and A. Suwannasri, "Sol–gel derived mesoporous titania nanoparticles: effects of calcination temperature and alcoholic solvent on the photocatalytic behavior," *Ceramics International*, vol. 38, no. 3, pp. 2233–2237, 2012.

[13] Y. P. Tsai, R. Doong, J. C. Yang, and Y. J. Wu, "Photo-reduction and adsorption in aqueous Cr(VI) solution by titanium dioxide, carbon nanotubes and their composite," *Journal of Chemical Technology and Biotechnology*, vol. 86, no. 7, pp. 949–956, 2011.

[14] H. Liu, X. Yu, and H. Yang, "The integrated photocatalytic removal of SO_2 and NO using Cu doped titanium dioxide supported by multi-walled carbon nanotubes," *Chemical Engineering Journal*, vol. 243, pp. 465–472, 2014.

[15] H. Liu, H. Zhang, and H. Yang, "Photocatalytic removal of nitric oxide by multi-walled carbon nanotubes-supported TiO_2," *Chinese Journal of Catalysis*, vol. 35, no. 1, pp. 66–77, 2014.

[16] Y. Koo, G. Littlejohn, B. Collins et al., "Synthesis and characterization of Ag–TiO_2–CNT nanoparticle composites with high photocatalytic activity under artificial light," *Composites Part B: Engineering*, vol. 57, pp. 105–111, 2014.

[17] S. S. Mali, C. A. Betty, P. N. Bhosale, and P. S. Patil, "Synthesis, characterization of hydrothermally grown MWCNT–TiO_2 photoelectrodes and their visible light absorption properties," *ECS Journal of Solid State Science and Technology*, vol. 1, no. 2, pp. M15–M23, 2012.

[18] T. A. Saleh, M. A. Gondal, Q. A. Drmosh, Z. H. Yamani, and A. al-yamani, "Enhancement in photocatalytic activity for acetaldehyde removal by embedding ZnO nano particles on multiwall carbon nanotubes," *Chemical Engineering Journal*, vol. 166, no. 1, pp. 407–412, 2011.

[19] N. Hintsho, L. Petrik, A. Nechaev, S. Titinchi, and P. Ndungu, "Photo-catalytic activity of titanium dioxide carbon nanotube

nano-composites modified with silver and palladium nanoparticles," *Applied Catalysis B: Environmental*, vol. 156-157, pp. 273–283, 2014.

[20] M. Barmala, A. Z. Moghadam, and M. R. Omidkhah, "Increased photo-catalytic removal of sulfur using titania/MWCNT composite," *Journal of Central South University*, vol. 23, no. 5, pp. 1066–1070, 2016.

[21] M. E. Simonsen and E. G. Sogaard, "Sol–gel reactions of titanium alkoxides and water: influence of pH and alkoxy group on cluster formation and properties of the resulting products," *Journal of Sol-Gel Science and Techology*, vol. 53, no. 3, pp. 485–497, 2010.

[22] H. Hossini, M. Safari, R. Rezaee, R. D. C. Soltani, O. Giahi, and Y. Zandsalimi, "Application of experimental design approach for optimization of the photocatalytic degradation of humic substances in aqueous solution using immobilized ZnO nanoparticles," *Journal of Advances in Environmental Health Research*, vol. 3, no. 3, pp. 154–163, 2015.

[23] K. M. Lee and H. SBA, "Simple response surface methodology: investigation on advance photocatalytic oxidation of 4-chlorophenoxyacetic acid using UV-active ZnO photocatalyst," *Maternité*, vol. 8, no. 12, pp. 339–354, 2015.

[24] M. Antonopoulou, A. Giannakas, and I. Konstantinou, "Simultaneous photocatalytic reduction of Cr(VI) and oxidation of benzoic acid in aqueous N-F-codoped TiO2 suspensions: optimization and modeling using the response surface methodology," *International Journal of Photoenergy*, vol. 2012, Article ID 520123, 10 pages, 2012.

[25] F. J. Zhang, M. L. Chen, K. Zhang, and W. C. Oh, "Visible light photoelectrocatalytic properties of novel yttrium treated carbon nanotube/titania composite electrodes," *Bulletin of the Korean Chemical Society*, vol. 31, no. 1, pp. 133–139, 2010.

[26] O. A. al-Harbi, E. M. A. Hamzawy, and M. M. Khan, "Effect of different concentrations of titanium oxide (TiO$_2$) on the crystallization behavior of Li$_2$O-Al$_2$O$_3$-SiO$_2$ glasses prepared from local raw materials," *Journal of Applied Sciences*, vol. 9, no. 16, pp. 2981–2986, 2009.

[27] X. Bin, Z. C. Yong, C. Bing, and F. H. Song, "Porous titanium treated by nitric acid with varied concentration and the bioactivity in vitro," *Journal of Inorganic Materials*, vol. 27, no. 5, pp. 555–560, 2012.

[28] N. A. Buang, F. Fadil, Z. A. Majid, and S. Shahir, "Characteristic of mild acid functionalized multiwalled carbon nanotubes towards high dispersion with low structural defects," *Digest Journal of Nanomaterials and Biostructures*, vol. 7, no. 1, pp. 33–39, 2012.

[29] D. Zhao, J. Zhang, J. Wang, W. Liang, and H. Li, "Photocatalytic oxidation desulfurization of diesel oil using Ti-containing zeolite," *Petroleum Science and Technology*, vol. 27, no. 1, pp. 1–11, 2009.

[30] Z. Juan, Z. Dishun, Y. Liyan, and L. Yongbo, "Photocatalytic oxidation dibenzothiophene using TS-1," *Chemical Engineering Journal*, vol. 156, no. 3, pp. 528–531, 2010.

19

Preparation and Physical and Photocatalytic Activity of a New Niobate Oxide Material Containing NbO$_4$ Tetrahedra

Hengkai Pan,[1,2] **Bei Wang,**[1,2] **Feng Zhang**(ID)**,**[1,2] **Weifeng Zhang,**[1,2] **and Guoqiang Li**(ID)[1,2]

[1]*Henan Key Laboratory of Photovoltaic Materials, Henan University, Kaifeng 475004, China*
[2]*Laboratory of Low-Dimensional Materials Science, Henan University, Kaifeng 475004, China*

Correspondence should be addressed to Feng Zhang; zhangfeng.home@163.com and Guoqiang Li; gqli1980@henu.edu.cn

Academic Editor: Leonardo Palmisano

The shape and connection type of MO$_x$ are critical to the physical and chemical properties. A series of new material Sr$_{2-x}$Na$_x$NbO$_4$ containing NbO$_4$ tetrahedra was prepared by controlling the ratio of SrCO$_3$ to sodium niobate under ambient air. With increasing the content of Sr in the sample, the MO$_x$ shape will change from NbO$_6$ octahedra to NbO$_4$ tetrahedra, which is confirmed by the Raman scattering spectra. With increasing the content of NbO$_4$ in the sample, the lattice parameter increases, optical band gap becomes larger, and the surface changes to be more active for oxygen adsorption, resulting in a higher photocatalytic activity.

1. Introduction

Complex metal oxide (AMO$_y$) is built upon the framework of MO$_x$ with sharing edges or corners and the inserted metal ion. The shape of MO$_x$ is diverse, such as octahedron and tetrahedron. The shape and connection types of MO$_x$ are critical to the physical and chemical properties [1–5]. Recently, theoretical predictions and experimental results confirmed that the TaS$_2$ built from the edge-sharing of TaS$_6$ trigonal prisms exhibits superior catalytic performance in comparison with that of TaS$_6$ octahedra [6, 7]. Those early studies suggested a potential way to improve the catalytic activity via tuning the shape of MO$_x$. The above hypothesis is rarely investigated in niobates, which is one kind of photocatalyst for water splitting and organic degradation, such as NaNbO$_3$, SrNb$_2$O$_6$, and Cs$_2$Nb$_4$O$_{11}$ [8–15]. Most of the niobate photocatalysts contain NbO$_6$ octahedra. For instance, NaNbO$_3$ contains the formwork of NbO$_6$ octahedra sharing corners with a slight distortion [16, 17]. The NbO$_6$ octahedra in SrNb$_2$O$_6$ share edges and corners simultaneously [14]. Cs$_2$Nb$_4$O$_{11}$ is the one that coexisted with NbO$_6$ octahedra and NbO$_4$ tetrahedra in the structure [12, 18]. The tetrahedral NbO$_4$ structure is rarely found in niobium oxide compounds

because the Nb^{5+} atom is usually too large to fit into an oxygen-anion tetrahedron and exists in the rare earth ANbO$_4$ (A = Y, Sm, and La) compounds [1, 2, 19]. To investigate the effect of MO$_x$ shape on photocatalytic activity, the challenge is preparing the sample containing NbO$_4$ tetrahedra.

Here, we reported a series of new material Sr$_{2-x}$Na$_x$NbO$_4$ containing NbO$_4$ tetrahedra, which was prepared from sodium niobate and SrCO$_3$ with different ratios under ambient air. The MO$_x$ shape will change from NbO$_6$ octahedra to NbO$_4$ tetrahedra. The Raman results identified that NbO$_4$ tetrahedra existed in the samples. The change from the NbO$_6$ octahedra to NbO$_4$ tetrahedra results in the different physical and chemical properties, such as optical band gap, adsorption property, and photocatalytic activity.

2. Experimental Section

2.1. Sample Preparation. Sodium niobate precursor was synthesized via a simple hydrothermal process. Typically, the prepared mixture of 50 mL of NaOH solution (8 M) and 2 g of Nb$_2$O$_5$ powder was poured into a 100 mL teflon-lined stainless steel autoclave. The hydrothermal reaction was performed in a drying oven at 120°C for 3 h [20–22]. After naturally cooling down to room temperature, we obtained the samples and rinsed the samples with deionized water and absolute ethanol to remove the residual unreacted precursor. The sodium niobate precursor was confirmed as the $Na_2Nb_2O_6 \bullet xH_2O$ from the XRD.

$SrCO_3$ and sodium niobate precursor were mixed in the mortar at different molar ratios and grounded for 15 min. Then, they were heated at 500°C for 3 h and calcined at 800°C for 10 h in a muffle furnace. After cooling, we got the samples. For simplicity, we denoted the samples prepared from the ratios of $SrCO_3$ to sodium niobate precursor of 1 : 2, 2 : 2, and 4 : 2 as A, B, and C, respectively. For comparison, the sample was also prepared by solid state reaction method using $NaNbO_3$ and $SrCO_3$ as starting materials at 900°C for 10 h.

2.2. Characterizations. The crystal structures were identified by X-ray diffraction (XRD). Raman scattering spectrum was measured using a laser Raman spectrophotometer at room temperature. The X-ray photoelectron spectra (XPS) were measured by a Kratos AXIS Ultra photoelectron spectroscope. The diffusion reflection spectrum was recorded with a UV-vis spectrophotometer (Shimadzu 2550) using $BaSO_4$ as the reference and transformed to the absorption spectra automatically. The temperature programmed desorption (TPD) was carried out on Chembet Pulsar TPD using He pretreatment at 300°C for 1 h. Inductively coupled plasma (ICP) was performed on Agilent 720/730 after pretreatment in mixture of HCl and HNO_3.

Photocatalytic activity for the decomposition of RhB in an aqueous solution was evaluated in the presence of samples A, B, and C under full arc light irradiation of Xe lamp as reported previously [23]. The initial concentration and pH value of RhB solution were about $2.5 \, mg \, L^{-1}$ and 4.5. Before illumination, the reagent was left in the dark for 30 min to achieve adsorption-desorption equilibrium. After adsorption of dye, the pH value is 6.7 for sample A, 6.6 for sample B, and 8.5 for sample C. The light emitted from the 300-W Xe lamp (the spectra is the same as that in our previous report [24]) directly irradiated on the solution in the reactor. The stirring was on during the reaction. The variation in the concentration of RhB was recorded by measuring the absorbance of the main peak in the UV-vis spectrum (UV-2550, Shimadzu) every 30 min.

3. Results and Discussions

All the samples show the similar XRD patterns with that of Sr_2NbO_4 (PDF#28-1245), as shown in Figure 1(a) [25]. A slight amount of $SrCO_3$ impurity was found in sample C due to too much $SrCO_3$ in the starting mixture. The compositions of samples were determined as $Sr_{0.6}Na_{1.4}NbO_4$, $Sr_{0.97}Na_{1.03}NbO_4$, and $Sr_{1.73}Na_{0.27}NbO_4$ for A, B, and C, from the ICP results. The total charge at A site will increase from 2.8 for sample A, 2.97 for sample B, and up to 3.73 for sample C. Moreover, the positions of diffraction peaks were successively shifted towards smaller 2θ with increasing the content of $SrCO_3$, as shown in Figure 1(b). Two factors, namely strains and lattice parameter changes, will contribute

to the shift of diffraction peaks [26]. The strains could be estimated by the following equations.

$$\frac{\beta \cos \theta}{\lambda} = \frac{1}{D} + \frac{\eta \sin \theta}{\lambda}, \qquad (1)$$

where β is the full width at half-maximum (FWHM) of the θ-2θ peak; θ is the diffraction angle; λ is the X-ray wavelength; η is the effective strain, and D is the crystallite size. The strain (η) is calculated from the slope, and the crystallite size (D) is calculated from the intercept of a plot of $\beta \cos \theta/\lambda$ against $\sin \theta/\lambda$. The calculated effective strain of samples A, B, and C are 1.15%, 0.87%, and 0.41%, respectively. However, the decrease in the effective strain is not reasonable. Kumar et al. found that the diffraction peak of $NaNbO_3/CdS$ core/shell particles shifted to larger 2θ in comparison with pristine $NaNbO_3$, where the effective strain of $NaNbO_3/CdS$ core/shell particles will decrease down to -0.76% from 0.89% for $NaNbO_3$ [26]. Our result is opposite to the above tendency, so we thought the shift of diffraction peak is not caused by the strains. According to the Bragg's law, we calculated the lattice parameter using Sr_2NbO_4 as the parent structure, as shown in Figure 1(c). It is clearly seen that the lattice parameter increases with the rise in the content of Sr and trend towards 4.11 Å that of Sr_2NbO_4.

Raman scattering is an effective method of investigating the changes at local structure. Figure 2(a) shows Raman scattering spectra of samples prepared from different ratios. The Raman band near $780-840 \, cm^{-1}$ appears in all the samples, whereas the band near $500-600 \, cm^{-1}$ only existed in sample A. The Raman band around $780-830 \, cm^{-1}$ is corresponding to the vibrational modes of a regular NbO_4 tetrahedron, and the band near $500-700 \, cm^{-1}$ is assigned as the symmetric stretching mode of the NbO_6 octahedra [1, 18, 19]. The correlation between the Raman wavenumbers for the stretching bands in niobate compounds and the Nb-O bond length has previously been established. We calculated the Nb-O bond length using the empirical equation $\nu/cm^{-1} = 29522 \times \exp(-1.9168R)$, where ν is the Nb-O stretching wavenumber, and R is the Nb-O bond length [13, 27]. The results are shown in Figure 2(b), and all of the Nb-O length satisfied the requirement of Nb-O bond length in NbO_4 tetrahedron of 1.83–1.93 Å, indicating that all the samples contain NbO_4 tetrahedron [1]. The precursor of sodium niobate only has the band near $500-600 \, cm^{-1}$, and the $YNbO_4$ contains the Raman band near $780-840 \, cm^{-1}$. Above results confirmed that sample A contains NbO_6 octahedran and NbO_4 tetrahedra, and samples B and C mainly contain NbO_4 tetrahedron. The formation mechanism will be discussed later. The Nb-O bond length increases with the rise in the content of Sr.

The optical properties of samples are displayed in Figure 3. The absorption edge of sample A locates at ~324 nm corresponding to ~3.8 eV, whereas that of samples B and C is ~310 nm corresponding to ~4.0 eV, implying that sample A has a narrower band gap in comparison with samples B and C. The increase of lattice distance usually results in the larger band gap due to the smaller expansion of the energy levels according to the principle of solid state physics.

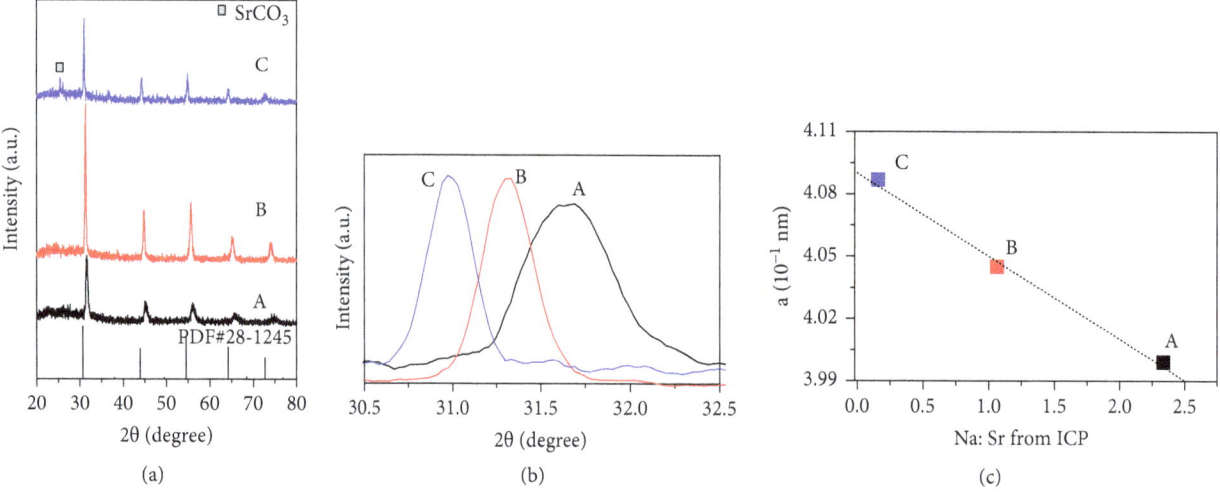

FIGURE 1: (a) XRD patterns, (b) enlarged view of main peak, (c) lattice parameter vs. Na : Sr ratio of samples prepared from different ratio of SrCO₃ to sodium niobate of (A) 1 : 2, (B) 2 : 2, and (C) 4 : 2. The standard XRD pattern of Sr_2NbO_4 (PDF#28-1245) was plotted at the bottom in Figure 1(a).

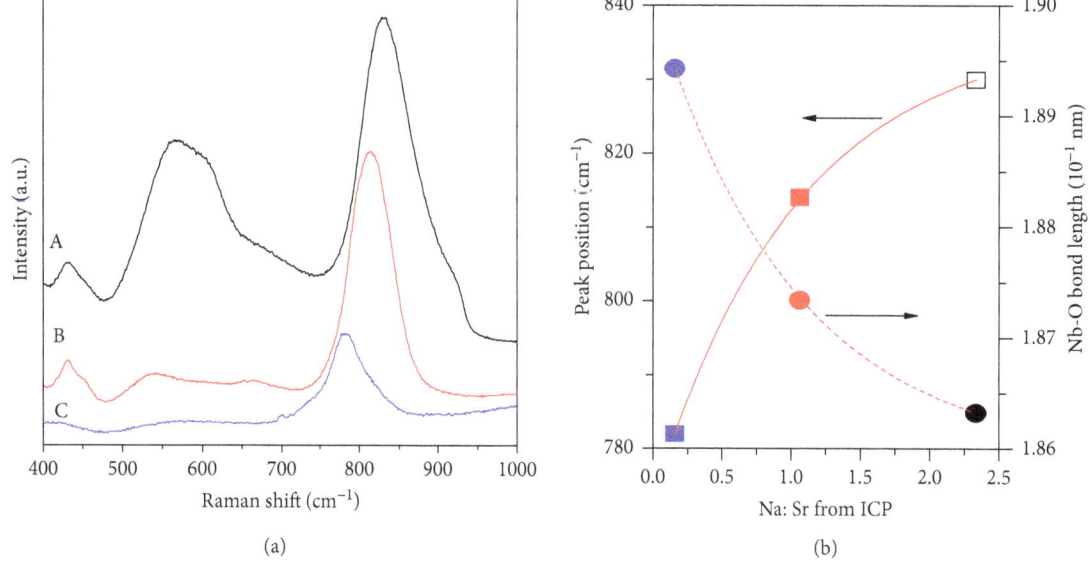

FIGURE 2: (a) Raman scattering spectra of samples prepared from different ratios and (b) plot of Raman peak position vs. Na : Sr ratio.

Consequently, it is reasonable to get the bigger optical band gap in samples B and C because they have larger lattice parameter. Samples B and C have the similar absorption edge and band gap, which is possibly attributed to their common properties of containing NbO_4 tetrahedra. Sample C has a tail around 310–500 nm, indicating that sample C possibly has more defects in comparison with sample B.

The O1s XPS lines obtained from samples are shown in Figure 4. The binding energies were calibrated to the C1s peak at 284.8 eV. The O1s spectra can be divided into two peaks at 530.0 and 531.4 eV, respectively. The peak at 530.0 eV represents the binding energy of lattice oxygen ions [28–31]. The peak at 531.4 eV was considered as the binding energy of oxygen defects [31] or surface oxygen, such as O_2^- [29, 30]. The ratio of O_{ads} to O_{latt} increases with the rise in the content of Sr. We used the TPD to investigate the property of

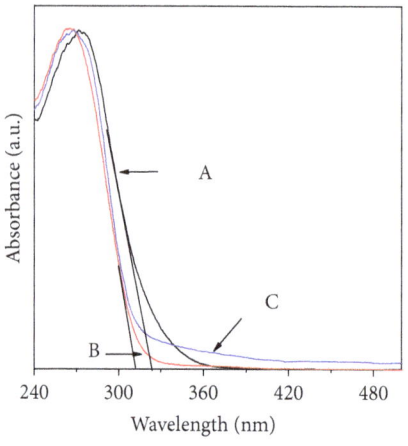

FIGURE 3: UV-vis spectra of samples prepared from different ratios.

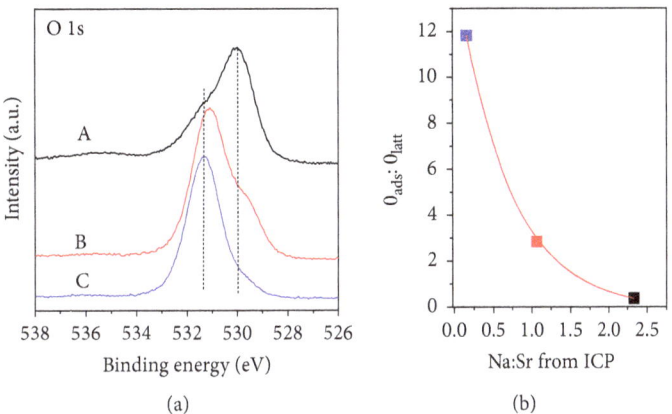

FIGURE 4: (a) O1s XPS lines obtained from samples and (b) the ratio of O_{ads} to O_{latt} change with Na : Sr ratio in the sample.

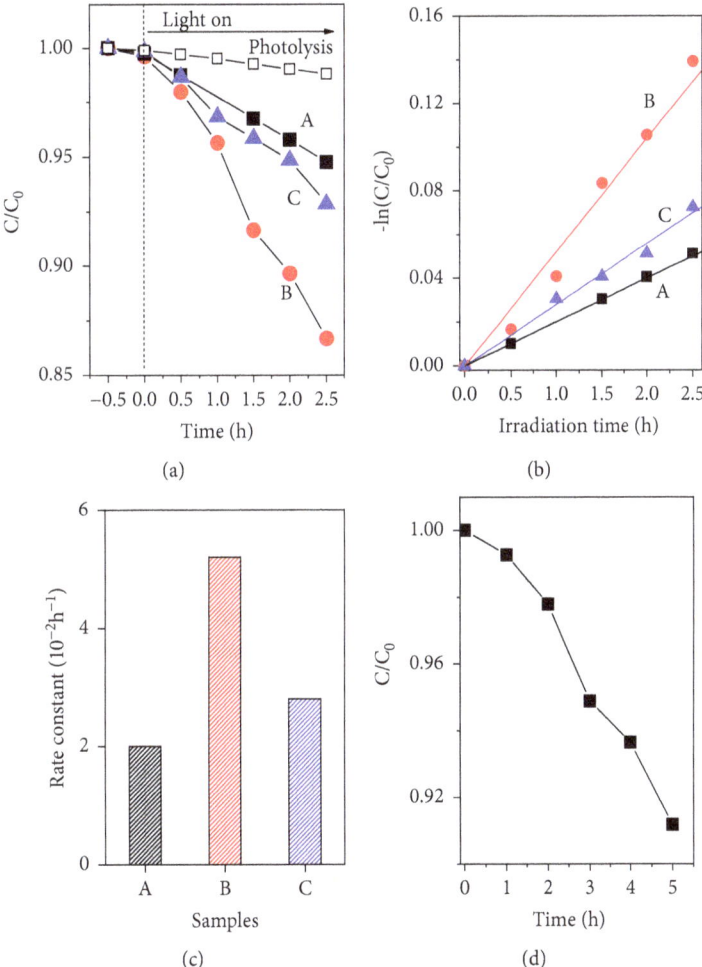

FIGURE 5: (a) RhB concentration varied with reaction time, (b) plot of $-\ln(C/C_0)$ vs. time, (c) the rate constant of all the samples, and (d) dichlorid phenol concentration varied with reaction time over sample B.

oxygen adsorption. The amount of adsorbed oxygen on samples is consistent with the XPS results.

Photocatalytic activity was evaluated by the photocatalytic RhB photodegradation under 300-W Xe lamp. Figure 5(a) shows the RhB concentration varied with reaction time. Obviously, after irradiation, the RhB concentration is decreased.

When the reagent solution is dilute, the reaction rate (r) can be expressed as $r = kC$, where k is the apparent rate constant, and C is the instantaneous concentration of the reactant. The plots of $-\ln(C/C_0)$ vs. time are shown in Figure 5(b), and the rate constants of RhB degradation are also displayed in Figure 5(c). k is 2.0×10^{-2}, 5.2×10^{-2}, and

$2.8 \times 10^{-2} \, h^{-1}$ for samples A, B, and C, respectively. Sample B exhibits 2.6 times higher activity than sample A. Dichlorid phenol (DCP), a colorless toxic pollutant, was used as a model pollutant, as shown in Figure 5(d). The DCP concentration was decreased over the sample B.

We will address the formation mechanism of NbO_4 tetrahedra and how to affect the photocatalytic activity. Usually the Sr_2NbO_4 could be synthesized under vacuum or reduced atmosphere [25]. In this work, we used sodium niobate as one of the starting materials and got $Sr_{2-x}Na_xNbO_4$ under ambient air. When sodium niobate reacts with $SrCO_3$, the Sr^{2+} will substitute Na^+. The Sr atom will give one excess charge and create an equivalent reduced atmosphere, resulting in the change of NbO_6 octahedra to NbO_4 tetrahedra. With increasing the content of Sr in the structure, the lattice parameter will be increased due to larger ionic radius of Sr^{2+} (1.18 Å) than Na^+ (1.02 Å) [16]. More amount of Sr inserted into the structure will cause the unbalance charge in A site (AMOx), leading to more defects generated in the sample, as that observed in sample C. Moreover, we could get the Na-doped Sr_2NbO_4 via the solid state reaction method using pseudoperovskite $NaNbO_3$ as the starting material, indicating that Na is very important to get $Sr_{2-x}Na_xNbO_4$, not the structure of sodium niobate. With increasing the content of NbO_4 in the sample, supported by the results of Raman spectra, the lattice parameter increases, optical band gap becomes larger, and the surface changes to be more active for oxygen adsorption, resulting in a higher photocatalytic activity. Above trends could be obtained from the results of samples A and B, supporting the catalytic property of MO_4 tetrahedron is superior to that of MO_6 octahedron. Due to more defects in sample C, its photocatalytic activity decreases in comparison with sample B.

4. Conclusions

We prepared a series of $Sr_{2-x}Na_xNbO_4$ photocatalytic materials containing NbO_4 tetrahedra by controlling the ratio of $SrCO_3$ to sodium niobate under ambient air. With increasing the content of NbO_4 in the sample, the lattice parameter increases, optical band gap becomes larger, and the surface changes to be more active for oxygen adsorption, resulting in a higher photocatalytic activity. The efficiency of this catalyst can also be due to the photosensibilisation of RhB and should be tested by using another organic molecule as model pollutant.

Conflicts of Interest

The authors declare that they have no conflicts of interest.

Acknowledgments

This work was supported by the Young Core Instructor Foundation from the Education Commission of Henan Province (2015GGJS-021), the Program for Science & Technology Innovation Talents in Universities of Henan Province, China (17HASTIT014), and Henan University funding (0000A40374).

References

[1] J. M. Jehng and I. Wachs, "Structural chemistry and Raman spectra of niobium oxides," *Chemistry of Materials*, vol. 3, no. 1, pp. 100–107, 1991.

[2] K. Nakajima, Y. Baba, R. Noma et al., "$Nb_2O_5 \cdot nH_2O$ as a heterogeneous catalyst with water-tolerant Lewis acid sites," *Journal of the American Chemical Society*, vol. 133, no. 12, pp. 4224–4227, 2011.

[3] M. M. Ahmad, E. S. Yousef, and E. S. Moustafa, "Dielectric properties of the ternary $TeO_2/Nb_2O_5/ZnO$ glasses," *Physica B: Condensed Matter*, vol. 371, no. 1, pp. 74–80, 2006.

[4] H. Chen, F. Zhang, W. Zhang, Y. Du, and G. Li, "Negative impact of surface Ti^{3+} defects on the photocatalytic hydrogen evolution activity of $SrTiO_3$," *Applied Physics Letters*, vol. 112, no. 1, article 013901, 2018.

[5] G. Q. Li, Z. G. Yi, H. T. Wang, C. H. Jia, and W. F. Zhang, "Factors impacted on anisotropic photocatalytic oxidization activity of ZnO: surface band bending, surface free energy and surface conductance," *Applied Catalysis B: Environmental*, vol. 158-159, pp. 280–285, 2014.

[6] D. N. Chirdon and Y. Wu, "Hydrogen evolution: not living on the edge," *Nature Energy*, vol. 2, no. 9, article 17132, 2017.

[7] Y. Liu, J. Wu, K. P. Hackenberg et al., "Self-optimizing, highly surface-active layered metal dichalcogenide catalysts for hydrogen evolution," *Nature Energy*, vol. 2, no. 9, article 17127, 2017.

[8] H. Wang, L. Zhang, Z. Chen et al., "Semiconductor heterojunction photocatalysts: design, construction, and photocatalytic performances," *Chemical Society Reviews*, vol. 43, no. 15, pp. 5234–5244, 2014.

[9] F. E. Osterloh, "Inorganic nanostructures for photoelectrochemical and photocatalytic water splitting," *Chemical Society Reviews*, vol. 42, no. 6, pp. 2294–2320, 2013.

[10] H. Shi, G. Chen, C. Zhang, and Z. Zou, "Polymeric g-C_3N_4 coupled with $NaNbO_3$ nanowires toward enhanced photocatalytic reduction of CO_2 into renewable fuel," *ACS Catalysis*, vol. 4, no. 10, pp. 3637–3643, 2014.

[11] Q. N. Yu, F. Zhang, G. Q. Li, and W. F. Zhang, "Preparation and photocatalytic activity of triangular pyramid $NaNbO_3$," *Applied Catalysis B: Environmental*, vol. 199, pp. 166–169, 2016.

[12] Y. Miseki, H. Kato, and A. Kudo, "Water splitting into H_2 and O_2 over $Cs_2Nb_4O_{11}$ photocatalyst," *Chemistry Letters*, vol. 34, no. 1, pp. 54-55, 2005.

[13] I.-S. Cho, S. Lee, J. H. Noh et al., "$SrNb_2O_6$ nanotubes with enhanced photocatalytic activity," *Journal of Materials Chemistry*, vol. 20, no. 19, article 3979, 2010.

[14] D. Chen and J. H. Ye, "Selective-synthesis of high-performance single-crystalline $Sr_2Nb_2O_7$ nanoribbon and $SrNb_2O_6$ nanorod photocatalysts," *Chemistry of Materials*, vol. 21, no. 11, pp. 2327–2333, 2009.

[15] Y. Miseki, H. Kato, and A. Kudo, "Water splitting into H_2 and O_2 over niobate and titanate photocatalysts with (111) plane-type layered perovskite structure," *Energy & Environmental Science*, vol. 2, no. 3, p. 306, 2009.

[16] G. Q. Li, T. Kako, D. F. Wang, Z. G. Zou, and J. Ye, "Composition dependence of the photophysical and photocatalytic properties of (AgNbO3)$_{1-x}$(NaNbO3)$_x$ solid solutions," *Journal of Solid State Chemistry*, vol. 180, no. 10, pp. 2845–2850, 2007.

[17] P. Li, S. X. Ouyang, G. C. Xi, T. Kako, and J. H. Ye, "The effects of crystal structure and electronic structure on photocatalytic H$_2$ evolution and CO$_2$ reduction over two phases of perovskite-structured NaNbO$_3$," *The Journal of Physical Chemistry C*, vol. 116, no. 14, pp. 7621–7628, 2012.

[18] J. Liu, E. P. Kharitonova, C. G. Duan, W. N. Mei, R. W. Smith, and J. R. Hardy, "Phase transition in single crystal Cs$_2$Nb$_4$O$_{11}$," *The Journal of Chemical Physics*, vol. 122, no. 14, article 144503, 2005.

[19] J. Pellicer-Porres, A. B. Garg, D. Vázquez-Socorro, D. Martínez-García, C. Popescu, and D. Errandonea, "Stability of the fergusonite phase in GdNbO$_4$ by high pressure XRD and Raman experiments," *Journal of Solid State Chemistry*, vol. 251, pp. 14–18, 2017.

[20] H. Zhu, Z. Zheng, X. Gao et al., "Structural evolution in a hydrothermal reaction between Nb$_2$O$_5$ and NaOH solution: - from Nb$_2$O5 grains to microporous Na$_2$Nb$_2$O$_6$·2/3H$_2$O fibers and NaNbO$_3$ cubes," *Journal of the American Chemical Society*, vol. 128, no. 7, pp. 2373–2384, 2006.

[21] H. Xu, M. Nyman, T. M. Nenoff, and A. Navrotsky, "Prototype Sandia octahedral molecular sieve (SOMS) Na$_2$Nb$_2$O$_6$·H$_2$O: - synthesis, structure and thermodynamic stability," *Chemistry of Materials*, vol. 16, no. 10, pp. 2034–2040, 2004.

[22] M. N. Iliev, M. L. F. Phillips, J. K. Meen, T. M. Nenoff, and J. Phys, "Raman spectroscopy of Na$_2$Nb$_2$O$_6$•H$_2$O and Na$_2$Nb$_{2-x}$MxO$_{6-x}$(OH)$_x$•H$_2$O (M = Ti, Hf) ion exchangers," *The Journal of Physical Chemistry B*, vol. 107, no. 51, pp. 14261–14264, 2003.

[23] G. Q. Li, N. Yang, W. L. Wang, and W. F. Zhang, "Synthesis, photophysical and photocatalytic properties of N-doped sodium niobate sensitized by carbon nitride," *Journal of Physical Chemistry C*, vol. 113, no. 33, pp. 14829–14833, 2009.

[24] W. L. Wang, G. Q. Li, N. Yang, and W. F. Zhang, "Composition dependence of surface photoelectric and photodegradation activities of silver antimonates with pyrochlore-like structure," *Materials Chemistry and Physics*, vol. 123, no. 1, pp. 322–325, 2010.

[25] K. Isawa and M. Nagano, "Synthesis and physical properties of niobium-based oxide, Sr$_{2-x}$La$_x$NbO4 (0⩽×<0.2)," *Physica C Superconductivity*, vol. 357–360, pp. 359–362, 2001.

[26] S. Kumar, S. Khanchandani, M. Thirumal, and A. K. Ganguli, "Achieving enhanced visible-light-driven photocatalysis using type-II NaNbO$_3$/CdS core/shell heterostructures," *ACS Applied Materials & Interfaces*, vol. 6, no. 15, pp. 13221–13233, 2014.

[27] F. D. Hardcastle and I. E. Wachs, "Determination of vanadium-oxygen bond distances and bond orders by Raman spectroscopy," *The Journal of Physical Chemistry*, vol. 95, no. 13, pp. 5031–5041, 1991.

[28] H. Tan, Z. Zhao, W. B. Zhu et al., "Oxygen vacancy enhanced photocatalytic activity of pervoskite SrTiO$_3$," *ACS Applied Materials & Interfaces*, vol. 6, no. 21, pp. 19184–19190, 2014.

[29] T. Kanagaraj and S. Thiripuranthagan, "Photocatalytic activities of novel SrTiO$_3$ – BiOBr heterojunction catalysts towards the degradation of reactive dyes," *Applied Catalysis B: Environmental*, vol. 207, pp. 218–232, 2017.

[30] K. Xie, N. Umezawa, N. Zhang, P. Reunchan, Y. Zhang, and J. Ye, "Self-doped SrTiO$_{3-\delta}$ photocatalyst with enhanced activity for artificial photosynthesis under visible light," *Energy & Environmental Science*, vol. 4, no. 10, article 4211, 2011.

[31] W. Yu, G. Ou, W. Si, L. Qi, and H. Wu, "Defective SrTiO$_3$ synthesized by arc-melting," *Chemical Communications*, vol. 51, no. 86, pp. 15685–15688, 2015.

20

Photocatalytic Activity under Simulated Sunlight of Bi-Modified TiO$_2$ Thin Films Obtained by Sol Gel

D. A. Solís-Casados ⓘ,[1] L. Escobar-Alarcón,[2] V. Alvarado-Pérez,[1] and E. Haro-Poniatowski[3]

[1]*Universidad Autónoma del Estado de México, Centro Conjunto de Investigación en Química Sustentable UAEMéx-UNAM, Toluca, MEX, Mexico*
[2]*Departamento de Física, Instituto Nacional de Investigaciones Nucleares, Apartado Postal 18-1027, 11801 Mexico City, Mexico*
[3]*Departamento de Física, Universidad Autónoma Metropolitana, Apartado Postal 55-532, 09340 Mexico City, Mexico*

Correspondence should be addressed to D. A. Solís-Casados; solis_casados@yahoo.com.mx

Academic Editor: Juan Rodriguez

The synthesis of Bi-modified TiO$_2$ thin films, with different Bi contents, is reported. The obtained materials were characterized by energy-dispersive X-ray spectroscopy (EDS), X-ray photoelectron spectroscopy (XPS), Raman spectroscopy (RS), X-ray diffraction (XRD), photoluminescence (PL), and diffuse reflectance spectroscopy (DRS), in order to obtain information on their chemical composition, vibrational features, and optical properties, respectively. Compositional characterization reveals that the bismuth content can be varied in an easy way from 0.5 to 25.4 at. %. Raman results show that the starting material corresponds to the anatase phase of crystalline TiO$_2$, and Bi addition promotes the formation of bismuth titanates, Bi$_2$Ti$_2$O$_7$ at Bi contents of 10.4 at. % and the Bi$_4$Ti$_3$O$_{12}$ at Bi contents of 21.5 and 25.4 at. %. Optical measurements reveal that the band gap narrows from 3.3 eV to values as low as 2.7 eV. The photocatalytic activity was tested in the degradation reaction of the Malachite Green carbinol base dye (MG) as a model molecule under simulated sunlight, where the most relevant result is that photocatalytic formulations containing bismuth showed higher catalytic activity than pure TiO$_2$. The higher photocatalytic activity of MG degradation of 67% reached by the photocatalytic formulation of 21.5 at. % of bismuth is attributed to the presence of the crystalline phase perovskite-type bismuth titanate, Bi$_4$Ti$_3$O$_{12}$.

1. Introduction

Pollution in wastewaters is one of the most important environmental topics nowadays due to the increasing necessity of human beings of clean water. Some dyes in wastewaters are considered pollutants and, in most cases, are considered toxic to humans and other living organisms, even when they are present in low quantities. Several processes have been proposed to remove or degrade these pollutants from wastewaters; particularly, the photocatalysis is currently considered a promising alternative to remove dyes from water in an efficient way. Photocatalysis is an advanced oxidation technology (AOT), based on physicochemical processes that

produce changes in the chemical structure of the organic compounds including their mineralization. AOT processes are based on the generation and the use of highly reactive oxidizing species, such as the hydroxyl (OH$^\bullet$), hydroperoxyl ($^\bullet$OOH), and superoxide radicals (O$_2{}^\bullet$), which are reactive sites towards degradation of organic compounds until their complete mineralization [1]. Among the photocatalytic materials, titanium dioxide (TiO$_2$) has been the most used because of its specific properties, such as resistance to chemical corrosion, nontoxic, and inexpensive, and its high photoactivity with UV radiation. However, TiO$_2$ has two important drawbacks; the first one is that it is activated only by ultraviolet light, due to its relatively high band gap energy of 3.2 eV for the anatase crystalline phase. The second one is the high recombination rate of the photogenerated electron-hole pairs that reduce its efficiency [2]. Several strategies have been proposed to improve the photocatalytic activity of TiO_2 such as doping it with metals and nonmetals, as well as the use of mixtures of the two main TiO_2 crystalline phases, anatase and rutile [3, 4]. Coupling of semiconductors has been also considered as an alternative route to develop high efficient photocatalytic materials that can compensate the disadvantages of the individual components inducing synergistic effects such as efficient charge separation, band gap narrowing, and consequently improvement of their photocatalytic performance. Therefore, the development of visible light-driven coupled photocatalysts is currently of great interest. Particularly, bismuth oxide (α-Bi_2O_3) has been reported as an efficient photocatalyst due to its unique structure and band gap energy close to 2.8 eV which makes it active in the visible region of the electromagnetic spectrum [5, 6]. Several studies have reported that the system formed by mixtures of TiO_2 and Bi_2O_3 can result in the formation of different crystalline phases of bismuth titanates ($Bi_4Ti_3O_{12}$ and $Bi_{12}TiO_{20}$) depending on the proportion of TiO_2 and Bi_2O_3 [7, 8]. It is worth mentioning that the photocatalysts based on $Bi_{12}TiO_{20}$ [9] and $Bi_4Ti_3O_{12}$ [10] crystals have been tested in the photodecolorization of methyl orange under UV irradiation showing high photocatalytic activity, similar in both cases. The aim of this work is to investigate the degradation of the Malachite Green dye (MG), carbinol base, under visible light irradiation, using a simulated sunlight source, in an attempt to correlate the photocatalytic activity of titanium oxide modified with different amounts of Bi_2O_3 with their physicochemical properties in particular; the band gap energy, the microstructure, and the electron-hole recombination rate.

2. Experimental

2.1. Bi-Modified TiO_2 Thin Films. The precursor solutions were prepared by the sol-gel technique. Titanium isopropoxide (Ti[OCH(CH_3)_2]_4, Aldrich 97%), nitric acid (HNO_3, Fermont 70%), 2-propanol (CH_3CHOHCH_3, Fermont 99.8%), bismuth nitrate pentahydrate (Bi(NO_3)_3·5H_2O, J.T. Baker) were used as precursors. A sol was prepared under environmental conditions mixing 10 mL of 2-propanol with 1 mL of titanium isopropoxide, stirred for 1 h. Bismuth nitrate pentahydrate was added slowly under stirring to obtain theoretical amounts of 0, 5, 30, 50, 70, and 80 wt. % of Bi_2O_3. 1 mL of nitric acid was added drop to drop to induce gelling as a variant of the sol-gel technique reported before [11]. The sol was sonicated in an ultrasonic bath during 5 min and aged for 12 h to obtain an incipient gelled solution. Afterwards, the precursor solution was deposited layer to layer by the spin coating technique onto borosilicate glass substrates (25 mm × 25 mm × 1 mm) functionalized with hydrofluoric acid (HF, 10% vol) to obtain homogeneous thin films. The spin coater was a KW-4A from Chemat Technology working at 1500 rpm under environmental conditions. Deposited thin films were thermally treated at 300°C during 1 h to eliminate organic residues, and

subsequently, the temperature was raised to 450°C at a heating rate of 3°C/min and maintained isothermally for 4 h to form a crystalline thin film.

2.2. Thin Film Characterization. Determination of the atomic bismuth content in the films was done by energy-dispersive X-ray spectroscopy (EDS) using a microprobe attached to a JEOL JSM 6510LV scanning electron microscope, and EDS analysis was carried out with an acceleration voltage of 15 kV; surface morphology was observed from micrographs obtained with the same microscope. The chemical bonding of the present elements was investigated by X-ray photoelectron spectroscopy (XPS). The wide and narrow XPS spectra were acquired using a JEOL JPS-9200 spectrometer. The adventitious carbon peak at 284.8 eV (1s) was used as the internal standard to compensate for sample charging. Raman spectroscopy (RS) was used to study the structural features of the films; spectra were acquired using an HR LabRam 800 spectrometer with an Olympus BX40 confocal microscope. A Nd:YAG laser beam (532 nm) was focused with a 50x objective onto the sample surface. A cooled CCD camera was used to record the spectra, and typically, an average of 50 accumulations of 10 seconds was done to improve the signal-to-noise ratio. The crystalline phases of the thin films were identified by the X-ray diffraction technique (XRD) with a Bruker D8 Advance Diffractometer using the Cu-Kα radiation line ($\lambda = 1.54$ Å); the diffraction patterns were recorded in steps of 0.05°. Diffuse reflectance spectroscopy (DRS) spectra were acquired on a PerkinElmer Lambda 35 spectrophotometer with an integration sphere with a resolution of ±1 nm; from the reflectance spectra, the Kubelka-Munk function was determined and the band gap energy was estimated [12, 13]. Photoluminescence spectra were acquired in a FluoroMax4, HORIBA Jobin Yvon spectrofluorometer, exciting the samples at 492 nm.

2.3. Photocatalytic Activity. The photocatalytic activity of the thin films was tested through the degradation of Malachite Green carbinol base dye ($C_{23}H_{26}N_2O$, Aldrich) contained in an aqueous solution (10 μmol/L). The reaction was carried out in a batch system in a borosilicate glass reactor, in which the thin film was introduced into a 25 mL of the MG solution; afterwards, the reaction system was stirred in dark condition in order to establish adsorption equilibrium between dye solution and photocatalyst. Thin films were activated by illumination with light emitted from a solar simulator SF150 of Sciencetech equipped with an AM1.0D filter which simulates the solar spectrum of direct light from the sun on the ground when the sun is at a zenith angle of 0°; the samples were irradiated with an average intensity of 60 mW/cm^2 keeping the distance between the liquid surface and the light source at 15 cm. The MG photodegradation was followed through the decrease of its characteristic absorption band peaking at 619 nm in the UV-Vis absorbance spectra. The spectra were obtained each 15 min in the first hour of reaction time, and afterwards, each 30 min during the second and third hours of reaction from the aliquots of 4 mL taken from reaction system and were returned to the reactor after each spectrum was taken. Absorbances obtained at each reaction time were

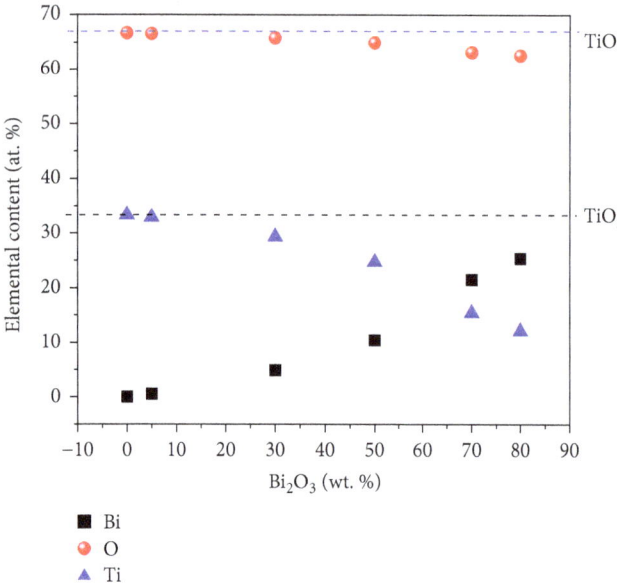

FIGURE 1: Atomic proportion of the thin films as a function of the Bi$_2$O$_3$ wt. %.

TABLE 1: Elemental chemical composition, determined by EDS, of the thin films as a function of the theoretical Bi$_2$O$_3$ (wt. %).

Chemical composition (at. %)			
Bi$_2$O$_3$ (wt. %)	EDS		
	Ti	O	Bi
0	33.3	66.7	0
5	32.9	66.6	0.5
30	29.3	65.8	4.9
50	24.7	64.9	10.4
70	15.4	63.1	21.5
80	12.1	62.5	25.4

correlated to dye concentrations through a calibration curve. A nonlinear least square data treatment was used to determine the values of the kinetic constant considering a pseudo-first order kinetic model. Total organic carbon (TOC) was determined in each solution after reaction as ppm of carbon, TOC was obtained by the combustion method, and the mineralization degree was calculated taking the TOC content in the MG solution, at the initial concentration of 10 μmol/L, 9.2 ppm of TOC as a reference.

3. Results and Discussion

3.1. Elemental Composition. The elemental chemical composition of the deposited thin films, as a function of the Bi$_2$O$_3$ wt. % used for their preparation, is shown in Figure 1 and Table 1 (EDS measurements). The results obtained from the EDS and XPS techniques showed good agreement following the same tendency. It is clearly observed that the film without Bi is almost stoichiometric TiO$_2$. When Bi is incorporated increasing the Bi$_2$O$_3$ load, the Bi content in the film increases monotonically from 0.5 to 25.4 at. % whereas the O content remains around 65 at. %. Simultaneously, the Ti content decreases from 33 to 12 at. %. It is worth noting that the elemental composition of the sample prepared using 70 wt. % of Bi$_2$O$_3$ (Bi = 21.5 at. %, Ti = 15.4 at. %, and O = 63.1 at. %) agrees very well with the composition of the Bi$_4$Ti$_3$O$_{12}$ titanate (Bi = 21.0 at. %, Ti = 15.8 at. %, and O = 63.2 at. %).

3.2. Raman Spectroscopy. Figure 2 shows the Raman spectra corresponding to the thin films containing different bismuth contents, from 0.0 to 25.4 at. %. For Bi contents lower than 4.9 at. %, the spectra consist of four characteristic bands located at 144, 396, 516, and 637 cm^{-1} attributed to the anatase crystalline phase of TiO$_2$ [14]. The inset in Figure 2

reveals that as the Bi content increases, the band at 144 cm^{-1} shifts to higher frequencies and its intensity diminishes, and at the same time, its FWHM becomes wider. These changes are attributed to structural disorder induced by the incorporation of bismuth into the TiO$_2$ lattice. When the Bi content reaches 10.4 at. %, low intensity peaks associated to the anatase crystalline phase remain and new features appear at 83, 105, 256, 401, and 530 cm^{-1}, these signals that appear shifted and broader can be assigned to the bismuth titanate Bi$_4$Ti$_3$O$_{12}$ [9]. These spectra resemble an amorphous material indicating a high degree of structural disorder. At the highest Bi content, the Raman spectrum is characterized by signals at 235, 276, 350, 540, 615, and 856 cm^{-1}; the presence of the Raman modes at 276, 540, and 856 cm^{-1} suggests the presence of the perovskite structure [9].

3.3. X-Ray Diffraction. Figure 3 shows the X-ray diffraction patterns from 20 to 40° of the samples with different Bi content. From Figure 3(a), the diffraction lines at 2θ = 25.3 and 37.8° characteristic of the TiO$_2$ in its anatase crystalline phase can be observed (JCPDS 89-4921). Small features of the anatase crystalline phase are observed as well (Figure 3(b)), in thin film containing 0.5 at. % of Bi. The film with bismuth content of 4.9 at. % shows diffraction lines at 2θ = 14.9, 28.7, 29.96, 32.3, 34.74, and 38.01° which are attributed to the bismuth titanate Bi$_2$Ti$_2$O$_7$ (JCPDS 32-0118) (Figure 3(c)). The film containing 10.4 at. % of Bi exhibits the same diffraction lines as is seen in Figure 3(d). Further increase in bismuth content up to 21.5 at. % shows new diffraction lines peaking at 2θ = 22.1, 23.3, 27.2, 29.46, 30.1, and 33.1° (Figure 3(e)), characteristics of the crystalline phase of bismuth titanate with molecular structure Bi$_4$Ti$_3$O$_{12}$ (JCPDS 35-0795) as is expected because of its elemental composition. From Figure 3(f), it can be observed that crystalline phase Bi$_4$Ti$_3$O$_{12}$ remains in the thin film with 25.4 at. % of bismuth content in good agreement with the Raman results.

3.4. X-Ray Photoelectron Spectroscopy. Shifts in the peak positions in photoelectron spectra are frequently used to determine the chemical state of elements if the shifts are large enough. Sometimes, peaks representing different chemical states are overlapped and a deconvolution procedure is required. Figure 4 shows the chemical shifts of the Ti 2p region in the XPS spectra of the Bi-modified TiO$_2$ thin films. Spectra were deconvoluted using Gaussian functions in order

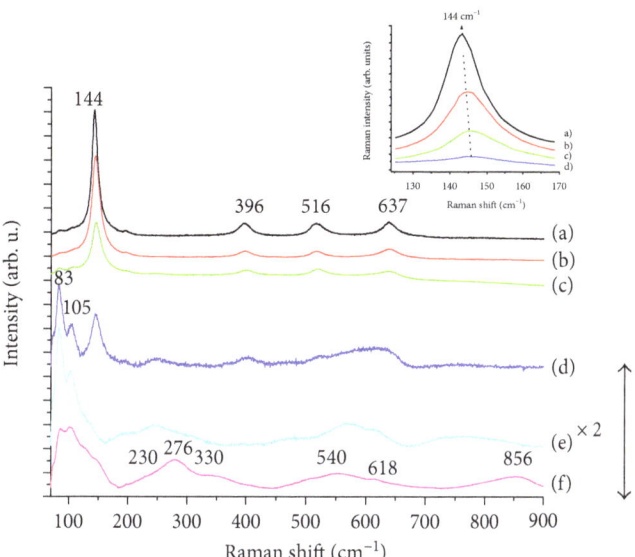

FIGURE 2: Raman spectra of the TiO_2 thin films modified with several bismuth contents of (a) 0, (b) 0.5, (c) 4.9, (d) 10.4, (e) 21.5, and (f) 25.4 at. %.

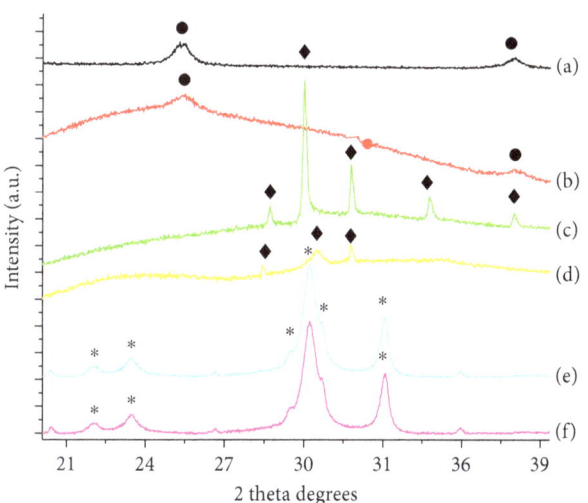

* 35-0795 $Bi_4Ti_3O_{12}$
♦ 32-0118 $Bi_2Ti_2O_7$
● 89-4921 TiO_2 anatase

FIGURE 3: X-ray patterns of the TiO_2 thin films modified with several bismuth contents (a) 0, (b) 0.5, (c) 4.9, (d) 10.4, (e) 21.5, and (f) 25.4 at. %.

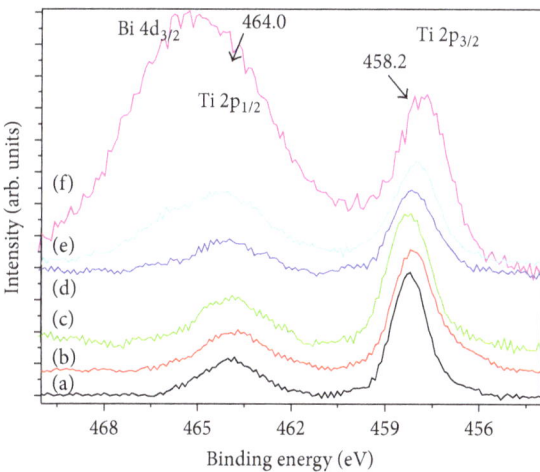

FIGURE 4: Ti 2p region spectra evolution as a function of bismuth content (a) 0, (b) 0.5, (c) 4.9, (d) 10.4, (e) 21.5, and (f) 25.4 at. %.

to obtain information about the interaction of the Ti atoms with the Bi and O atoms in the last atomic layer in the thin films as well as its chemical state (Figure 5). In Figure 5(a), two doublets can be observed; the first one with peaks located at 458.2 and 464.0 eV is attributed to the doublet of Ti-O bonds of the TiO_2 in its anatase phase, whereas the peaks at 456.7 and 462.9 eV could be attributed to a second doublet of Ti-O bonds in Ti_2O_3 (Figure 5(a)) [15]. The presence of Ti^{3+} due to oxygen vacancies has been reported before, resulting from the transfer of two electrons towards two adjacent Ti^{4+} to form Ti^{3+} on the surface. The spectrum of the Ti region in the sample with 10.4 at. % of bismuth (Figure 5(b)

shows peaks attributed to three main doublets: the first one, located at 456.7 and 462.3 eV, is attributed to the Ti-O bonds as in Ti_2O_3; the second one, at 457.9 and 463.9 eV, has a closed binding energy than the one reported by Wang and Ma for the compounds similar to bismuth titanates, so this peaks are attributed to the Ti-Bi-O bonds in the bismuth titanate [16]; finally, the third one, with peaks located at 458.3 and 464.1 eV, could be attributed to the Ti-O bonds in the anatase phase of TiO_2. Spectrum of the thin film with bismuth content of 25.4 at. % (Figure 5(c)) shows two doublets: the first doublet at 456.7 and 462.5 eV attributed to the Ti-O bonds in the Ti_2O_3 and the second doublet at 457.9 and 464.7 eV revealing the presence of Ti-Bi-O bonds as in the bismuth titanates $Bi_4Ti_3O_{12}$. Ti $2p_{1/2}$ photoemission is overlapped in a partial way by the Bi $4d_{3/2}$ core level peak [15, 16].

Figure 6 shows the Bi 4f photoelectron spectra of the thin films displaying a characteristic doublet located at

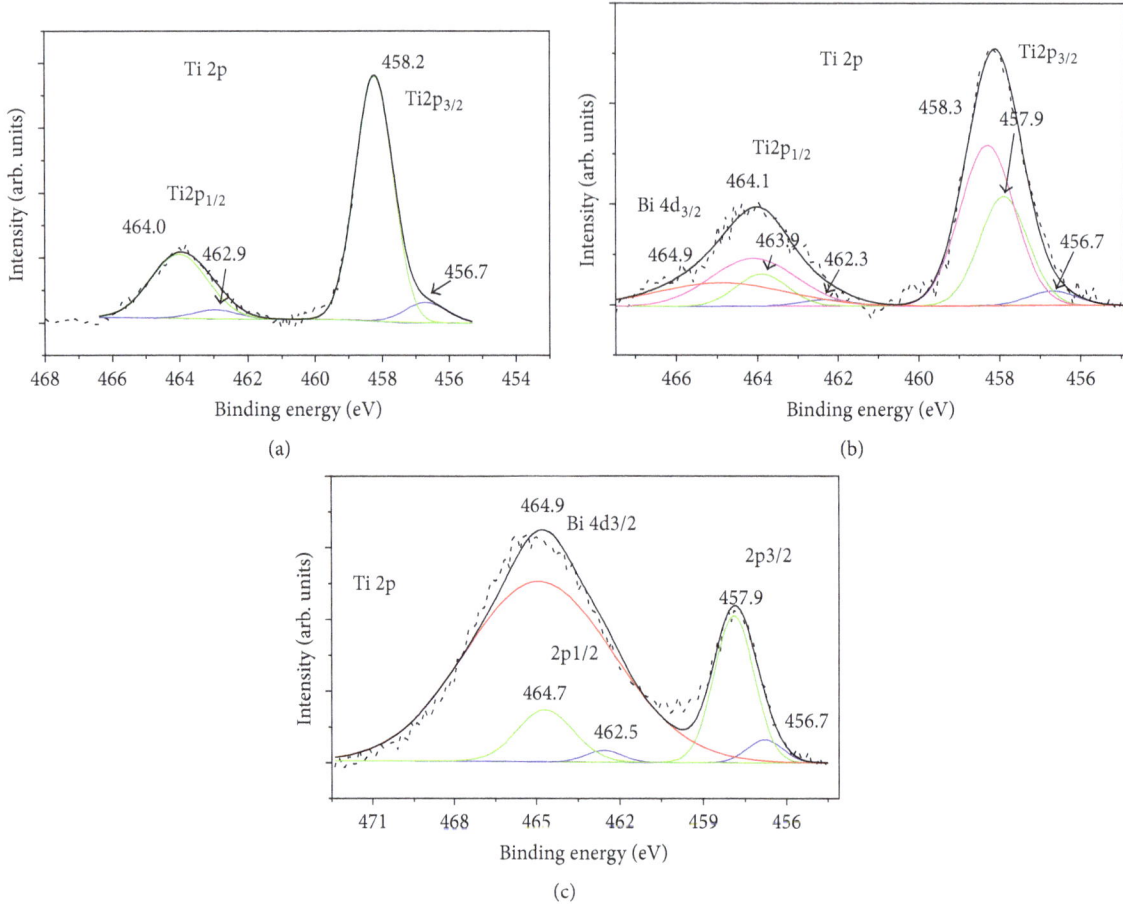

FIGURE 5: Gaussian deconvolution of the XPS spectra, Ti 2p region (a) 0 at. %, (b) 10.4 at. %, and (c) 25.4 at. % of bismuth.

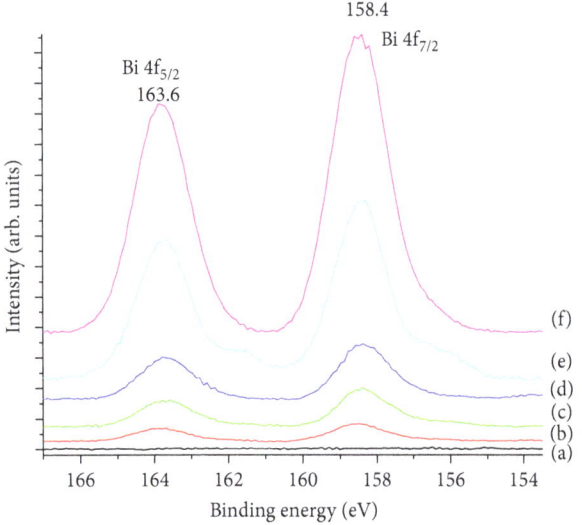

FIGURE 6: XPS spectra evolution of Bi 4f region as a function of bismuth content, (a) 0, (b) 0.5, (c) 4.9, (d) 10.4, (e) 21.5, and (f) 25.4 at. %.

158.4 and 163.6 eV. These signals were deconvoluted fitting two doublets: one of them located at 158.5 and 163.9 eV could be correlated to the Bi-Ti-O bonds in the layer $Bi_2Ti_3O_{10}{}^{2-}$ of the perovskite-type structure $Bi_4Ti_3O_{12}$

(Figure 7(a)), and the other doublet with peaks located at 156.8 and 161.9 eV could be assigned to the Bi-O bonds most probably as in the $(Bi_2O_2)^{2+}$ layer of the perovskite-type structure. These results suggest that the incorporation and further increase of Bismuth into the film changes the proportion of Bi-O bonds most probably in the $(Bi_2O_2)^{2+}$ layer, forming the perovskite structure at 21.5 at. % of Bi (Figure 7(b)). XPS spectrum of the sample with the highest bismuth content (25.4 at. %) increases the proportion of the doublet located at 158.4 and 163.8 eV correlated with the Bi-Ti-O bonds in the $Bi_4Ti_3O_{12}$, and the other doublet with peaks located at 156.6 and 162.4 eV assigned to the Bi-O bonds in the $(Bi_2O_2)^{2+}$ layer decreases probably due to starting of the Bi_2O_3 formation (Figure 7(c)). The line shape of the O 1s core level photoemission spectra of the thin films is shown in Figures 8(a)–8(f), and it is important to remark that two peaks are clearly observed. The first peak located at the low binding energy 529.8 eV can be assigned to the Ti-O bond while the second peak located at 532.5 eV can be attributed to the oxygen bonded to bismuth [17, 18]. Jovalekić et al. have reported that oxygen atom in a stronger Ti-O bond carries a higher effective negative charge than in a weaker Bi-O bond [17]. This correlation between the effective charge of oxygen atom and the binding energy of oxygen core electrons agrees with previous reports [17, 18].

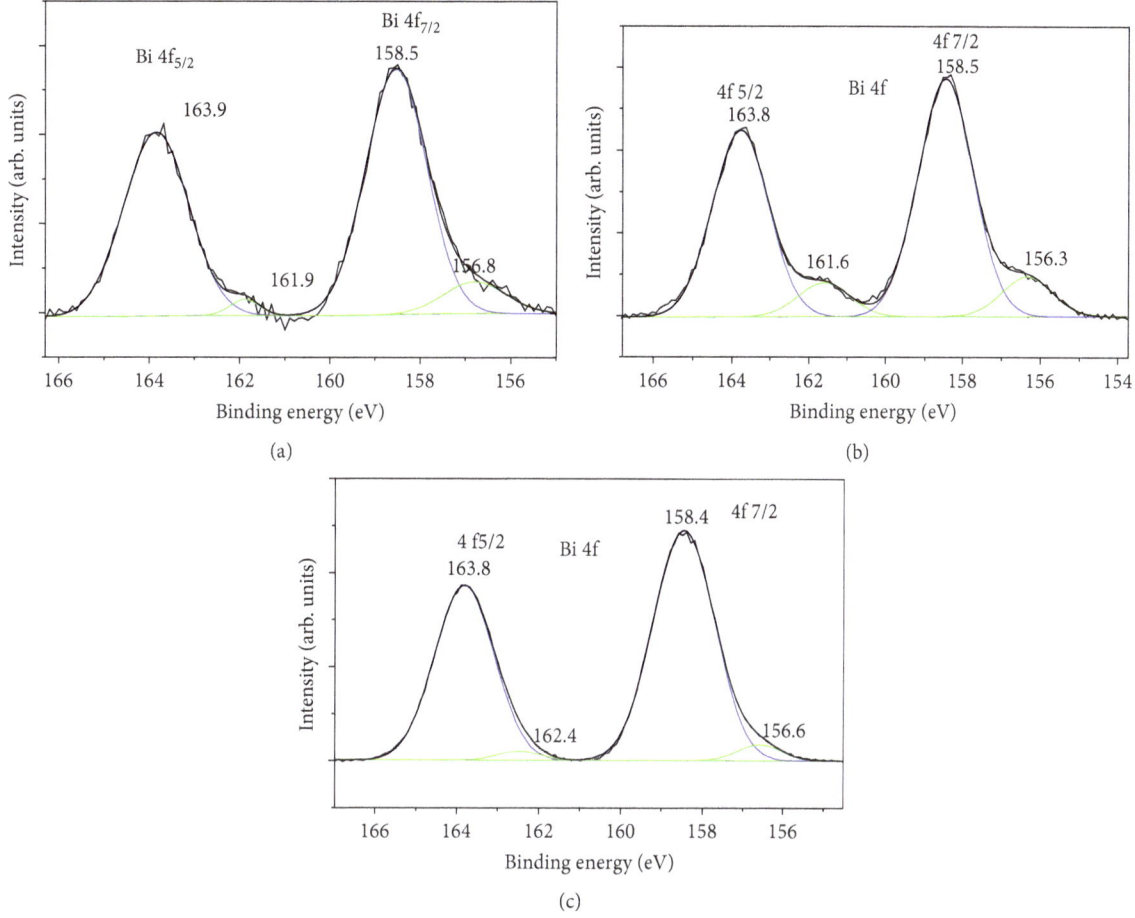

FIGURE 7: Gaussian deconvolution of the XPS spectra, Bi 4f7/2 region (a) 0.5, (b) 21.5, and (c) 25.4 at. % of bismuth.

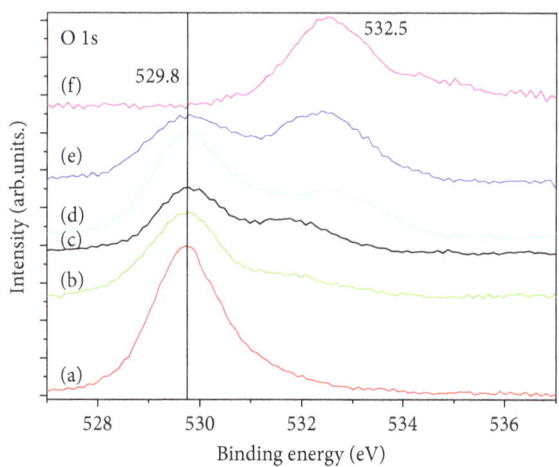

FIGURE 8: XPS spectra evolution of O 1s region as a function of bismuth content, (a) 0, (b) 0.5, (c) 4.9, (d) 10.4, (e) 21.5, and (f) 25.4 at. %.

TABLE 2: Effect of bismuth content on the band gap energy.

Bismuth content (at. %)	Band gap energy (eV)
0	3.3
0.5	3.3
4.9	3.2
10.4	3.1
21.5	2.9
25.4	2.7

3.5. Diffuse Reflectance Spectroscopy. Table 2 shows the optical band gap (Eg) values determined using the Kubelka-Munk method; this was done by transforming the reflectance spectra of the samples with different Bi contents to the Kubelka-Munk function, F(R), and then plotting $(F(R) E)^{1/2}$ versus E, considering a direct allowed transition band gap. The Eg values were obtained by a linear fit of the linear portions of the curve, determining its intersection with the photon energy axis [12]. The reflectance spectra of the samples and the $(F(R) E)^{1/2}$ versus E graphs as well as linear fits for estimating the band gap energy are shown in Figure 9. In these cases, the employed method allows the determination of the band gap values with good accuracy [13]. As it is observed in Table 2, the obtained results reveal that when Bi content increases in the thin films, the band gap narrows from 3.3 eV to values as low as 2.7 eV ($\lambda = 459$ nm); these low band gap values make this material potentially active under light illumination in the visible region of the electromagnetic spectrum.

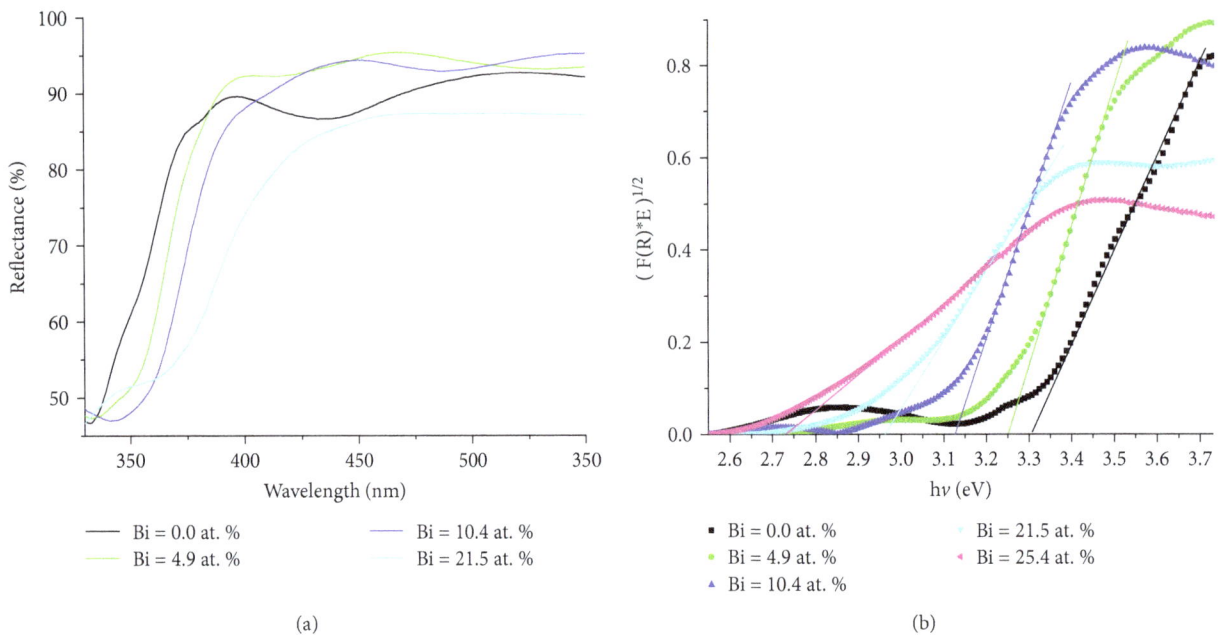

FIGURE 9: (a) Reflectance spectra and (b) $(F(R) E)^{1/2}$ versus E graphs of the prepared films.

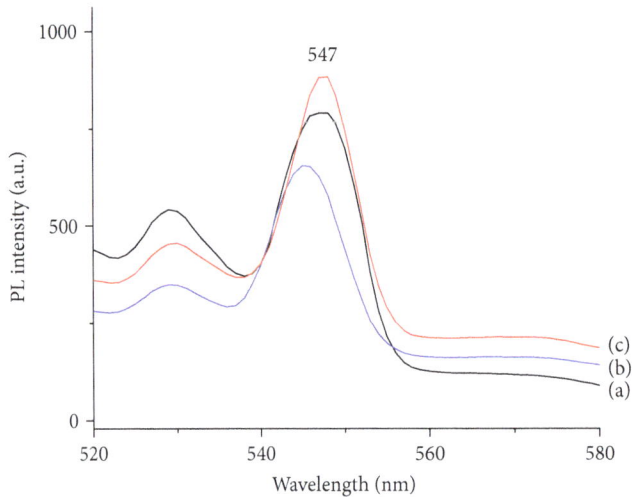

FIGURE 10: Photoluminescence spectra of (a) 0, (b) 21.5, and (c) 25.4 at. % of bismuth.

3.6. Photoluminescence Spectra.

Figure 10 shows the photoluminescence spectra of the TiO_2 and the Bi-modified TiO_2 thin films. The PL spectrum of the TiO_2 is characterized by an intense band peaking at 547 nm. This can be interpreted as a higher electron hole recombination rate in the TiO_2 thin film as the PL intensity is related directly to the electron-hole recombination rate [19]. The same band with similar intensity is seen for the photocatalyst with 25.4 at. % of Bi. The PL intensity of the spectrum corresponding to the photocatalyst with 21.5 at. % of Bi is approximately 20% lower than the PL intensity determined as the area under the curve of the TiO_2 film, indicating that this sample exhibits the lowest recombination rate.

3.7. Photocatalytic Activity.

To determine the photocatalytic activity, a photodegradation experiment of MG under simulated solar light was performed. Figure 11 shows the MG degradation degree as a function of the reaction time using thin films with different bismuth content. The MG degradation due to the photolysis process was close to 27% after 180 minutes of irradiation, and it was the lowest degradation degree as is seen in Figure 11. The TiO_2 film without bismuth behaves similarly to the photolysis process reaching a slightly higher conversion close to 34.7%. The film containing 0.5 at. % of bismuth shows a higher photocatalytic activity, of approximately 10% greater than the TiO_2 catalyst. Further increase in the Bi content, 4.9 at. %, improves in 34% the degradation degree. For a Bi content of 10.4 at. %, a decrease in the photoactivity is observed. The film with a bismuth content of 21.5 at. % improves significantly the photocatalytic activity reaching 64.6% of MG degradation after 180 min of irradiation time, 84% higher than the activity of the TiO_2

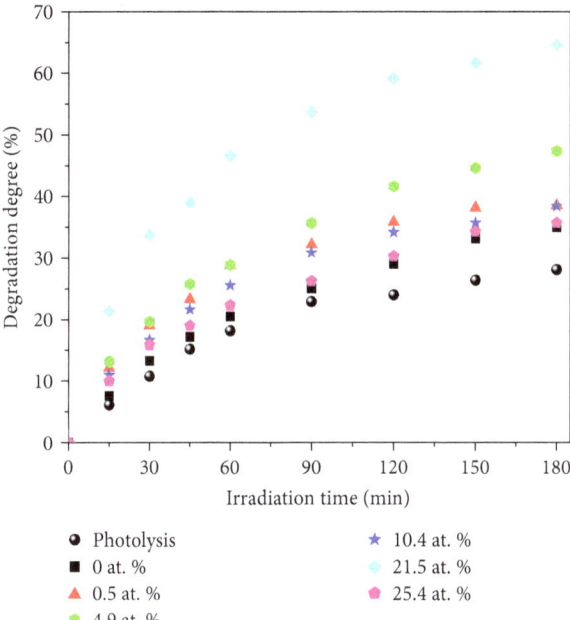

FIGURE 11: Photodegradation percent of MG dye using the Bi-modified TiO$_2$ thin films under solar simulated irradiation.

TABLE 3: Kinetic rate constant (k_{app}) for photocatalytic formulations as a function of bismuth content, determined using a least square data treatment [20].

Bismuth content (at. %)	k_{app} (min^{-1})
Uncatalyzed	0.0029 ± 0.00006
0	0.0038 ± 0.00009
0.5	0.0038 ± 0.00009
4.9	0.0048 ± 0.00007
10.4	0.0038 ± 0.00008
21.5	0.0063 ± 0.00009
25.4	0.0036 ± 0.00006

TABLE 4: Mineralization degree through TOC and photodegradation percent of Malachite Green dye through UV-VIS at 180 min of irradiation time, using a solar simulator as irradiation source.

Bismuth content (at. %)	TOC (ppm)	Mineralization degree (TOC)	Degradation degree (UV-Vis)
Reference	9.2	0	0
Uncatalyzed	8.4	9.1	28.1
0	5.9	35.8	34.9
0.5	6.3	31.7	38.5
4.9	5.0	45.2	47.4
10.4	5.7	38.2	38.4
21.5	4.1	55.8	64.6
25.4	5.8	37.5	35.6

film. This enhanced photocatalytic activity could be attributed to the presence of Bi$_4$Ti$_3$O$_{12}$ as the main crystalline phase in the photocatalytic formulation as well as to the lower recombination rate of the electron-hole pairs as was suggested by the PL results. In fact, the sample with the highest Bi content diminishes the degradation degree to values nearly to the obtained using TiO$_2$. In general terms, thin films containing bismuth exhibit a better photocatalytic activity than pure TiO$_2$ thin film.

Table 3 shows the values of the kinetic constant, k_{app} (min^{-1}), obtained from a fitting of the concentration as a function of the reaction time assuming a pseudo-first order expression using a least square data treatment with an acceptable precision [20]. These values agree with the photocatalytic degradation degree reached for each thin film. Additionally, the mineralization degree was followed by the quantification of the total organic carbon (TOC) through the reaction time. The degradation degrees determined by TOC and UV-Vis measurements are quite similar as it can be seen from Table 4. This indicates that the photodegradation process follows the route of mineralization of the organic dye tested. Further studies with reactive trap molecules indicate that mineralization of MG is mainly through the electron route, specifically by the O$_2^{\bullet}$ superoxide radicals.

4. Conclusions

Bi-modified TiO$_2$ thin films were prepared by the sol-gel and spin coating techniques. In this way, films with Bi contents from 0.5 to 25.4 at. % were obtained. The sol-gel technique induces the Bi incorporation into the TiO$_2$ lattice with the consequent formation of bismuth titanate, Bi$_4$Ti$_3$O$_{12}$, at higher Bi contents. Additionally, the Bi content in the thin films has a strong effect on the band gap energy which decreases from 3.3 eV to values as low as 2.7 eV making these materials potentially photoactive under solar radiation. This is due to the higher wavelength required to generate the electron-hole pairs increasing the absorption spectral window of these materials. However, this lower band gap energy can be responsible for a higher recombination rate of the photogenerated charge carriers. The improved photocatalytic activity leading to the MG degradation under simulated solar light can be attributed to the Bi$_4$Ti$_3$O$_{12}$ phase at a bismuth content of 21.5 at. %.

Additional Points

Highlights. The perovskite-type structure of bismuth titanate was obtained. The incorporation of Bi makes photocatalysts active with solar radiation. Bismuth titanate, Bi$_4$Ti$_3$O$_{12}$, enhances the photocatalytic activity.

Conflicts of Interest

The authors declare that there is no conflict of interests regarding the publication of this paper.

Acknowledgments

The authors thank CONACYT for the provided equipment through the CB-168827 and CB-240998 project. The authors would like to thank Dr. Uvaldo Hernández Balderas, M. en C. Alejandra Nuñez, Dra. Melina Tapia, M. en C. Lizbeth Triana, and LIA Citlalit Martinez Soto for their technical assistance.

References

[1] M. Yasmina, K. Mourad, S. H. Mohammed, and C. Khaoula, "Treatment heterogeneous photocatalysis; factors influencing the photocatalytic degradation by TiO_2," *Energy Procedia*, vol. 50, pp. 559–566, 2014.

[2] S. Bagwasi, Y. Niu, M. Nasir, B. Tian, and J. Zhang, "The study of visible light active bismuth modified nitrogen doped titanium dioxide photocatlysts: role of bismuth," *Applied Surface Science*, vol. 264, pp. 139–147, 2013.

[3] M. Pelaez, N. T. Nolan, S. C. Pillai et al., "A review on the visible light active titanium dioxide photocatalysts for environmental applications," *Applied Catalysis B: Environmental*, vol. 125, pp. 331–349, 2012.

[4] D. A. Solis-Casados, L. Escobar-Alarcón, L. M. Gómez-Oliván, E. Haro-Poniatowski, and T. Klimova, "Photodegradation of pharmaceutical drugs using Sn-modified TiO_2 powders under visible light irradiation," *Fuel*, vol. 198, pp. 3–10, 2017.

[5] II. Cheng, B. Huang, J. Lu et al., "Synergistic effect of crystal and electronic structures on the visible-light-driven photocatalytic performances of Bi_2O_3 polymorphs," *Physical Chemistry Chemical Physics*, vol. 12, no. 47, pp. 15468–15475, 2010.

[6] W. Raza, M. M. Haque, M. Muneer, T. Harada, and M. Matsumara, "Synthesis, characterization and photocatalytic performance of visible light induced bismuth oxide nanoparticle," *Journal of Alloys and Compounds*, vol. 648, pp. 641–650, 2015.

[7] H. Zhang, M. Lü, S. Liu et al., "Preparation and photocatalytic properties of sillenite $Bi_{12}TiO_{20}$ films," *Surface and Coatings Technology*, vol. 202, no. 20, pp. 4930–4934, 2008.

[8] X. Lin, P. Lv, Q. Guan, H. Li, H. Zhai, and C. Liu, "Bismuth titanate microspheres: directed synthesis and their visible light photocatalytic activity," *Applied Surface Science*, vol. 258, no. 18, pp. 7146–7153, 2012.

[9] C. Du, D. Li, Q. He et al., "Design and simple synthesis of composite $Bi_{12}TiO_{20}/Bi_4Ti_3O_{12}$ with a good photocatalytic quantum efficiency and high production of photo-generated hydroxyl radicals," *Physical Chemistry Chemical Physics*, vol. 18, no. 38, pp. 26530–26538, 2016.

[10] R. A. Golda, A. Marikani, and D. P. Padiyan, "Mechanical synthesis and characterization of $Bi_4Ti_3O_{12}$ nanopowders," *Ceramics International*, vol. 37, no. 8, pp. 3731–3735, 2011.

[11] D. A. Solís-Casados, L. Escobar-Alarcón, A. Arrieta-Castañeda, and E. Haro-Poniatowski, "Bismuth–titanium oxide nanopowders prepared by sol–gel method for photocatalytic applications," *Materials Chemistry and Physics*, vol. 172, pp. 11–19, 2016.

[12] A. Murphy, "Band-gap determination from diffuse reflectance measurements of semiconductor films, and application to photoelectrochemical water-splitting," *Solar Energy Materials and Solar Cells*, vol. 91, no. 14, pp. 1326–1337, 2007.

[13] R. López and R. Gómez, "Band-gap energy estimation from diffuse reflectance measurements on sol–gel and commercial TiO_2: a comparative study," *Journal of Sol-Gel Science and Technology*, vol. 61, no. 1, pp. 1–7, 2012.

[14] E. Haro-Poniatowski, R. Rodríguez-Talavera, H. M. de la Cruz, O. Cano-Corona, and R. Arroyo-Murillo, "Crystallization of nanosized titania particles prepared by the sol-gel process," *Journal of Materials Research*, vol. 9, no. 08, pp. 2102–2108, 1994.

[15] C. Jovalekic, M. Zdujic, and L. J. Atanasoska, "Surface analysis of bismuth titanate by Auger and X-ray photoelectron spectroscopy," *Journal of Alloys and Compounds*, vol. 469, no. 1-2, pp. 441–444, 2009.

[16] L. Wang and W. Ma, "$Bi_4Ti_3O_{12}$ synthesized by high temperature solid phase method and it's visible catalytic activity," *Procedia Environmental Sciences*, vol. 18, pp. 547–558, 2013.

[17] Č. Jovalekić, M. Pavlović, P. Osmokrović, and L. J. Atanasoska, "X-ray photoelectron spectroscopy study of $Bi_4Ti_3O_{12}$ ferroelectric ceramics," *Applied Physics Letters*, vol. 72, no. 9, pp. 1051–1053, 1998.

[18] Z. Hu, H. Gu, Y. Hu, Y. Zou, and D. Zhou, "Microstructural, Raman and XPS properties of single-crystalline $Bi_{3.15}Nd_{0.85}Ti_3O_{12}$ nanorods," *Materials Chemistry and Physics*, vol. 113, no. 1, pp. 42–45, 2009.

[19] P. Malathy, K. Vignesh, M. Rajarajan, and A. Suganthi, "Enhanced photocatalytic performance of transition metal doped Bi_2O_3 nanoparticles under visible light irradiation," *Ceramics International*, vol. 40, no. 1, pp. 101–107, 2014.

[20] G. Lente, *Deterministic Kinetics in Chemistry and Systems Biology*, Springer, London, UK, 2015.

21

Annual Optical Performance of a Solar CPC Photoreactor with Multiple Catalyst Support Configurations by a Multiscale Model

Manuel I. Peña-Cruz [ID],[1] **Patricio J. Valades-Pelayo,**[2,3] **Camilo A. Arancibia-Bulnes** [ID],[2] **Carlos A. Pineda-Arellano,**[1] **Iván Salgado-Tránsito,**[1] **and Fernando Martell-Chavez**[1]

[1]*CONACYT-Centro de Investigaciones en Óptica, A.C., Unidad Aguascalientes, Prol. Constitución 607, Frac. Reserva Loma Bonita, Aguascalientes, Aguascalientes 20200, Mexico*
[2]*Instituto de Energías Renovables-Universidad Nacional Autónoma de México, Privada Xochicalco s/n, A.P. 34, Col. Centro, Temixco, Morelos 62580, Mexico*
[3]*Depto. de Ingeniería de Procesos e Hidráulica, Universidad Autónoma Metropolitana-Iztapalapa, Av. San Rafael Atlixco No. 186, C.P. 09340, Mexico City, Mexico*

Correspondence should be addressed to Manuel I. Peña-Cruz; mipec@cio.mx

Academic Editor: K. R. Justin Thomas

In this work, the seasonal and yearly optical performance of supported catalyst CPC solar photocatalytic reactors has been theoretically analyzed. A detailed model for the optical response of the anatase catalyst films is utilized, based on the characteristic matrix method, together with Monte Carlo ray tracing simulations. The catalyst is supported over glass tubes contained inside a larger glass tube that functions as receiver of the CPC reflector. Arrangements with four, five, and six tubes are considered. Overall, the four-tube scenario presents the worst performance of all, followed by the five-tube case. In general, the six-tube configuration is better. Nevertheless, important differences can be observed depending on the specific arrangement of tubes. The six-tube case surpasses the absorption rate of all the other configurations when the distance between tubes is extended. This configuration exhibits 27% increased yearly energy absorption with respect to the reference case and 47% with respect to the worst case scenario.

1. Introduction

Photocatalysis has been a prolific scientific field due to the attractiveness of its applications. Among these are the degradation of various organic compounds in gases [1] and liquids [2], hydrogen production by water splitting [3], and CO2 reduction for hydrocarbon fuel production (Sarkar et al., 2016). As this process involves the generation of hydroxyl radicals through a photoinduced electrochemical reaction activated in the UV-Vis range, it is feasible to use solar energy as a renewable energy source, making photocatalysis an eco-friendly technology capable of mitigating environmental water pollution [4] or producing alternative fuels [5].

One of the main technologic challenges faced is the need to increase the efficiency of photocatalytic systems at larger scales [6]. In this regard, a key step for a suitable implementation on a large scale is a proper modeling strategy that can help in photoreactor design and optimization.

While most of the available literature in photoreactor modeling focuses on suspended photoreactors, their supported counterpart still lacks a unified approach [7]. In this regard, the use of pseudocontinuum approximations [8] or empirical models [9] is a common choice. Nonetheless, while empirical models are unsuitable for scale-up purposes, pseudocontinuum models are only applicable to certain support arrangements, which usually are scale-dependent [10].

In particular, solar photocatalytic reactors [11] are very attractive due to the positive impact of using renewable energy as the source of UV radiation for the reactions. A common configuration uses glass tubes as reactors [12], which are illuminated by a solar reflector, either to increase radiation intensity, by means of parabolic trough concentrators [13], or to improve the distribution of energy in the tube perimeter, by means of V-trough or CPC reflectors [14–17]. Even a modification of the standard CPC geometry to accommodate for a single flat support for the catalyst inside a tubular reactor has been proposed [18].

Most radiation transfer studies have focused on reactors with suspended catalyst [13, 15, 19, 20]. However, fixed catalyst reactors are also of interest [17]. Manassero et al. [21] have carried out experimental comparison and Monte Carlo modeling of suspended and fixed catalyst photoreactors. Recently, our research group proposed a multiscale approach [7] using different methods at the support and the reactor scale. The methods employed are the characteristic matrix method [22] and the Monte Carlo ray tracing method. This method allows studying the effect of relevant parameters, such as photocatalyst film thickness, support surface location, concentrator geometry, optical properties, and even irradiance conditions, hence giving valuable information on the design of supported photoreactors.

In this paper, the assessment of the year-round optical performance of a supported solar CPC photocatalytic reactor is carried out. This evaluation is achieved by coupling meteorological software and a ray tracing software to generate 3D incoming radiation profiles as input for the multiscale model presented by Valadés-Pelayo et al. [7]. Given that no studies in the literature consider the geometrical arrangement (number and disposition) of the tubes, six different arrays are considered. On this basis, this study identifies the best multitubular support layout which collects more radiative energy throughout the year and points out directions towards the improvement of multitubular support configurations.

2. Methodology

2.1. Ray Tracing on the CPC. Ray tracing simulations of the CPC concentrator were carried out to model its optical behavior at different times of the year. For this purpose, the Tonatiuh ray tracing software [23] was used in this work. Tonatiuh is a freely distributed, open-source code, based on the Monte Carlo method, which has been widely utilized for different applications. Among its attractive features, it allows the use of a scripting tool, which enables continuous simulation for relevant scenarios, as parameters are varied. The CPC geometry is already defined in this software as one of its basic concentrator types (see Figure 1).

The characteristics of the solar source need to be defined prior to the simulations. Solar irradiance data has been obtained from a typical meteorological year (TMY) provided by the commercial software Meteonorm®. This software allows obtaining hourly irradiance values through the year for a particular location, employing numerical algorithms coupled to preexisting data from weather stations. In particular, UVA irradiance values were utilized in this analysis, which correspond to the region of the spectrum where the anatase TiO_2 photocatalyst is active [7]. Four typical days (March 21, June 21, September 21, and December 21) were

FIGURE 1: CPC under Tonatiuh ray tracing simulation.

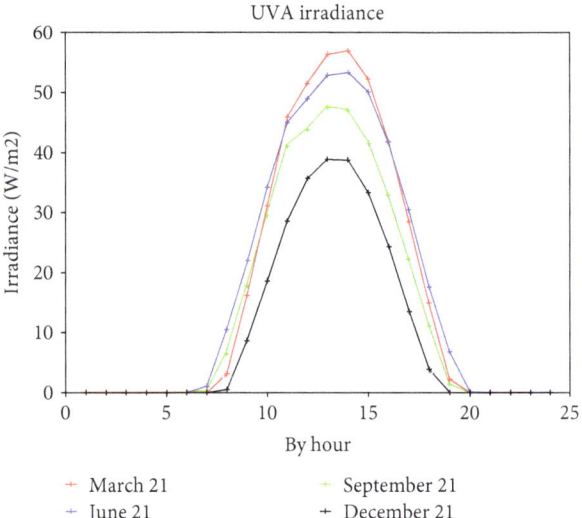

FIGURE 2: Typical UVA irradiance in the city of Aguascalientes, Mexico, as a function of hour, for the solstices and equinoxes.

chosen to present the results (see Figure 2), as they represent best the behavior of the angular position of the sun focusing on the CPC throughout the year. In this case, the local coordinates are defined as 21° 52′ 52″ N and 102° 17′ 29″ O, corresponding to the city of Aguascalientes, Mexico. We consider this analysis valuable for any particular location. Solar angles may change due to location, but the CPC must be oriented accordingly to the local latitude, providing similar results. UV irradiance may also vary from location to location, but the overall performance of the system must remain, as overall performance is modeled evenly by the amount of UV irradiance reaching the reactor but depends primarily on the geometrical disposition of the tubes for the energy absorption. However, for very high latitudes where winter days are very short, the energetic contribution of this season will be seriously reduced. On the other hand, the static nature of the collector does not allow taking advantage of the much longer summer days. Thus, the yearly energy absorption is reduced.

Commonly, nontracking CPC arrays are tilted to the latitude of the locality under study with the axis of the tubular receiver along the east-west direction, to optimize the yearly

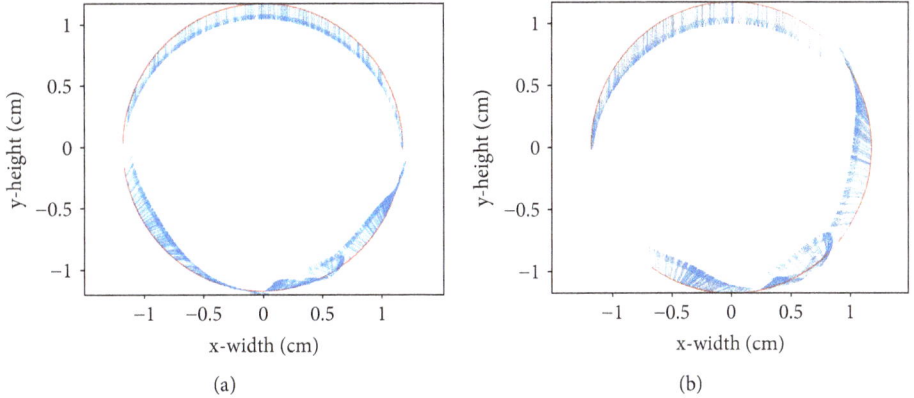

FIGURE 3: Photon vectors impinging on the receiver tube at 13:00 h, for (a) March 21 and (b) December 21.

collection of solar rays. This is also the configuration chosen for the present work.

A script is programmed in order to construct the virtual collector under design specifications, according to Salgado-Tránsito et al. [17]. In particular, the outer radius of the glass-tubular receiver is fixed at 1.175 cm. The radii of the inner TiO_2-coated glass tubes take the values 0.285, 0.342, and 0.4275 cm, according to the number of supporting tubes (6-5-4, respectively), in order to preserve in all cases the same total mass of catalyst inside the reactor. The length of all tubes is 70 cm, and anatase film thickness is fixed at 800 nm. All specular materials of the CPC are considered with ideal reflective properties, in order to evaluate solely the effect of the geometric disposition of the inner tubes in the receiver.

Sun position is calculated accordingly to the particular day and hour of interest. Then, a ray trace simulation with a million rays is performed. This is repeated for every working hour (from 9:00 to 17:00 h). The matrix of photons that reach the external part of the tubular receiver, resulting from these simulations, is stored for further processing by the optical model describing the inner part of the reactor.

The time-dependent solar position vector ($\mathbf{S_v}$) is used as an input for the ray trace simulations

$$\mathbf{S_v} = (\cos \gamma_s \sin \theta_z, -\sin \gamma_s \sin \theta_z, \cos \theta_z), \qquad (1)$$

where the zenith angle is calculated following Duffie and Beckman ([24]), as

$$\theta_z = \cos^{-1}(\cos \phi \cos \delta \cos \omega + \sin \phi \sin \delta), \qquad (2)$$

where ϕ is the site latitude, δ is the solar declination angle, and ω is the hour angle, which varies at a rate of 0.25°/min. Similarly, the azimuth angle is calculated as

$$\gamma_s = \operatorname{atan2}(\sin \phi \cos \delta \cos \omega - \cos \phi \sin \delta, \cos \delta \sin \omega), \quad (3)$$

where $\operatorname{atan2}(x, y)$ stands for the two-argument arctangent function, which is able to identify the appropriate quadrant of an angle whose tangent is given by y/x.

The Tonatiuh ray tracer does not deliver directly the photon vectors required in the next step of our simulation on its output file. Instead, it provides the history of collision points of every ray traced. As rays on a CPC can suffer several reflections before reaching the receiver, the direction vectors $\widehat{\mathbf{k}}$ must be obtained from the last two impacts of the ray, that with the tube $\mathbf{r_t}$, and the last reflection on the CPC $\mathbf{r_{lr}}$

$$\widehat{\mathbf{k}} = (\mathbf{r_t} - \mathbf{r_{lr}})|\mathbf{r_t} - \mathbf{r_{lr}}|^{-1}. \qquad (4)$$

Two plots with excerpts of the photon vectors impinging on the glass-tubular receiver are shown in Figure 3. These vectors are obtained for March 21 and December 21 at 13:00 h. The photon vector map is correlated directly to the day of the year, time, geometry (concentration factor), and tilt of the CPC. As exemplification, the low altitude of the sun on December 21 solstice (Figure 3(b)) produces an irregular distribution of photons, as compared to March 21 equinox (Figure 3(a)).

Ray vector maps such as the above are calculated for every working hour of the chosen days and are passed to the next step of the simulation, which considers propagation inside the receiver tube. A flowchart of the entire optical performance analysis has been resumed in Figure 4.

2.2. Radiation Absorption Model. The methodology allows modeling a number of important design parameters, such as incoming photon's trajectories, reactor geometry, photocatalyst layer thickness, and arrangement of the inner tubes. Key information required for the simulation radiation absorption inside the reactor, is the optical response of the absorber tubes, given in terms of their reflectance, transmittance, and absorptance coefficients. For the anatase film covering the absorber tubes, the characteristic matrix method (CMM) is used.

The CMM allows obtaining the optical response of thin film multilayers based on an electromagnetic description [22]. The application of this theory to glass-supported anatase films has been described in a previous publication [7], and only a brief description is presented here. The reflectance

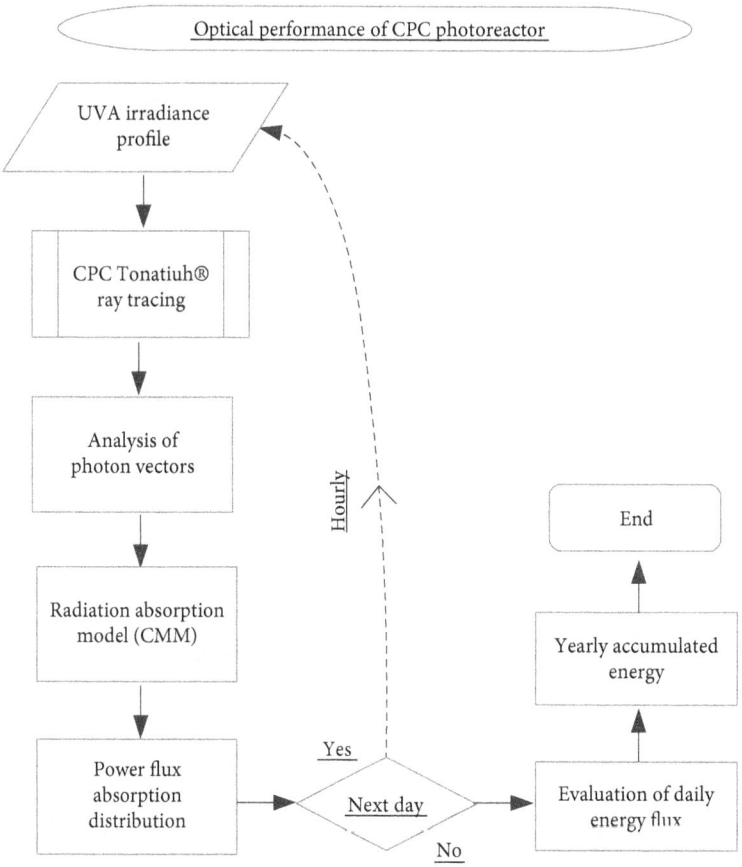

FIGURE 4: Flowchart: optical performance analysis for a CPC photoreactor.

R, transmitance T, and absorptance A of the films covering the tubes are expressed as

$$R = \left| \frac{\left(m'_{11} + m'_{12} p_f \right) p_0 - \left(m'_{21} + m'_{22} p_f \right)}{\left(m'_{11} + m'_{12} p_f \right) p_0 + \left(m'_{21} + m'_{22} p_f \right)} \right|^2,$$

$$T = \left| \frac{2 p_0}{\left(m'_{11} + m'_{12} p_f \right) p_0 + \left(m'_{21} + m'_{22} p_f \right)} \right|^2, \quad (5)$$

$$A = 1 - R - T,$$

where the components of the characteristic matrix m'_{ij} and the coefficients p_0 and p_f depend all on the optical properties of the media involved [7]. The refractive index of glass has been obtained from Rubin [25], the optical properties of the TiO_2 film have been obtained following Viseu et al. (2001), and a value of 1.33 has been considered for the refractive index of water.

For a single interface, for instance, glass/water or glass/air, the above equations reduce to Fresnel formulae [24] for reflection and transmission of radiation. This is the model applied to determine the properties of the receiver tube that encloses the reaction space where the absorber tubes are contained.

Once the involved optical coefficients have been obtained, the next step is to consider the propagation of each of these rays inside the reactor volume by a Monte Carlo method. For this, the set of ray incident in the outer part of the glass receiver tube (described in the previous section) is used. The core part of this process is the determination of all possible collisions of the ray with the different surfaces contained in the volume, i.e., the absorber tubes, or the walls of the receiver tube itself. At each step, once it is determined which surface is hit by the ray, it must be decided if it is absorbed, transmitted, or reflected, comparing the optical coefficients to a random ρ

$$0 \leq \rho \leq R_s, \text{reflection},$$

$$R_s < \rho \leq R_s + T_s, \text{transmission}, \quad (6)$$

$$R_s + T_s < \rho \leq 1, \text{absorption},$$

where the subscript "s" indicates that the coefficients of the surface in question are considered. At the beginning of the propagation of each ray, a random number is generated and compared to the reflectivity of the outer wall of the receiver tube, to determine if the ray is reflected or enters the reaction space. In case the ray is reflected, its propagation is terminated, and a new ray is propagated; otherwise, the ray is considered to enter the reactor volume and its propagation is continued. If at any point a ray is transmitted to the outside of the reactor, it is counted as lost and a new ray is generated. If it is reflected on an inner surface, the new propagation direction is obtained from the vector form of the law of reflection

$$\hat{\mathbf{r}} = \hat{\mathbf{i}} - 2(\hat{\mathbf{n}} \cdot \hat{\mathbf{i}})\hat{\mathbf{n}}, \tag{7}$$

where $\hat{\mathbf{r}}$, $\hat{\mathbf{i}}$, and $\hat{\mathbf{n}}$ are the reflected, incident, and normal unit vectors. After the new direction is obtained, the next intersection of the ray with a wall is sought.

On the other hand, if a ray is absorbed in an area element, a count is registered for this element, and a new ray is propagated. Typically, a million rays are propagated in each run. After all rays have been propagated, the radiative power absorbed at each surface element can be obtained by the following equation:

$$Q_{(i,j)} = \frac{P_{ray}}{A_{i,j}} \sum N_{i,j}, \tag{8}$$

where $A_{i,j}$ correspond to the area of the element (i, j), $N_{i,j}$ is the number of rays absorbed, and P_{ray} is the power assigned to each ray or power per photon. The P_{ray} value is obtained as a function of the solar irradiance energy distribution, and the total number of rays impinging the receiver. The area elements are obtained by dividing the perimeter of each tube into equal angular segments. Each element spans the whole length of the tube (longitudinal symmetry of the problem is assumed) and the whole thickness of the film.

The analyzed quantities are the distribution of power absorbed in the perimeter of the tubes per unit area, the total power absorbed by each tube, and the yearly energy absorbed by each tube configuration.

2.3. Analyzed Cases. Traditionally, fixed thin film analysis considers only the amount of film catalyst deposited on the supported tubes as the design parameter of importance. The geometrical arrangement of the tubes is commonly chosen arbitrary or selected due to a simple symmetric disposition. In this work, six configurations are compared in order to obtain the most effective design. For the first scenario, a four-tube square configuration is studied (see Figure 5(a)). For a second scenario, a five-tube configuration in a pentagram disposition is analyzed (see Figure 5(b)). This disposition is the reference case, being the configuration initially built and tested by Salgado-Tránsito et al. [17] for the degradation of the pesticide carbaryl. The third scenario consists of six tubes, with a pentagram disposition similar to the second scenario, but with smaller diameter, and adds a center tube (Figure 5(c)). The fourth case preserves the number of the tubes from the latter, but it considers a displacement of the surrounding tubes away from the centered one, closed to the receiver tube (Figure 5(d)). The fifth scenario does the opposite, moving the surrounding tubes closer to the center (Figure 5(e)). Finally, the sixth case preserves the geometry of the third case but with a rotation of the whole tube configuration by $\pi/5$ (Figure 5(f)).

All configurations are compared considering the same solar irradiance conditions, as well as the same CPC design (concentration factor and tubular receiver diameter). It is worth noticing that the total catalyst film area available for absorption is kept constant between the different configurations, by suitable choice of the radius of the tubes. Moreover, the total mass of TiO_2 considered is the same in all cases as thickness is not varied either. The only variations lie in the number, radius, and angular disposition of the absorber tubes.

3. Results

In this section, results obtained from annual performance analysis are discussed. The absorption rate is calculated for every scenario, absorber tube arrangement, date, and hour. The absorption parameter $Q_{(i,j)}$ is integrated to obtain average absorption rates for every tube, as well as for every configuration as a whole.

Figure 6 shows the absorption rate for the outer wall of the 5 tubes of configuration 2, for the 21st of June, at 13:00 hours. Beam and diffuse radiation results are presented separately, to see the effect of the different angular composition of incident radiation; while beam radiation comes from a single direction (approximately $23°$ from the normal to the aperture, in the plane of the collector cross section), diffuse radiation comes from every possible direction on a hemisphere around the aperture normal. It can be seen that for beam solar radiation, the upper tubes (i.e., tubes 1, 2, and 3 of Figure 4(b)) have a higher absorption rate on the upper zones (from $0°$ to $180°$), with a peak around $90°$, while the bottom tubes (tubes 4 and 5) have a wider collection angle due to the intrinsic downside reflections of the CPC geometry (involute section). There is however, asymmetry, because radiation is coming more inclined towards tube 4, resulting in lateral reflections to this tube ($180°$), while tube 5 receives more radiation from the section of the involute closest to the receiver tube bottom. It is also noticeable that tube 2 and tube 3 have smaller absorption on its bottom part (from $200°$ to $320°$) due to the effect of "shadowing" from the bottom tubes. A small amount that impinges tube 3 is due to lateral reflections of the CPC. These results are consistent with the distribution of incoming radiation at the walls of the reactor (receiver tube).

Meanwhile, for diffuse radiation, the situation is very different for each tube. Tube 1 receives radiation mainly from the right, tube 3 mainly from the left, and tube 2 (the top tube) both from left and right. The asymmetries observed are attributable to partial ground reflection of the diffuse component. The bottom tubes are illuminated from the bottom left and bottom right.

A similar analysis can be carried out for every configuration. In particular, Figure 7 compares results of configurations 2 and 3. In the latter, a sixth tube has been added at the center of the reaction space, and the diameter of the tubes has been decreased, to preserve the same total catalyst mass. For the 6-tube configuration, it is noticeable that although the tubes are smaller in comparison with the 5-tube case, the absorption rate for the external tubes remains almost

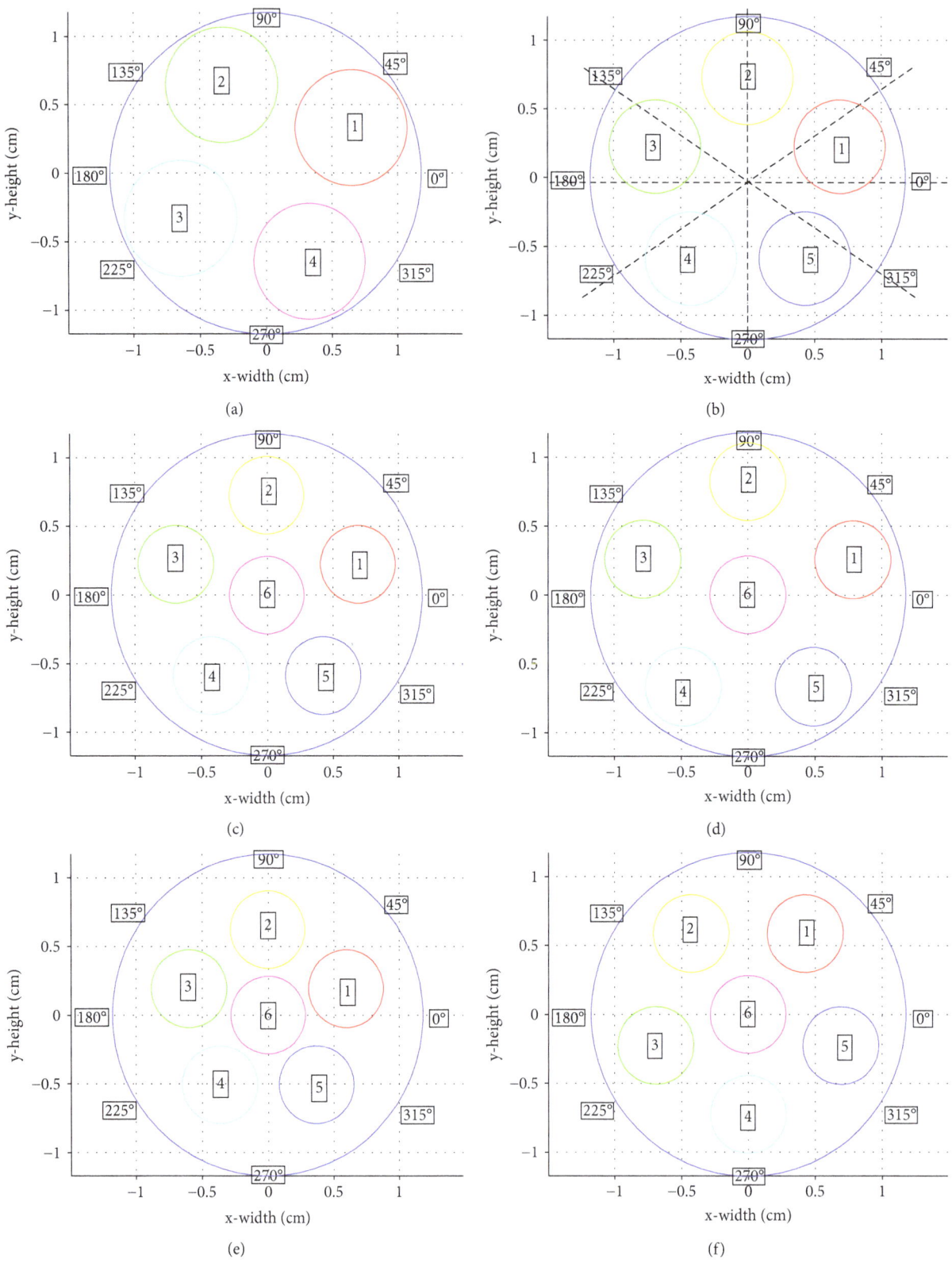

FIGURE 5: Diagram of the multitubular configurations analyzed: (a) four tubes in a square; (b) five tubes in pentagram disposition; (c) six tubes in a pentagram disposition with central tube; (d) six tubes with increased distance; (e) six tubes with decreased distance; (f) six tubes rotated by $\pi/5$.

the same. Due to the inclusion of a central tube (tube 6), the configuration is capable of capturing solar rays that pass along the spaces between the tubes in configuration 2 and

that were going to be lost through the receiver wall. Furthermore, the bottom tubes (4 and 5) are less "shadowed" by the upper tubes and capture a greater amount of photons in its

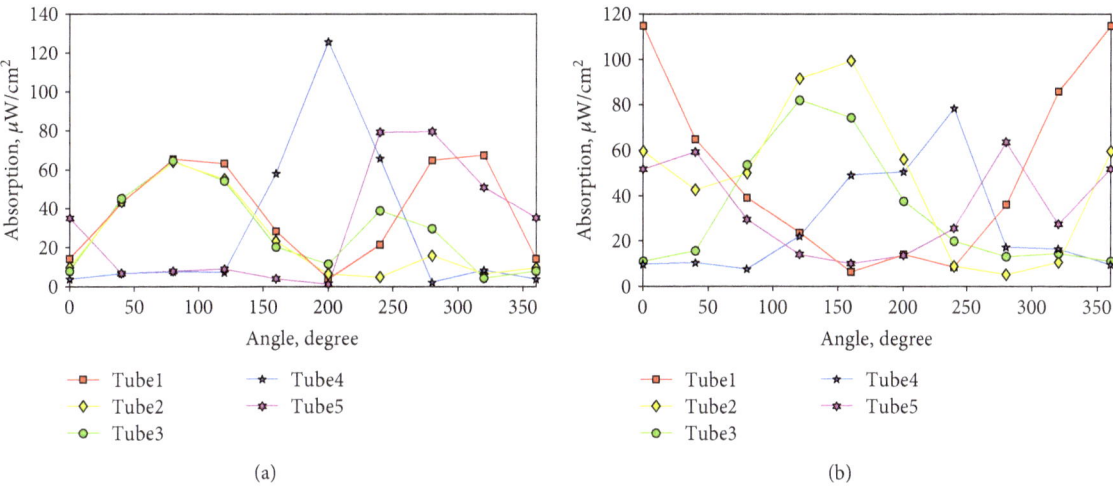

FIGURE 6: Power flux absorption distribution for the 5 tubes in configuration 2. June 21st at 13:00 hours. Beam (a) and diffuse (b) radiation contributions.

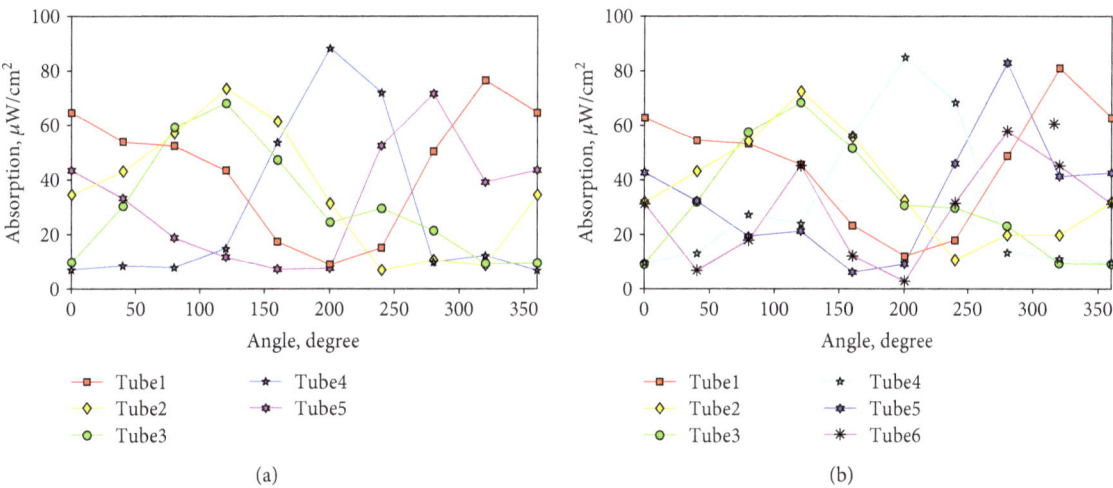

FIGURE 7: Power flux absorption distribution for the tubes in configurations 2 (a) and 3 (b), for combined beam and diffuse irradiance. June 21st at 13:00 hours.

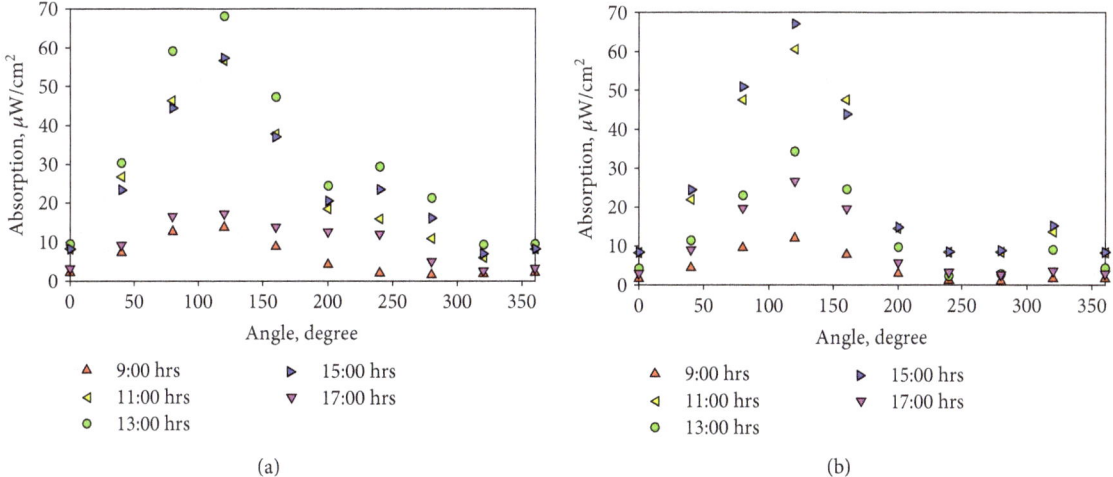

FIGURE 8: Power flux absorption distribution for tube 3, in configuration 2, for different hours. June 21st (left) and December 21st (right).

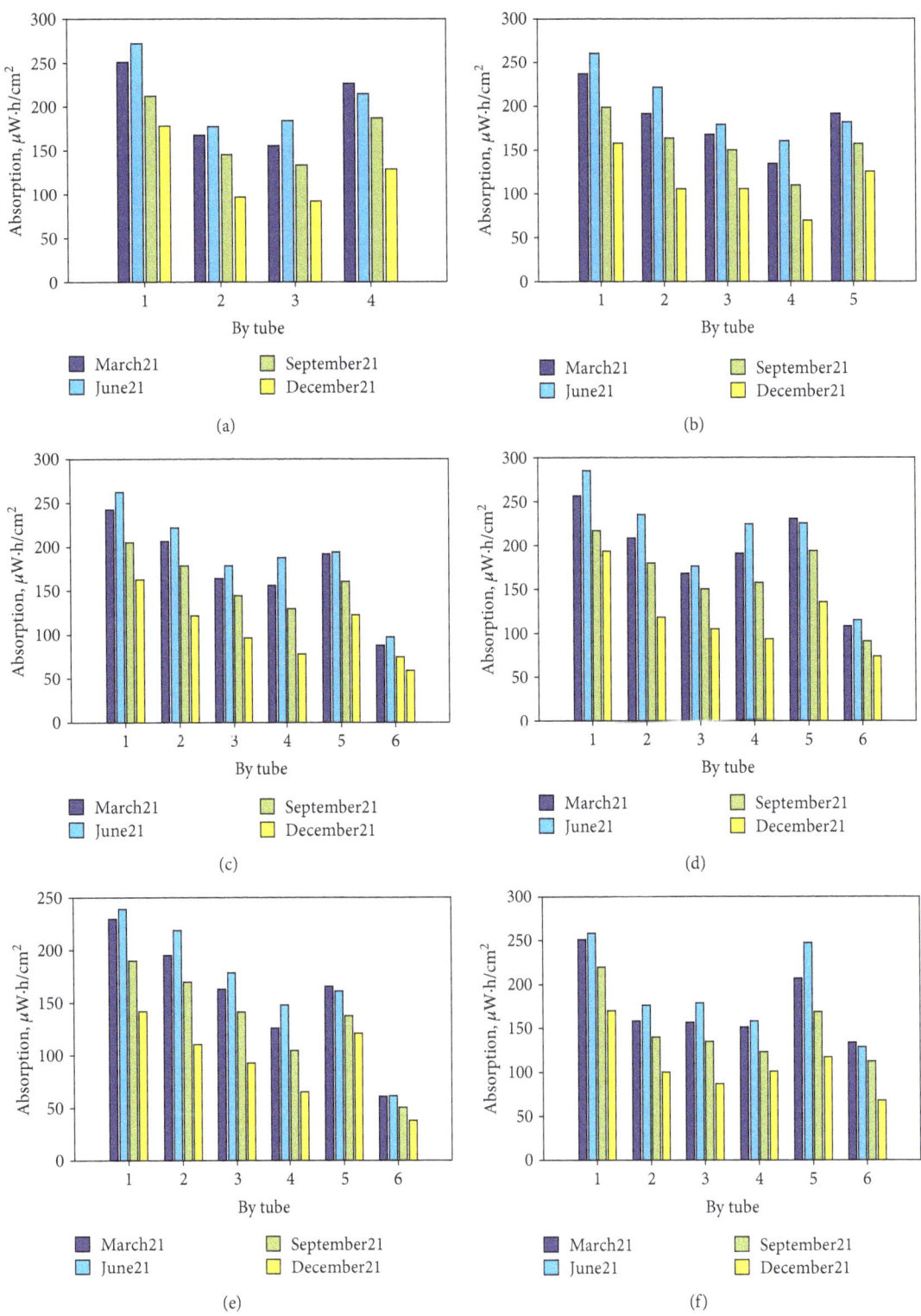

FIGURE 9: Average energy absorption for the solstices and equinoxes, for each tube in the different configurations. Panels (a) to (f) correspond to cases 1 to 6, respectively.

upper side (0° to 180°). Thus, an increase in performance is expected from the configurations with 6 tubes.

In Figure 8, the results are presented for different hours of the day for a single tube (tube 3), both for the 21st of June and the 21st of December, for combined beam and diffuse contributions. It is immediately clear that the absorption distribution maintains the same shape during the day, changing only in height as the incident solar irradiance changes with the hour. Small variations in the morning and afternoon can be appreciated due to a steeper angle of sunlight entering the CPC aperture. The reason for this preserved shape of the curves along the day is that the angle of beam radiation keeps more or less constant in the plane perpendicular to the receiver tube axis, because of the east-west orientation of this receiver. Thus, the angle changes occur mainly in the plane that contains the tube axis and the collector aperture normal, producing mainly a change in the collected power but not in the absorption distribution.

In both cases of Figure 7, the main contribution of absorbed radiation takes place at the top half of the tube. There is a secondary peak that displaces from the left bottom part of the tube for the summer and to the right bottom part for the winter. This is easy to understand, considering that the incidence angle to the normal of the collector is symmetrical for both dates, at 23.45° to the right and left, respectively.

For Figure 9, the flux absorption is averaged over the perimeter of each tube. The graph compares the average power absorption of each tube in each of the configurations, for the solstices and equinoxes, which are the more representative days of the year. The first feature that can be noted is the clear seasonal differences, attributable mainly to the variations on the duration of day along the year. Note however that the two equinoxes, that have the same sunlight duration, have very different absorbed power levels for different tubes. On the average, the power absorbed is less in September than in March, because autumn is the rainy season (see Figure 2). However, this is not true in every case, because the incident power in different tubes for different times of the year is also influenced by the distribution of incident rays on the tube walls.

Let us make a remark regarding units: if one wishes to compare the present results with those of Valadés-Pelayo et al. [7], it should be taken into account that the quantities presented are slightly different. In that paper, the power absorbed by the tubes was expressed in W/m^3, which refers to the power absorbed by the catalyst film per unit volume. However, we have preferred here to express results in terms of energy absorbed per film outer surface area $\mu Wh/cm^2$. The rationale being that charge separation, recombination, and degradation reactions take place at the catalyst surface; hence, radiation absorption per unit surface, not per unit volume, is a measure that better represents the behavior of the reactor. However, both are valid manners of expressing the same physical situation.

Yearly mean absorbed power flux is used as comparison parameter between configurations in Figure 10. It is obtained by averaging the absorbed power flux of each tube over all working hours of the year. It can be noticed that the configuration that presents the greatest absorption rate through the year is case 4, the 6-tube-far configuration. This configuration preserves the advantages of using a central tube, but higher photon energy is captured by bringing the tubes closer to the external receiver, intercepting a greater fraction of the rays reflected by the CPC. The 6-tube-far configuration exceeds up to 30% in the performance absorption rate from the 4-tube configuration (case 1) and up to 13% more efficiency versus the 6-tube configuration 3.

Finally, in Table 1, the total energy absorbed over one year is presented for the different configurations. Clearly, configuration 4 is the winner, with 13% improvement with respect to the initially proposed geometry (configuration 2).

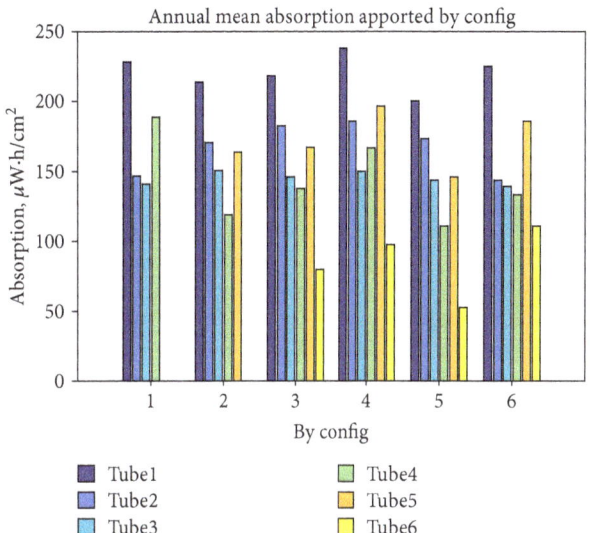

FIGURE 10: Comparison of energy absorption averaged over the whole year, for each tube of all the configurations. Config: 1 (Figure 9(a)); 2 (Figure 9(b)); 3 (Figure 9(c)); 4 (Figure 9(d)); 5 (Figure 9(e)); 6 (Figure 9(f)).

TABLE 1: Total annual absorbed energy for the different configurations.

Configuration	Annual absorbed energy (W·h)
(1) 4 tubes	48.4
(2) 5 tubes	44.8
(3) 6 tubes	47.0
(4) 6-tube-far	50.8
(5) 6-tube-near	43.1
(6) 6-tube-rot $\pi/5$	44.0

However, we can see that case 1 also shows a better performance. In both cases, this has to do with the reduction of losses through the tube gaps. However, from the point of view of better distribution of the catalyst over the reaction space, configuration 4 is clearly advantageous. The absorption by this improved configuration amounts approximately to 56% of the solar UV energy incident in the CPC collector during the entire year. This is quite a good performance, if we compare this with values typical of suspended catalyst reactors [19], where the limitations to absorption are imposed mainly by backscattering of radiation. In the present fixed catalyst configuration, reflection and transmission of radiation on the catalyst film produce also losses to the outside of the reactor.

4. Conclusions

In this study, the seasonal and yearly optical performance of CPC solar photocatalytic reactor with supported catalyst has been theoretically analyzed. A detailed model for the optical response of the anatase catalyst films has been utilized, together with Monte Carlo ray tracing simulations. The catalyst is supported over glass tubes contained inside a larger glass tube that functions as receiver of the CPC reflector. Different configurations of catalyst support have been considered in terms of number and arrangement of the supporting tubes. Arrangements with four, five, and six tubes have been considered. The radius of the tubes is adjusted as their number increases, in order to have the same surface area in all compared cases. The case of five tubes, proposed in a previous work on semiempirical grounds, serves as a reference case for comparison. Important seasonal dependence of the response of the different tubes on each array is observed. Overall, the five-tube scenario presents the worst performance of all, followed by some of the 6-tube cases, because they leave free trajectories for solar rays to cross the reactor. In general, the six-tube configurations are better than the originally proposed, but nevertheless important differences can be observed depending on the specific arrangement of tubes. Another interesting result is that the six-tube case and the rotated six-tube case present almost the same absorption capacity, proving that symmetrical rotations have very little effect on global performance. As can be seen, the six-tube case surpasses the absorption rate of all the other configurations in every working day of the year when the distance between tubes is extended to cover a larger cross section. This configuration exhibits 13% increased yearly energy absorption with respect to the reference. As a main conclusion, it can be said that the development of a modeling tool for evaluating rigorously the annual performance optical of CPC's photocatalytic reactors provides valuable insights to improve the design of the system.

Conflicts of Interest

The authors declare that they have no conflicts of interest.

Acknowledgments

The authors acknowledge CONACYT for the financial support for the project Problemas Nacionales 2015-01-1651: "Diseño y Construcción de Potabilizador Integral Solar de Agua para Comunidades Rurales". The authors also acknowledge the support from Fondo Sectorial CONACYT-SENER-Sustentabilidad Energetica through Grant 207450 and Centro Mexicano de Innovacion en Energia Solar (CeMIE-Sol), within strategic project no. 10, "Combustibles Solares y Procesos Industriales" (COSOLPi). Patricio Valades-Pelayo also acknowledges a postdoctoral fellowship by DGAPA-UNAM and a "Retención" scholarship by CONACYT.

References

[1] C. S. Lugo-Vega, B. Serrano-Rosales, and H. de Lasa, "Immobilized particle coating for optimum photon and TiO2 utilization in scaled air treatment photo reactors," *Applied Catalysis B: Environmental*, vol. 198, pp. 211–223, 2016.

[2] A. Pinedo, M. López, E. Leyva, B. Zermeño, B. Serrano, and E. Moctezuma, "Photocatalytic decomposition of metoprolol and its intermediate organic reaction products: kinetics and degradation pathway," *International Journal of Chemical Reactor Engineering*, vol. 14, no. 3, pp. 809–820, 2016.

[3] S. Escobedo, B. Serrano, A. Calzada, J. Moreira, and H. de Lasa, "Hydrogen production using a platinum modified TiO_2 photocatalyst and an organic scavenger. Kinetic modeling," *Fuel*, vol. 181, pp. 438–449, 2016.

[4] U. J. Rajput, H. Alhadrami, F. al-Hazmi, Q. Guo, and J. Yang, "Initial investigations of a combined photo-assisted water cleaner and thermal collector," *Renewable Energy*, vol. 113, pp. 235–247, 2017.

[5] K. Nakata and A. Fujishima, "TiO_2 photocatalysis: design and applications," *Journal of Photochemistry and Photobiology C: Photochemistry Reviews*, vol. 13, no. 3, pp. 169–189, 2012.

[6] J. Marugán, R. van Grieken, A. E. Cassano, and O. M. Alfano, "Quantum efficiency of cyanide photooxidation with TiO_2/SiO_2 catalysts: multivariate analysis by experimental design," *Catalysis Today*, vol. 129, no. 1-2, pp. 143–151, 2007.

[7] P. J. Valadés-Pelayo, C. A. Arancibia-Bulnes, I. Salgado-Tránsito, H. I. Villafán-Vidales, M. I. Peña-Cruz, and A. E. Jiménez-González, "Effect of photocatalyst film geometry on radiation absorption in a solar reactor, a multiscale approach," *Chemical Engineering Science*, vol. 161, pp. 24–35, 2017.

[8] V. Loddo, S. Yurdakal, G. Palmisano et al., "Selective photocatalytic oxidation of 4-methoxybenzyl alcohol to p-anisaldehyde in organic-free water in a continuous annular fixed bed reactor," *International Journal of Chemical Reactor Engineering*, vol. 5, no. 1, 2007.

[9] G. E. Imoberdorf, G. Vella, A. Sclafani, L. Rizzuti, O. M. Alfano, and A. E. Cassano, "Radiation model of a TiO_2-coated, quartz wool, packed-bed photocatalytic reactor," *AICHE Journal*, vol. 56, no. 4, pp. 1030–1044, 2010.

[10] G. E. Imoberdorf, F. Taghipour, M. Keshmiri, and M. Mohseni, "Predictive radiation field modeling for fluidized bed photocatalytic reactors," *Chemical Engineering Science*, vol. 63, no. 16, pp. 4228–4238, 2008.

[11] S. Malato, J. Blanco, D. C. Alarcón, M. I. Maldonado, P. Fernández-Ibáñez, and W. Gernjak, "Photocatalytic decontamination and disinfection of water with solar collectors," *Catalysis Today*, vol. 122, no. 1-2, pp. 137–149, 2007.

[12] E. R. Bandala, C. A. Arancibia-Bulnes, S. L. Orozco, and C. A. Estrada, "Solar photoreactors comparison based on oxalic acid photocatalytic degradation," *Solar Energy*, vol. 77, no. 5, pp. 503–512, 2004.

[13] C. A. Arancibia-Bulnes and S. A. Cuevas, "Modeling of the radiation field in a parabolic trough solar photocatalytic reactor," *Solar Energy*, vol. 76, no. 5, pp. 615–622, 2004.

[14] J. Blanco, S. Malato, P. Fernández et al., "Compound parabolic concentrator technology development to commercial solar detoxification applications," *Solar Energy*, vol. 67, no. 4-6, pp. 317–330, 1999.

[15] J. Colina-Márquez, F. Machuca-Martínez, and G. Li Puma, "Photocatalytic mineralization of commercial herbicides in a pilot-scale solar CPC reactor: photoreactor modeling and reaction kinetics constants independent of radiation field," *Environmental Science and Technology*, vol. 43, no. 23, pp. 8953–8960, 2009.

[16] S. Malato Rodríguez, J. Blanco Gálvez, M. I. Maldonado Rubio et al., "Engineering of solar photocatalytic collectors," *Solar Energy*, vol. 77, no. 5, pp. 513–524, 2004.

[17] I. Salgado-Tránsito, A. E. Jiménez-González, M. L. Ramón-García, C. A. Pineda-Arellano, and C. A. Estrada-Gasca, "Design of a novel CPC collector for the photodegradation of carbaryl pesticides as a function of the solar concentration ratio," *Solar Energy*, vol. 115, no. 0, pp. 537–551, 2015.

[18] J. Chaves and M. C. Pereira, "New CPC solar collector for planar absorbers immersed in dielectrics. Application to the treatment of contaminated water," *Journal of Solar Energy Engineering*, vol. 129, no. 1, pp. 16–21, 2007.

[19] C. A. Arancibia-Bulnes, E. R. Bandala, and C. A. Estrada, "Radiation absorption and rate constants for carbaryl photocatalytic degradation in a solar collector," *Catalysis Today*, vol. 76, no. 2-4, pp. 149–159, 2002.

[20] J. Colina-Márquez, F. Machuca-Martínez, and G. L. Puma, "Radiation absorption and optimization of solar photocatalytic reactors for environmental applications," *Environmental Science & Technology*, vol. 44, no. 13, pp. 5112–5120, 2010.

[21] A. Manassero, M. L. Satuf, and O. M. Alfano, "Photocatalytic reactors with suspended and immobilized TiO2: comparative efficiency evaluation," *Chemical Engineering Journal*, vol. 326, pp. 29–36, 2017.

[22] M. Born and E. Wolf, *Principles of Optics: Electromagnetic Theory of Propagation, Interference and Diffraction of Light*, Cambridge University Press, 1999.

[23] M. Blanco, A. Mutuberria, A. Monreal, and R. Albert, "Results of the empirical validation of Tonatiuh at Mini-Pegase CNRS-PROMES facility," in *SolarPACES 2011 Granada*, Spain, 2011.

[24] J. A. Duffie and W. A. Beckman, *Solar Engineering of Thermal Processes 4th Edition*, Wiley, 2013.

[25] M. Rubin, "Optical properties of soda lime silica glasses," *Solar Energy Materials*, vol. 12, no. 4, pp. 275–288, 1985.

22

Synthesis and Characterization of TiO_2/SiO_2 Monoliths as Photocatalysts on Methanol Oxidation

Rigoberto Regalado-Raya,[1] **Rubí Romero-Romero** ⓘ,[2] **Osmín Avilés-García** ⓘ,[1] **and Jaime Espino-Valencia** ⓘ[1]

[1]*Facultad de Ingeniería Química, Universidad Michoacana de San Nicolás de Hidalgo, Edif. V1, Ciudad Universitaria, 58060 Morelia, Michoacán, Mexico*
[2]*km 14.5 Carretera Toluca-Atlacomulco, San Cayetano, Piedras Blancas, Centro Conjunto de Investigación en Química Sustentable UAEMéx-UNAM, Toluca, Estado de México, Mexico*

Correspondence should be addressed to Jaime Espino-Valencia; jespinova@yahoo.com.mx

Academic Editor: Juan M. Rodriguez

Photocatalytic materials based on silica-titania (SiO_2-TiO_2) were synthesized by sol-gel and dip-coating method. TEOS and titanium butoxide were used as precursors of the silica-titania, respectively. A thin film with anatase phase was obtained on the surface of the support. The effect of variables as dispersion mechanism, immersion time, and number of treatment cycles were studied. The materials were characterized using X-ray diffraction, scanning electron microscopy, energy dispersion scanning, and N_2 adsorption-desorption. The highest crystallinity of TiO_2 on silica, high specific surface area in TiO_2-SiO_2 materials, and thin film formation were obtained by using a stirring plate and minimum immersion time. The so synthesized catalyst allowed the production of formaldehyde from the photocatalyzed methanol oxidation in a packed-bed reactor.

1. Introduction

Among the advanced oxidation processes (AOPs), heterogeneous photocatalysis has been widely applied in the degradation of organic compounds, hydrogen production from water, reduction of heavy metals, and selective oxidation reactions [1–4]. Selective oxidation of alcohols to aldehydes by photocatalysis is an attractive route because it can be carried out under mild conditions (room temperature, atmospheric pressure, and neutral pH) [5, 6]; particularly, formaldehyde is a highly important intermediate compound in the chemical industry because of its use in the synthesis of adhesives, fertilizers, pyridines, drugs, polyols, and dyes among others [7, 8]. The industrial production of formaldehyde is performed from the oxidation and dehydrogenation of methanol with iron molybdate and silver catalysts, respectively [9, 10].

The most studied semiconductor in the field of heterogeneous photocatalysis is titanium dioxide (TiO_2) due to its high oxidative capacity, nontoxicity, low cost, high chemical and physical stability, corrosion resistance, and chemical inertness [11, 12]. TiO_2 has been widely used in both powders and thin films [13, 14]. When it is used as suspended particles, a separation step such as filtration or centrifugation is required at the end of the process. For this reason, the immobilization of the particles in an appropriate substrate has attracted great interest [15].

There are several methods for the preparation of TiO_2 thin films, and the physicochemical properties strongly depend on the selected method [16–19]. Among all the techniques, the dip-coating sol-gel method is the most widely used because of its good homogeneity, simplicity, low cost, and low temperature during the process [20]. On the other hand, the nature and type of substrate should not be left aside. Amorphous SiO_2 has shown to be an excellent support due to its mechanical properties, good adsorption capacity, and high surface area [21, 22].

This work aims to elucidate the effect on morphological, textural, and structural properties of three important variables involved in the synthesis of TiO_2 coated SiO_2 monoliths. These variables are the mechanism to disperse Ti alkoxide species in an appropriate medium allowing the hydrolysis-condensation processes, dipping time, and number of coating cycles (dip + drying + calcination). In order to evaluate the photocatalytic activity of the synthesized catalysts, the photooxidation of methanol was carried out in a bench-scale continuous-flow packed-bed reactor. This reaction was elected because of its industrial importance and because it is a consecutive reaction, whose selectivity towards intermediate compounds may help to prove the application of TiO_2 in selective oxidation processes instead of the organic compounds mineralization. This process could offer the advantage being performed at mild temperature and pressure conditions unlike other existing processes that occur at relatively high temperatures.

2. Experimental

2.1. Preparation of SiO2 Monoliths.
Silica dioxide monoliths were synthesized by sol-gel method. The synthesis of the silica support was performed using tetraethyl orthosilicate (TEOS) $[Si(OC_2H_5)_4]$ as alkoxide precursor of the Si sol.

First, the ethanol was added into a beaker and it was maintained under continuous stirring until the temperature reached 60°C. At this point, alkoxide was added and mixed into the beaker for 15 min. After this time, a water and nitric acid solution (1 : 0.0012 molar ratio) was added and the mixture was kept under stirring and keeping the temperature constant for one hour. The molar ratio of water : ethanol : TEOS was of 16 : 4 : 1, respectively.

After that, 2.5 ml of the resulting sol were poured into one container to begin the aging process. This was repeated several times to obtain various monoliths. The container lids were previously drilled to allow solvent diffusion. During this step, alkoxide groups are removed by acid- or base-catalyzed hydrolysis reactions, and link networks O-Si-O are formed in subsequent condensation reactions involving hydroxyl groups [23, 24]. Depending upon the water : alkoxide molar ratio R, pH, temperature, and solvent, condensation leads to different polymeric structures such as linear, entangled chains, clusters, and colloidal particles [25].

The obtained monoliths were then dried from room temperature to 100°C during 14 hours with a slow heating profile to eliminate the solvent. The drying was performed in an Isotemp Vacuum Oven programmable stove model 282 A. The drying treatment was slow to lead the formation of open pores. The drying profile was as follows: 1 h at 40°C, 2.5 h at 50°C, 13 h at 60°C, 2.5 h at 70°C, 3.5 h at 80°C, 2.5 h at 90°C, and 27 h at 100°C. This procedure was performed in order to keep the structure, since a fast drying profile could cause a structure collapse causing cracking of the monolith.

Finally, to provide the monoliths with the appropriate structural and mechanical properties, they were calcined from room temperature (25°C) to 550°C for 6 h at a heating rate of 2.5°C/min using a Jelrus muffle with 2 steps. Amorphous silica compounds without a defined crystalline phase are found at this temperature.

2.2. TiO2 Synthesis.
Ethanol, water, titanium butoxide, and diethanolamine (basic catalyst) were used to obtain Ti sols via sol-gel method. Titanium butoxide $(Ti[O(CH_2)_3CH_3]_4)$ was dispersed in the ethanol. Immediately, a diethanolamine and water solution was dropped into the volume. It is necessary to maintain a 1 : 1 alkoxide : water molar ratio. Once the solution addition was completed, the agitation was maintained for two hours. After this time, the solution was aged for further two hours without any stirring. Diethanolamine was elected because of its low reactivity during sol-gel process, this makes the hydrolysis reactions slow by favoring the thin film formation [26].

Finally, SiO_2 monoliths were immersed into the Ti sol obtained. In this process, the studied variables were the dispersion mechanism (mechanic or ultrasound) and the residence time into the Ti sol. For the former, a stirring plate and an ultrasonic cleaner were utilized as agitation media. Regarding residence time in the Ti sol, this variable was studied at three levels (half, one, and three hours) for each stirring medium. The number of cycles (immersion-drying-calcination) was also studied in order to establish its relationship with the amount of titania on SiO_2.

The monoliths coated with titanium species were dried at room temperature for 24 h and calcined under an air flow at 550°C for 5 hours.

2.3. Catalysts Characterization.
A Bruker Advance 8 diffractometer was employed to carry out the X-ray diffraction analysis and determine the presence of anatase in the synthesized catalysts. The patterns were obtained using $CuK\alpha$ radiation at 20 kV and 20 mA. Data were collected over 2θ range of 5–50° with a step of 0.5°/min.

A JEOL JSM-6510LV electron microscope coupled with an energy dispersive X-ray spectrometer was employed to observe the surface morphology of the prepared catalysts and to perform elemental analysis of the catalysts.

Autosorb-1 Quantachrome sorption equipment was employed to determine the specific surface area and average pore diameter of the synthesized samples by using liquid nitrogen (77 K). The pore size distributions and the specific surface areas of the materials were estimated by Dubinin-Astakhov (DA) and Brunauer Emmett–Teller (BET) methodologies, respectively.

2.4. Bench-Scale Photocatalytic Reactor.
The photooxidation of methanol was performed using a bench-scale continuous packed-bed reactor. An eight-watt UV lamp emitting 254 nm waves was placed right in the center of the reactor. 30 monoliths constituted the catalytic bed. Compressed air was used as carrier gas to help methanol to flow through the packed-bed reactor. The air flow was constant at 50 ml/min. The methanol liquid was heated at 65°C in order to vaporize it. The reactor set-up is depicted in Figure 1.

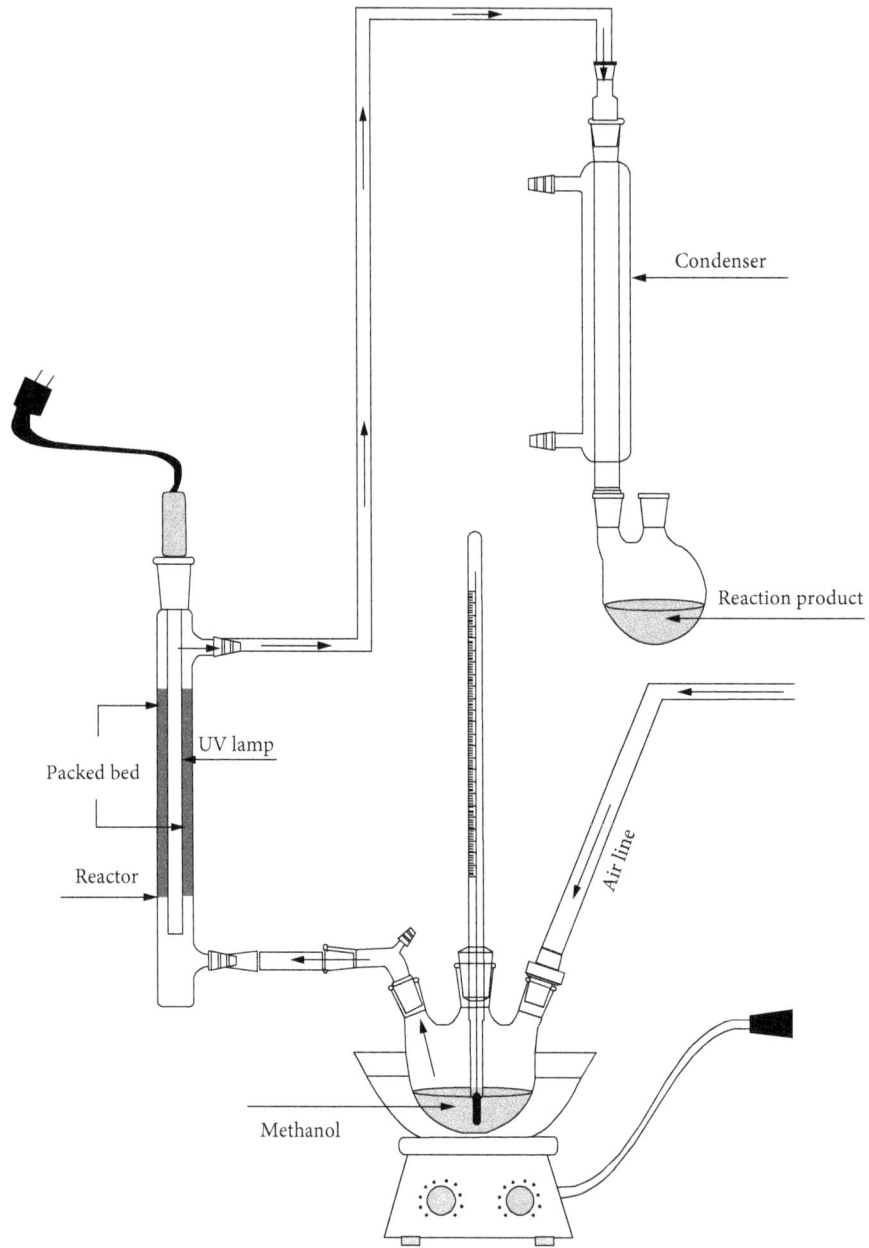

FIGURE 1: Bench-scale photocatalytic reactor set-up.

Identification of formaldehyde by photooxidation of methanol was carried out according to the following methodology: a proper container with a 2,4-dinitrophenylhydrazine (2,4-dnph) solution was placed instead of the condenser (see Figure 1), in such a way that the exit stream was directly bubbled into the solution. This was conducted with all synthesized materials. If there was formaldehyde in such a stream, then a precipitate was observed. This precipitate was the 2,4-dinitrophenylhydrazone, which is the product of the reaction between the 2,4-dnph and the aldehyde as depicted in Figure 2 [27].

The 2,4-dnph solution was prepared as follows: 2 ml of concentrated sulfuric acid was mixed under stirring with 0.4 gr of 2,4-dnph and 3 ml of water until total dissolution

appears. At this point, 10 ml of ethanol at 95% are added to the solution.

The quantitative analysis of formaldehyde after photooxidation of methanol was verified by collecting the condensed reaction product in a container at 4°C and analyzed in a Varian GC 3800 using a 52 CP WAX column (30 m × 0.320 mm).

3. Results and Discussion

SiO_2 monoliths with a diameter of 15 mm and thickness of 1 mm approximately were obtained following the methodology described in the previous section. Figure 3 shows such monoliths. It can be observed that they are totally transparent. This is expected to allow an excellent light transmittance

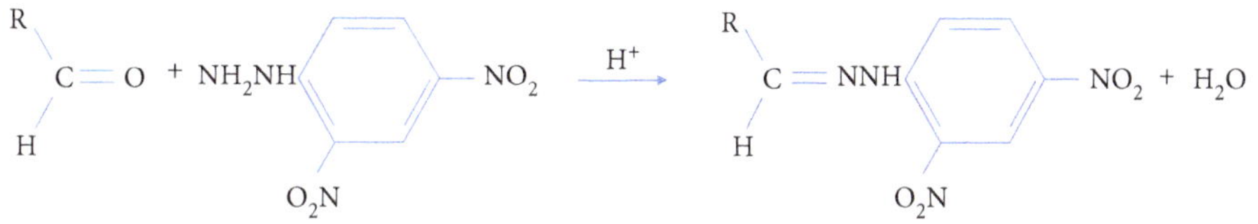

FIGURE 2: 2,4-Dinitrophenylhydrazine general reaction with aldehyde functional groups.

FIGURE 3: SiO$_2$ monoliths obtained by sol-gel technique.

TABLE 1: Nomenclature of synthesized materials (B by ultrasound bath and P by stirring plate).

Sample name	Number of immersion-drying-calcination cycles	Immersion time (hours)
3ST12B	3	1/2
3ST12P	3	1/2
3ST1B	3	1
3ST1P	3	1
3ST3B	3	3
3ST3P	3	3
4ST1P	4	1

and an appropriated use in photocatalytic reactions using UV light.

In total, 8 samples were characterized in order to decide at what conditions the monoliths for the methanol photooxidation should be synthesized. From the 8 monoliths, 7 were coated with TiO$_2$ while the left one was only SiO$_2$. The nomenclature used to name the samples is explained in Table 1.

Figure 4 shows the diffraction patterns with respect to the agitation mechanism and number of immersion-drying-calcination cycles. The obtained crystalline phase is mainly anatase, and this is expected because of the calcination temperature (550°C) [28]. Usually, when the calcination temperature is increased to more than 550°C, the anatase phase is observed to gradually change into a rutile phase with a larger particle size that results unfavorable for photocatalytic degradation reactions [29].

The average crystallite size of samples was estimated using the Scherrer's equation through the full width at half maximum of the anatase (101) peak (see Table 2). Based on these results, it can be observed that the agitation mechanism has a significant effect on crystallinity since the monoliths coated in the ultrasound bath exhibit the smallest crystalline size (Figure 4(d)) than that of prepared with mechanical stirring (Figure 4(c)) to the same number of cycles (3 cycles) and residence time in the sol (3 hours). The diffraction pattern for commercial Degussa P-25 is included on the top right corner of Figure 4 for reference purposes. It can also be observed that the obtained SiO$_2$ is amorphous (Figure 4(e)). As can be seen in Table 2, the number of treatment cycles decreases crystal growth. In addition, the increase in immersion time for samples with 3 treatment cycles decreased the crystallinity. The sample with the largest crystalline size corresponds

to 3 treatment cycles and half an hour of immersion (sample 3ST12P). It is worth mentioning that more than 3 cycles were unsuccessfully attempted since monoliths got broken, with the exception of those with 1 hour immersion time. In this case, monoliths did not stand the fifth cycle.

Regarding samples 3ST1B and 3ST12B, these were discarded because the ultrasound influenced the structure to an extent that the SiO$_2$ monolith was broken. This may be ascribed to the vibrational movements caused by ultrasonic, causing the structure to become weaker. This is a consequence of the immersion under ultrasound presence during TiO$_2$/SiO$_2$ monoliths preparation as well as the decrease in crystallinity.

Figure 5 (3ST12P sample) shows the surface morphology of TiO$_2$ film obtained after 3 cycles of immersion-drying-calcination treatments. The residence time was half an hour for each of the immersion processes. The film exhibits a homogeneous morphology with a lineal growing up with almost totally flat surface. The EDS analysis shows the Ti presence. Figure 6 (3ST1P sample) presents the surface of the TiO$_2$ film obtained after 3 cycles of immersion-heating-calcination treatments and one hour of residence time for each immersion process.

Figure 7 (3ST3P sample) illustrates the characteristic morphology of a TiO$_2$ film after 3 treatment cycles and 3 hours of immersion under mechanical stirring. The comparison of Figures 7 and 5 make it clear that the amount of material deposited on the surface increases with immersion time. This increase is not related to the crystalline growth of anatase on the surface as has been evidenced by XRD analysis. The clusters presented in Figure 7 can be ascribed to the time given to the monolith in the sol where hydrolysis and condensation reactions are occurring, so the longer the immersion time, the larger the agglomerate.

By comparing Figures 5–7, it can be observed that the residence time has a significant effect on the morphology of

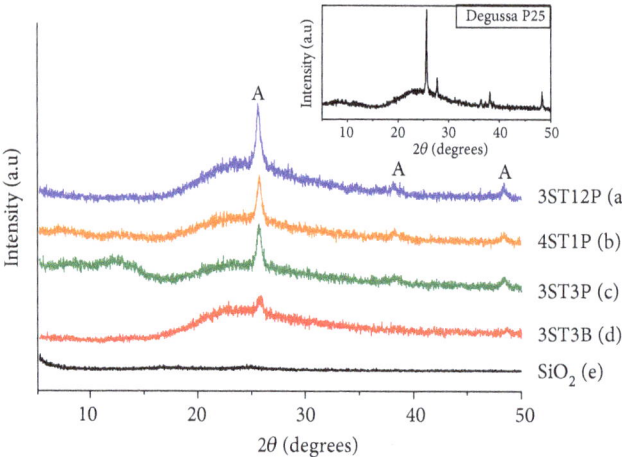

FIGURE 4: XRD patterns of TiO_2/SiO_2 and pure SiO_2 samples.

TABLE 2: Average crystallite size of the synthesized materials in Figure 4.

Sample	Average crystallite size (nm)
3ST12P	19.4
4ST1P	17.0
3ST3P	16.3
3ST3B	13.5

the coating. A time higher than half an hour (Figures 6 and 7) promotes the appearance of microcracks, and therefore the film loses homogeneity. These microcracks may be ascribed to a stress effect during the drying-calcination treatment due to thermal shrinkage and expansion phenomena. The stress is caused by chemical reactions during the drying and thermal expansion coefficients difference between the support $(5 \times 10^{-7}$°C) and TiO_2 film $(2.1-2.8 \times 10^{-6}$°C) [30]. It seems that the effect of this phenomenon becomes stronger when the amount of TiO_2 increases due to the drying and calcination process. The crystallinity of the sample shown in Figure 5 may be related to the absence of cracking since with the growth of the crystal and densification of the film, the compressive stresses are reduced.

Figure 8 (4ST1P sample) shows an image of a TiO_2 film after 4 treatment cycles and 1 hour of immersion under mechanical stirring. The comparison of the EDS analysis of this with that in Figure 6 confirms that the amount of TiO_2 is a direct function of the number of dip coating/heat treatment cycles and that the extent and frequency of the microcracks increase with the number of cycles [31, 32].

The final percentage in weight gained of TiO_2 by the SiO_2 monoliths is shown in Table 3. It can be observed that the increase in immersion time (samples 3ST12P and 3ST1P) favors the amount of TiO_2 deposited on the monoliths. On the other hand, the increase in the number of treatment cycles (samples 3ST1P and 4ST1P) decreases the final percentage of weight gained of TiO_2.

Table 4 shows the specific surface area and average pore diameter of synthesized materials. All samples presented type

I isotherms and average pore sizes of 18 Å, which according to the IUPAC classification corresponds to materials with microporous texture. It can be seen that the pure SiO_2 monolith presented the highest surface area $(339 \text{ m}^2/g)$. As the immersion time of the monoliths increases, the surface area decreases by about 50% (3ST3P sample). This decrease can be attributed to the amount of TiO_2 on the SiO_2 surface. Although the surface area decreases, the pore size distribution and the mean diameter are maintained.

It can be said that the synthesis conditions in which better crystallinity of the anatase phase is obtained; the highest specific surface area as well as a better uniformity of the formed film are with half an hour immersion, 3 cycles and stirring plate as dispersion mechanism. Therefore, these conditions were used to synthesize 30 monoliths to pack the bed reactor in order to perform the photocatalytic oxidation of methanol.

The minimum air flow rate to carry the methanol gas through the reactor was established as 50 ml/min. Two sets of experiments by triplicate were performed; one set without catalyst and the other one with the catalyst. In the former case, no formaldehyde was detected by the employed analysis method (2,4-dnph), and therefore the production of formaldehyde by photolysis was discarded. In the latter set of experiments, formaldehyde was identified. The analytical technique for the qualitative analysis of formaldehyde with a 2,4-dnph solution was carried out. The formation of micelles as precipitates is due to the formation of 2,4-dinitrophenylhydrazone indicating the presence of formaldehyde during the photocatalytic reaction. A precipitate indicating the presence of the aldehyde was only observed with the material 3ST12P. This does not mean that the other materials did not have photoactivity but that this could be so high that total methanol oxidation rather than selective oxidation was attained. To determine the amount of formaldehyde formed during photooxidation of methanol, the first condensed reaction product was evaluated by gas chromatography (GC). The final concentration of produced formaldehyde corresponds to a value of 457 micromol/L (13.7 mg/L). This result is superior to those obtained with other titania-silica systems [33].

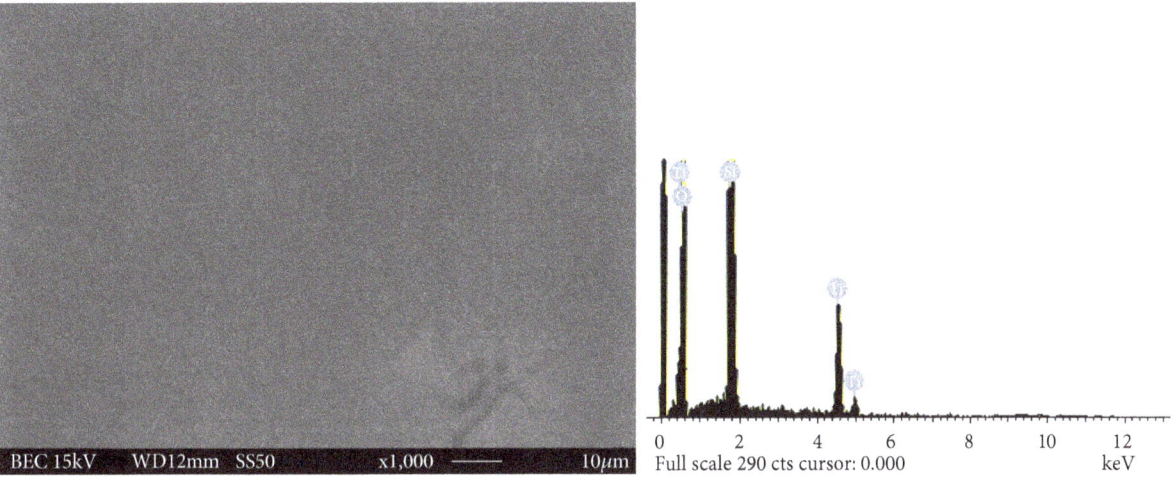

FIGURE 5: SEM image of TiO$_2$ film morphology obtained after 3 treatment cycles using a stirring plate and half an hour of immersion.

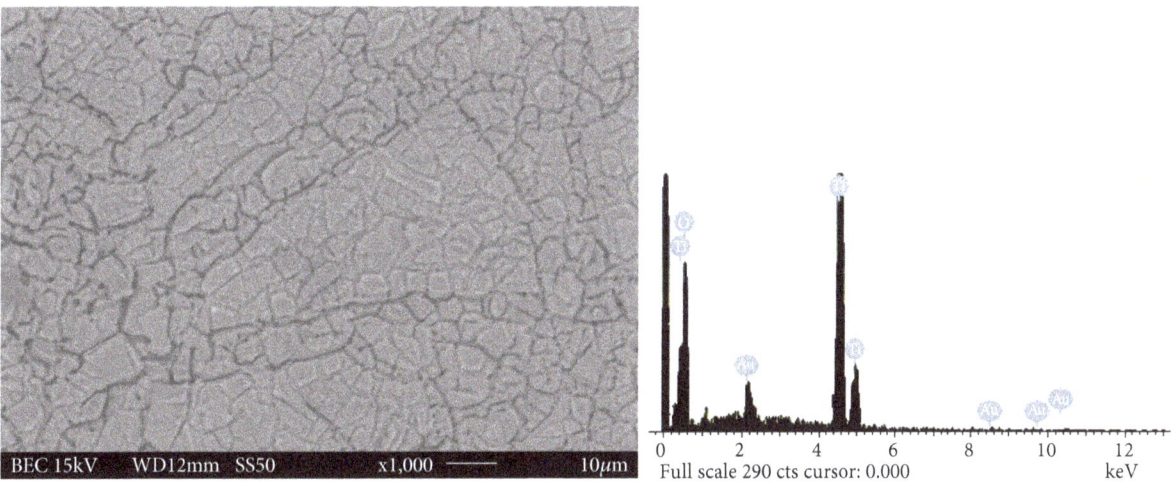

FIGURE 6: SEM image after 3 treatment cycles using stirring plate and 1 hour of immersion.

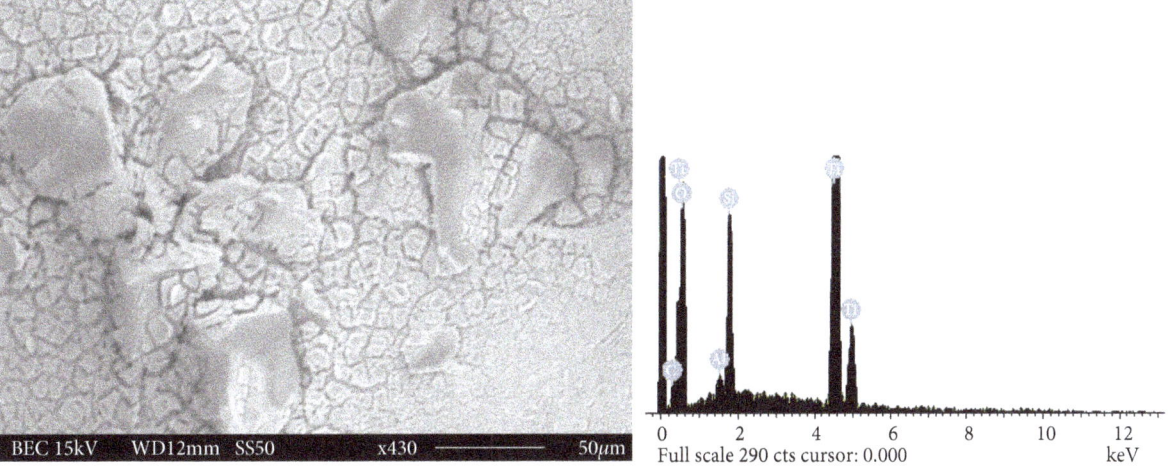

FIGURE 7: SEM image after 3 treatment cycles using stirring plate and 3 hours of immersion.

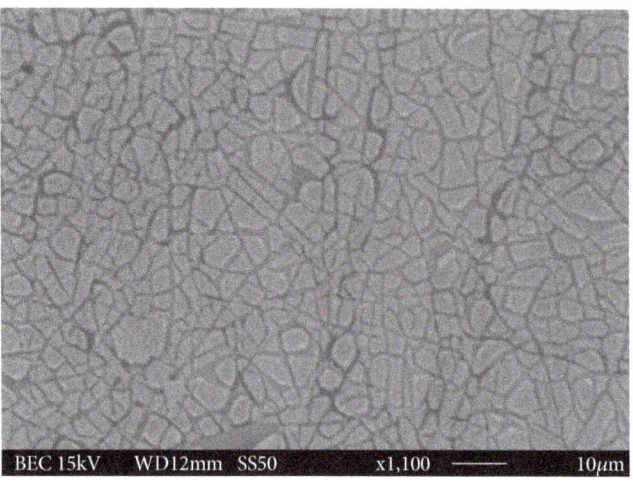 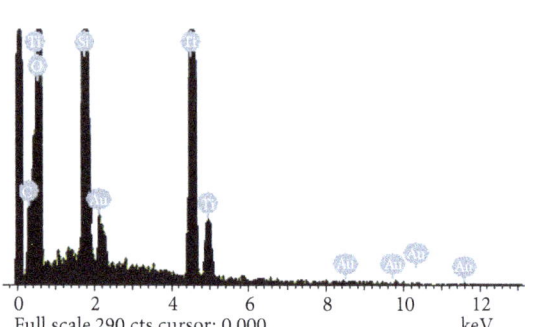

FIGURE 8: SEM image of TiO$_2$ film after 4 treatment cycles and 1 hour of immersion under mechanical stirring.

TABLE 3: Final percentage in weight gained of TiO$_2$ on the SiO$_2$ monoliths.

Sample	Percentage in weight gained of TiO$_2$ (%)
3ST12P	14.3
3ST1P	18.5
4ST1P	17.4

TABLE 4: Specific surface area and average pore size of the synthesized materials.

Sample	Specific surface area (m^2 g^{-1})	Average pore size (Å)
SiO$_2$	339	18
3ST12P	241	18
3ST1P	210	18
3ST3P	170	18
4ST1P	240	18

4. Conclusions

SiO$_2$ monoliths coated with thin films of TiO$_2$ anatase phase were successfully prepared by using a dip-coating sol-gel method. The dispersion mechanism, the immersion time, and the number of dip-coating cycles of the SiO$_2$ monoliths into the Ti sol were found to affect both the morphology and crystallinity of the TiO$_2$ deposit. SiO$_2$ monoliths coated with crackle-free TiO$_2$ films were obtained after three dip-coating cycles, with a dip time of 30 minutes. It can also be concluded that mechanical stirring should be preferred over ultrasound dispersion since the former favors the structural stability of the monolith and increases the film crystallinity, while the ultrasound dispersion method leads to monolithic structure breakage and also increases film crackles. Immersion time diminished both TiO$_2$ film and homogeneity. Immersion time and number of cycles also affect the surface area and deposit crystallinity. The surface area of the SiO$_2$-TiO$_2$ materials was decreased when the immersion time

increased, which is related to the amount of TiO$_2$ on the SiO$_2$ surface. The highest anatase phase crystallinity and specific surface area were obtained after 3 dip-coating cycles and half an hour of immersion under mechanical stirring. Under these preparation conditions, the attained surface area was 241 m^2/g and the crystallite size was 19.4 nm. The weight percentage gained by SiO$_2$ monoliths was 14.3% (TiO$_2$ film).

A formaldehyde concentration of 13.7 mg/L was attained at mild conditions of pressure and temperature in a continuous flow reactor packed with SiO$_2$ monoliths coated with TiO$_2$ anatase films prepared with 3 dip-coating cycles and 0.5 hours of immersion time.

Conflicts of Interest

The authors declare that there is no conflict of interests.

Acknowledgments

The authors are grateful to UAEMex for the financial support through project 4373/2017/CI and to CONACYT (project 269093). R. Regalado would like to thank CONACYT for the financial support and to CCIQS from UAEM for the granted support.

References

[1] Z. Li, S. Cong, and Y. Xu, "Brookite vs anatase TiO$_2$ in the photocatalytic activity for organic degradation in water," *ACS Catalysis*, vol. 4, no. 9, pp. 3273–3280, 2014.

[2] H. Park, C. D. Vecitis, W. Choi, O. Weres, and M. R. Hoffmann, "Solar-powered production of molecular hydrogen from water," *The Journal of Physical Chemistry C*, vol. 112, no. 4, pp. 885–889, 2008.

[3] J. Wang, Z. Bian, J. Zhu, and H. Li, "Ordered mesoporous TiO$_2$ with exposed (001) facets and enhanced activity in photocatalytic selective oxidation of alcohols," *Journal of Materials Chemistry A*, vol. 1, no. 4, pp. 1296–1302, 2013.

[4] M. Pera-Titus, V. García-Molina, M. A. Baños, J. Giménez, and S. Esplugas, "Degradation of chlorophenols by means of

advanced oxidation processes: a general review," *Applied Catalysis B: Environmental*, vol. 47, no. 4, pp. 219–256, 2004.

[5] A. Tanaka, K. Hashimoto, and H. Kominami, "Selective photocatalytic oxidation of aromatic alcohols to aldehydes in an aqueous suspension of gold nanoparticles supported on cerium(iv) oxide under irradiation of green light," *Chemical Communications*, vol. 47, no. 37, pp. 10446–10448, 2011.

[6] Y. Zhang, Z. R. Tang, X. Fu, and Y. J. Xu, "Engineering the unique 2D mat of graphene to achieve graphene-TiO_2 nanocomposite for photocatalytic selective transformation: what advantage does graphene have over its forebear carbon Nanotube?," *ACS Nano*, vol. 5, no. 9, pp. 7426–7435, 2011.

[7] G. Reuss, W. Disteldorf, O. Grundler, and A. Hilt, "Formaldehyde," in *Ullmann's Encyclopedia of Industrial Chemistry*, I. F. Ullmann, W. Gerhartz, Y. S. Yamamoto, F. T. Campbell, R. Pfefferkorn, and J. F. Rounsaville, Eds., VCH, Deerfield Beach, FL, USA, 1985.

[8] H. R. Gerberich, A. L. Stautzenberger, and W. C. Hopkins, "Formaldehyde," in *Kirk-Othmer Encyclopaedia of Chemical Technology*, pp. 231–250, John Wiley and Sons, New York, 1980.

[9] E. Cao and A. Gavriilidis, "Oxidative dehydrogenation of methanol in a microstructured reactor," *Catalysis Today*, vol. 110, no. 1-2, pp. 154–163, 2005.

[10] K. I. Ivanov and D. Y. Dimitrov, "Deactivation of an industrial iron-molybdate catalyst for methanol oxidation," *Catalysis Today*, vol. 154, no. 3-4, pp. 250–255, 2010.

[11] K. Nakata, T. Ochiai, T. Murakami, and A. Fujishima, "Photoenergy conversion with TiO_2 photocatalysis: new materials and recent applications," *Electrochimica Acta*, vol. 84, pp. 103–111, 2012.

[12] K. Nakata and A. Fujishima, "TiO_2 photocatalysis: design and applications," *Journal of Photochemistry and Photobiology C: Photochemistry Reviews*, vol. 13, no. 3, pp. 169–189, 2012.

[13] S. Shamaila, A. K. L. Sajjad, F. Chen, and J. Zhang, "Synthesis and characterization of mesoporous-TiO_2 with enhanced photocatalytic activity for the degradation of chloro-phenol," *Materials Research Bulletin*, vol. 45, no. 10, pp. 1375–1382, 2010.

[14] C. P. Lin, H. Chen, A. Nakaruk, P. Koshy, and C. C. Sorrell, "Effect of annealing temperature on the photocatalytic activity of TiO_2 thin films," *Energy Procedia*, vol. 34, pp. 627–636, 2013.

[15] W. Dai, X. Wang, P. Liu, Y. Xu, G. Li, and X. Fu, "Effects of electron transfer between TiO_2 films and conducting substrates on the photocatalytic oxidation of organic pollutants," *The Journal of Physical Chemistry B*, vol. 110, no. 27, pp. 13470–13476, 2006.

[16] Y. Cui, J. Sun, Z. Hu et al., "Synthesis, phase transition and optical properties of nanocrystalline titanium dioxide films deposited by plasma assisted reactive pulsed laser deposition," *Surface and Coatings Technology*, vol. 231, pp. 180–184, 2013.

[17] X. Zhao, M. Liu, and Y. Zhu, "Fabrication of porous TiO_2 film via hydrothermal method and its photocatalytic performances," *Thin Solid Films*, vol. 515, no. 18, pp. 7127–7134, 2007.

[18] D. Li, M. Carette, A. Granier, J. P. Landesman, and A. Goullet, "In situ spectroscopic ellipsometry study of TiO_2 films deposited by plasma enhanced chemical vapour deposition," *Applied Surface Science*, vol. 283, pp. 234–239, 2013.

[19] A. Arunachalam, S. Dhanapandian, C. Manoharan, and R. Sridhar, "Characterization of sprayed TiO_2 on ITO substrates for solar cell applications," *Spectrochimica Acta Part A: Molecular and Biomolecular Spectroscopy*, vol. 149, pp. 904–912, 2015.

[20] A. Y. Shan, T. I. M. Ghazi, and S. A. Rashid, "Immobilisation of titanium dioxide onto supporting materials in heterogeneous photocatalysis: a review," *Applied Catalysis A: General*, vol. 389, no. 1-2, pp. 1–8, 2010.

[21] M. A. L. Vargas, M. Casanova, A. Trovarelli, and G. Busca, "An IR study of thermally stable V_2O_5-WO_3-TiO_2 SCR catalysts modified with silica and rare-earths (Ce, Tb, Er)," *Applied Catalysis B: Environmental*, vol. 75, no. 3-4, pp. 303–311, 2007.

[22] W. Chang, L. Yan, Bin Liu, and R. Sun, "Photocatalyic activity of double pore structure TiO_2/SiO_2 monoliths," *Ceramics International*, vol. 43, no. 8, pp. 5881–5886, 2017.

[23] C. J. Brinker, "Hydrolysis and condensation of silicates: effects on structure," *Journal of Non-Crystalline Solids*, vol. 100, no. 1-3, pp. 31–50, 1988.

[24] G. Andrade-Espinosa, V. Escobar-Barrios, and R. Rangel-Mendez, "Synthesis and characterization of silica xerogels obtained via fast sol–gel process," *Colloid and Polymer Science*, vol. 288, no. 18, pp. 1697–1704, 2010.

[25] M. Ahmad, J. R. Jones, and L. L. Hench, "Fabricating sol–gel glass monoliths with controlled nanoporosity," *Biomedical Materials*, vol. 2, no. 1, pp. 6–10, 2007.

[26] Y. Farhang Ghoje Biglu and E. Taheri-Nassaj, "Investigation of phase separation of nano-crystalline anatase from TiO2-SiO2 thin film," *Ceramics International*, vol. 39, no. 3, pp. 2511–2518, 2013.

[27] S. Uchiyama, Y. Inaba, and N. Kunugita, "Derivatization of carbonyl compounds with 2,4-dinitrophenylhydrazine and their subsequent determination by high-performance liquid chromatography," *Journal of Chromatography B*, vol. 879, no. 17-18, pp. 1282–1289, 2011.

[28] M. S. Lee, S. S. Park, G. D. Lee, C. S. Ju, and S. S. Hong, "Synthesis of TiO_2 particles by reverse microemulsion method using nonionic surfactants with different hydrophilic and hydrophobic group and their photocatalytic activity," *Catalysis Today*, vol. 101, no. 3-4, pp. 283–290, 2005.

[29] R. Kaplan, B. Erjavec, J. Dražić, J. Grdadolnik, and A. Pintar, "Simple synthesis of anatase/rutile/brookite TiO_2 nanocomposite with superior mineralization potential for photocatalytic degradation of water pollutants," *Applied Catalysis B: Environmental*, vol. 181, pp. 465–474, 2016.

[30] Z. Fu, U. Eckstein, A. Dellert, and A. Roosen, "In situ study of mass loss, shrinkage and stress development during drying of cast colloidal films," *Journal of the European Ceramic Society*, vol. 35, no. 10, pp. 2883–2893, 2015.

[31] N. Arconada, A. Durán, S. Suárez et al., "Synthesis and photocatalytic properties of dense and porous TiO_2-anatase thin films prepared by sol–gel," *Applied Catalysis B: Environmental*, vol. 86, no. 1-2, pp. 1–7, 2009.

[32] C. M. Malengreaux, G. M. L. Léonard, S. L. Pirard et al., "How to modify the photocatalytic activity of TiO_2 thin films through their roughness by using additives. A relation between kinetics, morphology and synthesis," *Chemical Engineering Journal*, vol. 243, pp. 537–548, 2014.

[33] J. M. Stokke, D. W. Mazyck, C. Y. Wu, and R. Sheahan, "Photocatalytic oxidation of methanol using silica-titania composites in a packed-bed reactor," *Environmental Progress*, vol. 25, no. 4, pp. 312–318, 2006.

Permissions

All chapters in this book were first published in IJP, by Hindawi Publishing Corporation; hereby published with permission under the Creative Commons Attribution License or equivalent. Every chapter published in this book has been scrutinized by our experts. Their significance has been extensively debated. The topics covered herein carry significant findings which will fuel the growth of the discipline. They may even be implemented as practical applications or may be referred to as a beginning point for another development.

The contributors of this book come from diverse backgrounds, making this book a truly international effort. This book will bring forth new frontiers with its revolutionizing research information and detailed analysis of the nascent developments around the world.

We would like to thank all the contributing authors for lending their expertise to make the book truly unique. They have played a crucial role in the development of this book. Without their invaluable contributions this book wouldn't have been possible. They have made vital efforts to compile up to date information on the varied aspects of this subject to make this book a valuable addition to the collection of many professionals and students.

This book was conceptualized with the vision of imparting up-to-date information and advanced data in this field. To ensure the same, a matchless editorial board was set up. Every individual on the board went through rigorous rounds of assessment to prove their worth. After which they invested a large part of their time researching and compiling the most relevant data for our readers.

The editorial board has been involved in producing this book since its inception. They have spent rigorous hours researching and exploring the diverse topics which have resulted in the successful publishing of this book. They have passed on their knowledge of decades through this book. To expedite this challenging task, the publisher supported the team at every step. A small team of assistant editors was also appointed to further simplify the editing procedure and attain best results for the readers.

Apart from the editorial board, the designing team has also invested a significant amount of their time in understanding the subject and creating the most relevant covers. They scrutinized every image to scout for the most suitable representation of the subject and create an appropriate cover for the book.

The publishing team has been an ardent support to the editorial, designing and production team. Their endless efforts to recruit the best for this project, has resulted in the accomplishment of this book. They are a veteran in the field of academics and their pool of knowledge is as vast as their experience in printing. Their expertise and guidance has proved useful at every step. Their uncompromising quality standards have made this book an exceptional effort. Their encouragement from time to time has been an inspiration for everyone.

The publisher and the editorial board hope that this book will prove to be a valuable piece of knowledge for researchers, students, practitioners and scholars across the globe.

List of Contributors

Ivana Šagud and Irena Škorić
Department of Organic Chemistry, Faculty of Chemical Engineering and Technology, University of Zagreb, Marulićev trg 19, 10000 Zagreb, Croatia

Julio César González-Torres, Enrique Poulain, Raúl García-Cruz and Oscar Olvera-Neria
Área de Física Atómica Molecular Aplicada (FAMA), CBI, Universidad Autónoma Metropolitana-Azcapotzalco, Av. San Pablo 180, Col. Reynosa Tamaulipas, 02200 Ciudad de México, Mexico

Víctor Domínguez-Soria
Área de Química Aplicada, CBI, Universidad Autónoma Metropolitana-Azcapotzalco, Av. San Pablo 180, Col. Reynosa Tamaulipas, 02200 Ciudad de México, Mexico

Asim Abas and Shanpeng Wang
School of Physical Science and Technology, Lanzhou University, Lanzhou 730000, China

Guomei Tang
School of Physical Science and Technology, Lanzhou University, Lanzhou 730000, China
School of Mathematics and Computer Science Institute, Northwest Minzu University, Lanzhou, Gansu 730030, China
State Key Laboratory of Advanced Processing and Recycling of Non-Ferrous Metals, Lanzhou University of Technology, Lanzhou 730050, China

Armando Vázquez, Isabel Lázaro, Roel Cruz and Israel Rodríguez-Torres
Instituto de Metalurgia, Facultad de Ingeniería, Universidad Autónoma de San Luis Potosí, Av. Sierra Leona 550, 78210 San Luis Potosí, SLP, Mexico

Lucía Alvarado
Departamento de Ingeniería en Minas, Metalurgia y Geología, Universidad de Guanajuato, Ex. Hacienda de San Matías s/n Fracc. San Javier, 36025 Guanajuato, GTO, Mexico

José Luis Nava
Departamento de Ingeniería Geomática e Hidráulica, Universidad de Guanajuato, Av. Juárez 77, 36000 Guanajuato, GTO, Mexico

André E. Nogueira
Brazilian Nanotechnology National Laboratory (LNNano), Brazilian Center for Research in Energy and Materials (CNPEM), Zip Code 13083-970 Campinas, São Paulo, Brazil

National Laboratory of Nanotechnology for Agrobusiness (LNNA), EMBRAPA-Brazilian Agricultural Research Corporation, Rua XV de Novembro, 1452, São Carlos, SP 13560-970, Brazil

Lucas S. Ribeiro and Emerson R. Camargo
Interdisciplinary Laboratory of Electrochemistry and Ceramics (LIEC), Department of Chemistry, Federal University of São Carlos, Rod. Washington Luis km 235, CP 676 São Carlos, SP 13565-905, Brazil

Luiz F. Gorup
FACET-Department of Chemistry, Federal University of Grande Dourados, Dourados, Mato Grosso do Sul 79804-970, Brazil

Gelson T. S. T. Silva, Fernando F. B. Silva and Caue Ribeiro
National Laboratory of Nanotechnology for Agrobusiness (LNNA), EMBRAPA-Brazilian Agricultural Research Corporation, Rua XV de Novembro, 1452, São Carlos, SP 13560-970, Brazil

Julieta Cabrera, Alcides López, Pilar García, Dante Ríos and Juan M. Rodriguez
Universidad Nacional de Ingeniería, Av. TúpacAmaru s/n, Rimac, Lima, Peru

Dwight Acosta
Instituto de Física, Universidad Nacional Autónoma de México, 20364 Ciudad de México, Mexico

Roberto J. Candal
Instituto de Investigación e Ingeniería Ambiental, CONICET, Universidad Nacional de San Martín, Campus Miguelete, 25 de Mayo y Francia, 1650 San Martín, Provincia de Buenos Aires, Argentina

Claudia Marchi
Centro de Microscopias Avanzadas, FCEyN, Universidad de Buenos Aires, Ciudad Universitaria, 1428 Buenos Aires, Argentina

Maria E. Manríquez, Jin An Wang, Lifang Chen, Jose Salmones, Julio González-García and Carmen Reza
ESIQIE, Instituto Politécnico Nacional, Av. Instituto Politécnico Nacional s/n, Col. Zacatenco, 07738 Mexico City, Mexico

Luis Enrique Noreña
Departamento de Ciencias Básicas, Universidad Autónoma Metropolitana-Azcapotzalco, Av. San Pablo 180, Col. Reynosa-Tamaulipas, 02200 Mexico City, Mexico

Francisco Tzompantzi
Departamento Química, Universidad Autónoma Metropolitana-Iztapalapa, San Rafael Atlixco 186, 09340 Mexico City, Mexico

José G. Hernández Cortez
GDMyPQ, Eje Lázaro Cárdenas 152, Instituto Mexicano del Petróleo, 07730 Mexico City, Mexico

Liqun Ye and Haiquan Xie
College of Chemistry and Pharmaceutical Engineering, Nanyang Normal University, Nanyang, China

Carlos Díaz-Uribe, Jose Viloria, Lorraine Cervantes and William Vallejo
Grupo de Fotoquímica y Fotobiología, Universidad del Atlántico, Barranquilla, Colombia

Karen Navarro
Grupo de Fotoquímica y Fotobiología, Universidad del Atlántico, Barranquilla, Colombia
Universidad Nacional de Córdoba, Cordoba, Argentina

Eduard Romero
Departamento de Química, Universidad Nacional de Colombia, Bogotá, Colombia

Cesar Quiñones
Institución Universitaria Politécnico Gran Colombiano, Bogotá, Colombia

Shuisheng Wu, Nianyuan Tan, Donghui Lan and Bing Yi
College of Chemistry and Chemical Engineering, Hunan Institute of Engineering, Xiangtan 411104, China

Maryam Jami, Ralf Dillert and Yanpeng Suo
Institut für Technische Chemie, Leibniz Universität Hannover, Callinstraße 3, 30167 Hannover, Germany

Detlef W. Bahnemann
Institut für Technische Chemie, Leibniz Universität Hannover, Callinstraße 3, 30167 Hannover, Germany
Laboratory "Photoactive Nanocomposite Materials", Saint Petersburg State University, Ulyanovskaya Str. 1, Peterhof, Saint Petersburg 198504, Russia

Michael Wark
Technische Chemie, Universität Oldenburg, Carl-von-Ossietzky Str. 9-11, 26111 Oldenburg, Germany

O. Alvarado-Rolon and A. Ramírez-Serrano
Facultad de Química, Universidad Autónoma del Estado de México, Paseo Colon esq. Paseo Tollocan s/n, 50120 Toluca, MEX, Mexico

R. Natividad, R. Romero and L. Hurtado
Facultad de Química, Centro Conjunto de Investigación en Química Sustentable UAEM-UNAM, Universidad Autónoma del Estado de México, Carretera Toluca–Atlacomulco, Km 14.5, Unidad San Cayetano, 50200 Toluca, MEX, Mexico

Yu Niu, Fuying Li, Qiyou Wu, Peijing Xu and Renzhang Wang
Fujian Provincial Collaborative Innovation Center for Clean Coal Gasification, Technology College of Resources and Chemical Engineering, Sanming University, Sanming 365004, China

Kai Yang
School of Metallurgy and Chemical Engineering, Jiangxi University of Science and Technology, Ganzhou 341000, China

Laila M. Al-Harbi
Chemistry Department, Faculty of Science, King Abdulaziz University, Jeddah 21589, Saudi Arabia

Ghaida H. Munshi
Chemistry Department, Faculty of Science, King Abdulaziz University, Jeddah 21589, Saudi Arabia
Chemistry Department, Faculty of Science, University of Jeddah, Jeddah 23218, Saudi Arabia

Amal M. Ibrahim
Chemistry Department, Faculty of Science, University of Jeddah, Jeddah 23218, Saudi Arabia
Physical Chemistry Department, National Research Centre, Cairo 12622, Egypt

Tzayam Pérez
Departamento de Ing. Química, Noria Alta, Universidad de Guanajuato, 36050 Guanajuato, GTO, Mexico

José L. Nava
Departamento de Ingeniería Geomática e Hidráulica, Av. Juárez 77, Zona Centro, Universidad de Guanajuato, 36000 Guanajuato, GTO, Mexico

Fengmin Wu, Wenlu Duan, Mei Li and Hang Xu
School of Chemical Engineering and Pharmaceutics, Henan University of Science and Technology, Luoyang 471023, China

Jian-hui Xu, Xiao-mei Lv, Shan-hong Lan and Liu Peng
School of Environmental and Civil Engineering, Dongguan University of Technology, Dongguan, Guangdong 523808, China

Bin-bin Ding and Chao-lin Li
Environmental Science and Engineering Center, Harbin Institute of Technology Shenzhen Graduate School, Shenzhen, Guangdong 518055, China

Thammasak Rojviroon
Department of Civil Engineering, Faculty of Engineering, Rajamangala University of Technology Thanyaburi, Pathum Thani 12110, Thailand

Sanya Sirivithayapakorn
Department of Environmental Engineering, Faculty of Engineering, Kasetsart University, Bangkok 10900, Thailand

Molood Barmala
Department of Chemical Engineering, Dezful Branch, Islamic Azad University, Dezful, Iran

Mohammad Behnood
Department of Petroleum and Chemical Engineering, Science and Research Branch, Islamic Azad University, Tehran, Iran

Hengkai Pan, Bei Wang, Feng Zhang, Weifeng Zhang and Guoqiang Li
Henan Key Laboratory of Photovoltaic Materials, Henan University, Kaifeng 475004, China
Laboratory of Low-Dimensional Materials Science, Henan University, Kaifeng 475004, China

D. A. Solís-Casados and V. Alvarado-Pérez
Universidad Autónoma del Estado de México, Centro Conjunto de Investigación en Química Sustentable UAEMéx-UNAM, Toluca, MEX, Mexico

L. Escobar-Alarcón
Departamento de Física, Instituto Nacional de Investigaciones Nucleares, Apartado Postal 18-1027, 11801 Mexico City, Mexico

E. Haro-Poniatowski
Departamento de Física, Universidad Autónoma Metropolitana, Apartado Postal 55-532, 09340 Mexico City, Mexico

Manuel I. Peña-Cruz, Carlos A. Pineda-Arellano, Iván Salgado-Tránsito and Fernando Martell-Chavez
CONACYT-Centro de Investigaciones en Óptica, A.C., Unidad Aguascalientes, Prol. Constitución 607, Frac. Reserva Loma Bonita, Aguascalientes, Aguascalientes 20200, Mexico

Camilo A. Arancibia-Bulnes
Instituto de Energías Renovables-Universidad Nacional Autónoma de México, Privada Xochicalco s/n, A.P. 34, Col. Centro, Temixco, Morelos 62580, Mexico

Patricio J. Valades-Pelayo
Instituto de Energías Renovables-Universidad Nacional Autónoma de México, Privada Xochicalco s/n, A.P. 34, Col. Centro, Temixco, Morelos 62580, Mexico
Depto. de Ingeniería de Procesos e Hidráulica, Universidad Autónoma Metropolitana-Iztapalapa, Av. San Rafael Atlixco No. 186, C.P. 09340, Mexico City, Mexico

Rigoberto Regalado-Raya, Osmín Avilés-García and Jaime Espino-Valencia
Facultad de Ingeniería Química, Universidad Michoacana de San Nicolás de Hidalgo, Edif. V1, Ciudad Universitaria, 58060 Morelia, Michoacán, Mexico

Rubí Romero-Romero
km 14.5 Carretera Toluca-Atlacomulco, San Cayetano, Piedras Blancas, Centro Conjunto de Investigación en Química Sustentable UAEMéx-UNAM, Toluca, Estado de México, Mexico

Index

www.ingramcontent.com/pod-product-compliance
Lightning Source LLC
Chambersburg PA
CBHW050446200326
41458CB00014B/5077